CHEMICAL EVOLUTION:

SELF-ORGANIZATION OF THE MACROMOLECULES OF LIFE

Studies in
Chemical Evolution and the Origin of Life
Series Editor: Cyril Ponnamperuma
Published Volumes and Volumes in Preparation

Prebiological Self Organization of Matter
Cyril Ponnamperuma and Frederick R. Eirich (Eds.)

Trieste Conferences on
Chemical Evolution and the Origin of Life

Chemical Evolution: Origin of Life
Proceedings of the First Trieste Conference October 1992
Cyril Ponnamperuma and Julian Chela-Flores (Eds.)

Chemical Evolution: Self Organization of the Macromolecules of Life
Proceedings of the Second Trieste Conference October 1993
Julian Chela-Flores, Mohindra Chadha, Alicia Negrón-Mendoza, and Tairo Oshima (Eds.)

INTERNATIONAL CENTRE FOR THEORETICAL PHYSICS

INTERNATIONAL ATOMIC ENERGY AGENCY
UNITED NATIONS EDUCATIONAL, SCIENTIFIC AND CULTURAL ORGANIZATION

CHEMICAL EVOLUTION: SELF-ORGANIZATION OF THE MACROMOLECULES OF LIFE

Edited by

Julian Chela-Flores
International Centre for Theoretical Physics
Trieste, Italy
and
Instituto Internacional de Estudios Avanzados
Caracas, Venezuela

Mohindra Chadha
Bhabha Atomic Research Center
Bombay, India

Alicia Negrón-Mendoza
Universidad Nacional Autonoma de Mexico
Instituto de Ciencias Nucleares
Mexico City, Mexico

Tairo Oshima
Tokyo Institute of Technology
Yokohama, Japan

Proceedings of
The Trieste Conference on Chemical Evolution and the Origin of Life
25–29 October 1993

A. DEEPAK Publishing **1995**
A Division of Science and Technology Corporation
Hampton, Virginia USA

Published by
A. DEEPAK Publishing
A Division of Science and Technology Corporation
101 Research Drive
Hampton, Virginia 23666-1340 USA

Library of Congress Cataloging-in-Publication Data

Trieste Conference on Chemical Evolution and the Origin of Life (1993)
Chemical evolution—self—organization of the macromolecules of life : proceedings of the Trieste Conference on Chemical Evolution and the Origin of Life, 25–29 October 1993 / edited by Julian Chela Flores . . . [et al.].
p. cm.
Includes bibliographical references.
ISBN 0–937194–32–8 (hc)
1. Molecular evolution—Congresses. 2. Life—Origin—Congresses.
I. Chela Flores, Julian. II. Title.
QH325.T74 1993
577—dc20 95–19705
CIP

Printed in the United States of America

Dedicated to

CYRIL PONNAMPERUMA

on the occasion of his seventieth birthday

FOREWORD

Since this meeting fell at a time which was a landmark in the long and illustrious career of Professor Cyril Ponnamperuma, some of his former colleagues suggested that the event be held in his honour. Not only was he celebrating almost of four decades of a productive scientific career, but he was also assuming the first Directorship of the World Center for Sustainable Development at the University of Maryland, College Park, USA, in addition to his present position as Professor of Chemistry and Director of the Laboratory of Chemical Evolution.

CYRIL PONNAMPERUMA

A Biographical Sketch

Cyril Ponnamperuma is a native of Sri Lanka (Ceylon). He came to the United States in 1959, and in 1967 he became a naturalized citizen.

His early education was in Sri Lanka and India where he received a baccalaureate in philosophy. He then proceeded to London where he obtained a B.Sc. (Honors) degree in Chemistry at Birkbeck College, University of London, in 1959. During this time, he had the privilege of being associated with Professor J. D. Bernal, a pioneer in the field of the origin of life. After his studies at the University of London, Ponnamperuma attended the University of California, Berkeley, where he received his doctorate in Chemistry in 1962 under the direction of Nobel Laureate Professor Melvin Calvin.

In 1962, he was awarded a National Academy of Sciences Resident Associateship, tenable with NASA at the Ames Research Center. In 1963, he joined NASA's Exobiology Division and became Chief of the Chemical Evolution Branch. The primary goal of his laboratory was the study of the origin of life. When the Apollo program was established, he was selected as a principal investigator for organic analysis. He has been closely involved with NASA in the Viking and Voyager programs. He has served as member of both SESAC (Space and Earth Sciences Advisory Council) and LSAC (Life Sciences Advisory Council) of NASA. He is a member of the Governor's Advisory Council for Education for the State of Virginia.

Ponnamperuma has been associated with many universities in the United States and abroad. He was on the visiting faculty of Stanford University, the University of Nijmegen in the Netherlands, and the Sorbonne. The Indian Atomic Energy Commission appointed him a Distinguished Visiting Professor in 1967. In 1970, the USSR Academy of Sciences invited him to visit the Soviet Union as a guest lecturer. In 1970 and 1971, he was Director of the UNESCO Program for the Development of Basic Research in Sri Lanka. In 1982, the Chinese Academy of Sciences invited him to lecture in the People's Republic of China.

Since July, 1971, he has been at the University of Maryland as Professor of Chemistry and Director of the Laboratory of Chemical Evolution. Among his teaching responsibilities are a graduate course on Chemical Evolution and an interdisciplinary undergraduate course on Life in the Universe. In 1976, has was a Phi Beta Kappa visiting scholar. In 1978, he was named a Distinguished Professor of the University of Maryland. In 1980, the International Society for the Study of the Origin of Life awarded him the first A. I. Oparin Gold Medal for the "best sustained program" on the origin of life. He is a D.Sc. (Honoris Causa) of the University of Sri Lanka (1978), the University of Puget Sound (1982), the University of Peradeniya (1984), and the University of Sri Jayawardenepura (1985). In 1990, The President of Sri Lanka awarded him the "Vidya Jothi" (Luminary of Science) medal for his services to science and Sri Lanka. In 1991, the government of France conferred on him the title of "Chevalier de Lettres et des Artes" for promoting international understanding. In 1991, the University of Maryland, awarded him the first Distinguished International Science Award for a scholarly career combined with extraordinary services to the international community. In 1993, the Russian Academy of Creative Arts awarded him the first Harold Urey Prize for his outstanding contributions to the study of the origin of life.

The author of over four hundred publications related to chemical evolution and the origin of life, he has written and edited sixteen books on the subject, including the *Origins of Life* which has been translated into many languages. He is on the editorial board of the *Journal of Molecular Evolution* and for over a decade was editor-in-chief of the international journal *Origins of Life*.

Ponnamperuma is a member of the American Chemical Society, a Fellow of the Chemical Society, London, a Fellow of the Royal Institute of Chemistry, London, and a foreign Fellow of the Indian National Science Academy. He was president of the International Society for the Study of the Origin of Life, chairman of the ACS Committee for International Affairs and served on the Advisory Board of the AAAS Council of International Activities. He chaired the Future Actions Committee of CHEMRAWN (Chemical Research Applied to World Needs) in 1982. He is a member of the Cosmos Club, a fellow of the Explorers Club, and Chairman of the Board of the Sri Lanka Overseas Foundation. In 1984, he was appointed Science and Technology Advisor to the President of Sri Lanka. He was Director of the Institute of Fundamental Studies (1984-1991), Sri Lanka, and Director of the Arthur C. Clarke Centre for Modern Technologies (1985-1987) also located in Sri Lanka. He functions as Science and Technology Adviser to the President of Sri Lanka while maintaining his academic position and an active research program at the University of Maryland. In 1985, he was elected a Fellow of the Third World Academy of Sciences and was appointed the Chairman of the Global Frontiers of Science Committee. He is Vice President of the Third World Network of Scientific Organizations and President of the Third World Foundation of North America. He was also appointed Director General of the Network of 20 international centers for Sustainable Development to be established in the developing world by the Third World Academy of Sciences.

CONTENTS

Foreword .... *vi*
Biographical Sketch of Cyril Ponnamperuma .... *vii*
Preface .... *xi*
Photograph of Participants .... *xii*

PART I: Chemical Aspects of Self-Organization

The Physicochemical Origins of the Genetic Code 3
Cyril Ponnamperuma, Mitchell K. Hobish and Nalinie Wickramasinghe

Role of Nitriles and Other Reactive Molecules in Chemical Evolution 19
Mohindra S. Chadha

Formation of Amino Acid Precursors by Cosmic Radiation in Primitive Terrestrial and Extraterrestrial Environments 37
Kensei Kobayashi, Takeshi Saito, and Tairo Oshima

Hydrogen Cyanide Polymers: Prebiotic Agents for the Simultaneous Origin of Proteins and Nucleic Acids 41
Clifford N. Matthews

A Plausible Route for the Synthesis of Bio-Organic Compounds from Inorganic Carbon on the Primitive Earth. An Overview 49
G. Albarrán, A. Negrón-Mendoza, S. Ramos-Bernal, and C.H. Collins

PART II: Geophysical Aspects of Self-Organization

Early Terrestrial Life: Problems of the Oldest Record 65
Manfred Schidlowski

PART III: Biochemical Aspects of Self-Organization

Energy, Matter and Self-Organization in the Early Molecular Evolution of Bioenergetic Systems 83
Herrick Baltscheffsky and Margareta Baltscheffsky

The Evolution of Organisation and the Perils of Errors in the Origin of Life 91
Clas Blomberg

On Attempts to Create Life-Mimicking Cells 107
F.R. Eirich

Computational Support for Origins of Life Research: A Personal View 119
Mitchell K. Hobish

Membrane Phase Separations, Asymmetry and Implications in the Origin of Life 131
Michael O. Eze

General Crystal in Prebiotic Context 139
I. Simon

Self-Ordering and Polymerization in Biological Macromolecules and Symmetry Breaking .. 145
Giuseppe Vitiello

PART IV: Biophysical Aspects of Self-Organization

Return to Dichotomy: *Bacteria and Archaea* 155
A. Yamagishi and T. Oshima

RNA: Genotype and Phenotype 159
Peter F. Stadler

Gradual Rise of Cellular Translation 177
M. Rizzotti

Molecular Relics from Chemical Evolution and the Origin of Life 185
Julian Chela-Flores

Designing a Biosphere for Mars 201
Robert H. Haynes and Christopher P. McKay

Unexpected *In-Vitro* Intron Splicing of Common Bean Chloroplast trnL (UAA) Gene and Pseudogene by T7 RNA Polymerase 213
O. Carelse and M.V. Mubumbila

On Dyson's Model of the Origins of Life and Possible Experimental Verification 219
J.N. Islam and M.K. Pasha

The Role of Information Processing in the Evolution of Complex Life Forms 225
K. Tahir Shah

PART V: Chirality and Self-Organization

The Weak Force and the Origin of Life and Self-Organization 237
Alexandra J. MacDermott

Tunneling, Chirality and Cold Prebiotic Evolution 251
Vitalii I. Goldanskii

Finding the Necessary Condition and Scope of L-Amino Acid Surviving Under Beta Electron Irradiation 277
W.Q. Wang, J.L. Wu, X.M. Pan, and L.F. Luo

Chiral Interaction in Molecular Systems 283
G. Gilat

Cosmological Sources of Molecular Chirality 295
J. Chela-Flores and N. Kumar

PART VI: Celebration of Cyril Ponnamperuma's 70th Birthday

Photograph of the award of a plaque to Professor Cyril Ponnamperuma 303

On the Occasion of the Ponnamperumafest:

Message for Cyril Ponnamperuma from Abdus Salam 304

A Letter from Francois Raulin 305

A Comprehensive Bibliography of Cyril Ponnamperuma 307

List of Participants 336

PREFACE

An attempt to describe the role of replication processes of self-organization of biological macromolecules implies obtaining insights into evidences and understanding provided by a diversity of disciplines. Recent advances in several areas of human knowledge, and the development of new methodologies make it possible to study the different stages of the long history of evolution of organic matter. This background made it possible to present a coherent account of some aspects of molecular and biological evolution.

This volume gathers together the papers presented at the Conference on Self-Organization of the Macromolecules of Life, held in Trieste in October 1993. This was inserted in a programme that includes evidence of early life from the oldest known fossils in the geological record as well as the prior events of chemical evolution and self-organization, including the question of the chirality of protein amino acids.

Many topics related to the main theme of the conference are still at a speculative and uncertain stage, but this situation need not be discouraging as it offers opportunities for future research.

This book is organized in five sections corresponding to chemical, geological, biochemical, and biophysical aspects of self-organization, concluding with a section on chirality. These subjects were covered by 13 speakers from Austria, Germany, India, Italy, Japan, Sri Lanka, Sweden, United Kingdom, USA and Venezuela. In other plenary and discussion sessions 21 contributions were presented from participants from Bangladesh, Canada, France, Germany, Hungary, India, Israel, Mexico, Nigeria, People's Republic of China, the Russian Federation, United Kingdom, USA and Zimbabwe.

The Conference of Self-Organization of the Macromolecules of Life was generously supported by the International Centre for Theoretical Physics, International Centre for Science and High Technology, the European Commission, and UNESCO.

Mohindra Chadha, Julian Chela-Flores,
Alicia Negron-Mendoza, and Tairo Oshima

PARTICIPANTS

1) M. Schidlowski, 2) V.I. Goldanski, 3) M.S. Chadha, 4) J. Chela-Flores, 5) C. Ponnamperuma, 6) T. Oshima, 7) H. Baltscheffsky, 8) K.T. Shah, 9) Mrs. Schidlowski, 10) W. Wang, 11) W.Q. Wang, 12) O. Carelse, 13) A. Negron-Mendoza, 14) M.K. Pasha, 15) K. Rao, 16) Observer, 17) Mrs. Blomberg, 18) T. Saito, 19) R. Mistretta, 20) I.S. Kulaev, 21) F.R. Eirich, 22) H. Stavliotis, 23) O. Famurewa, 24) O. Tacheeni Scott, 25) N. Kumar, 26) M.O. Eze, 27) C. Blomberg, 28) B.E. Prieur, 29) C. Matthews, 30) A. Figureau, 31)W. Thiemann, 32) A. MacDermott, 33) G. Vitiello, 34) S. Rauch-Wojciechoswki, 35) A. Alstein, 36) A. Allegrini, 37) A. Rizzotti, 38) P. Stadler, 39) G. Longo, 40) G. Lenaz, 41) C. Cosmovici, 42) J. Doskocil, 43) I. Simon, 44) G. Gilat, 45) B. Heinz, 46) M.K. Hobish, 47) G. De Sarrazin, 48) Ninel Valderrama

PART I
CHEMICAL ASPECTS OF SELF-ORGANIZATION

THE PHYSICOCHEMICAL ORIGINS OF THE GENETIC CODE

Cyril Ponnamperuma, Mitchell K. Hobish,
and Nalinie Wickramasinghe
The Laboratory of Chemical Evolution,
University of Maryland
College Park, MD 29642

ABSTRACT

A study of the association of homocodonic amino acids and selected heterocodonic amino acids with selected nucleotides in aqueous solution was undertaken to examine a possible physical basis for the origin of codon assignments. These interactions were studied using 1H nuclear magnetic resonance spectroscopy (NMR).

The strongest association of all the homocodonic amino acids were with their respective anticodonic nucleotide sequences. The strength of association was seen to increase with increase in the chain length of the anticodonic nucleotide. The association of these amino acids with different phosphate esters of nucleotides suggests that a definite isomeric structure is required for association with a specified amino acid; the 5'-mononucleotides and (3'-5')-linked dinucleotides are the favored geometries for strong associations. Use of heterocodonic amino acids and nonprotein amino acids supports these findings. We conclude that there is at least a physicochemical, anticodonic contribution to the origin of the genetic code.

1. INTRODUCTION

The origin of nucleic acid-directed protein biosynthesis is one of the principal problems in the study of the origin on living systems, for it was a key circumstance in the transition from chemical to biological evolution. Although mechanisms of contemporary protein biosynthesis have been elucidated (Watson, 1975), the origin of this process remains unknown. The central question is: How did this mechanism of translation from information coded in one class of linear chain molecules (nucleic acids) into an entirely different class of molecules (proteins) arise? The mechanism may be at least partially evident in the so-called genetic code.

Speculations about the origin of the genetic code began even before the code was deciphered (Pauling and Delbruck, 1950; Gamow, 1954; Nirenberg and Matthei, 1961). At one extreme of these speculations is the "frozen accident" theory, which states that the code had as its basis a chance event which resulted in a random assignment of amino acids to codons at an early stage of evolution, followed by fixation of these at a later stage (Crick, 1968; Eigen, 1971). At the other extreme lies the "direct interaction" theory which states that the codon assignments are based on preferential physical and chemical interactions between amino acids and component nucleotides of the genetic code (Woese, 1967). Between these views is a virtual continuum of theories (Orgel, 1968; Jukes, 1973; Jukes, 1977).

While a frozen accident origin of the genetic code, though strictly irrefutable, cannot be substantiated in the laboratory as it is not possible to find a basis for experimentation. The direct interaction theory is, however, amenable to experimental verification, as it may be presumed that traces of those early selective interactions are even now discernible, though they may not play any role in the contemporary system. Since the genetic code (Fig. 1) relates a single amino acid to a sequence of three nucleotides (codon or the anticodon), that occurs along a strand of mRNA or tRNA, it may be supposed that amino acids and their codonic or anticodonic triplets could have been among the earliest groupings able to exercise some degree of chemical discrimination. Perhaps such discrimination could take place at the level of monomers or dimers. We chose to investigate this possibility by measuring the degree of interaction between selected amino acids and nucleotides in aqueous solution. Similar studies have been carried out previously, and have yielded the following general conclusions:

1.1 The interaction of amino acids with nucleotides (or derivatives thereof) are selective;

1.2 The interactions are relatively weak, but data are reproducible within and between laboratories;

1.3 Binding data have been obtained only for a few amino acids, most notably the aromatic acids (e.g., phenylalanine and tryptophan);

1.4 NMR is the most appropriate method for further investigation, due to its high sensitivity.

First base	second base: U	C	A	G	Third base
U	Phe	Ser	Tyr	Cys	U
	Phe	Ser	Tyr	Cys	C
	Leu	Ser	C.T.*	C.T.*	A
	Leu	Ser	C.T.*	Trp	G
C	Leu	Pro	His	Arg	U
	Leu	Pro	His	Arg	C
	Leu	Pro	Gln	Arg	A
	Leu	Pro	Gln	Arg	G
A	Ile	Thr	Asn	Ser	U
	Ile	Thr	Asn	Ser	C
	Ile	Thr	Lys	Arg	A
	Met	Thr	Lys	Arg	G
G	Val	Ala	Asp	Gly	U
	Val	Ala	Asp	Gly	C
	Val	Ala	Glu	Gly	A
	Val	Ala	Glu	Gly	G

* C.T. stands for chain terminator

Figure 1. The Genetic Code

With these clues in mind, this study was designed to answer the following questions to a degree of detail not heretofore available.

1.5 Are amino acid-nucleotide interactions preferentially codonic or anticodonic?

1.6 Would any such preference be enhanced by increasing the chain length of the nucleotides?

1.7 Are some geometries of nucleotides preferred over others when challenged by amino acids?

1.8 Are some geometries of amino acids preferred over others when challenged by nucleotides?

2. EXPERIMENTAL PROCEDURES

NMR was the method of choice in this investigation into the associations between amino acids and nucleotides. The utility of NMR as a probe of weak molecular associations in solution resides in the sensitivity that many nuclei display towards different magnetic environments (Kowalsky, 1962; Kowalsky and Cohn, 1964; Jardetzky and Roberts, 1981). The NMR signals of the heterocyclic ring protons of nucleic acid bases are shifted in the presence of a molecule with which the base may interact (Wagner and Lawaczeck, 1972; Reuben, 1978). The extent of this shift depends on the relative concentrations of the interacting groups (Fontaine et al., 1977). Information on the strength of the association can be obtained from analysis of such data. This principle is applied to the experiments described here, where the changes in the chemical shifts of several of the base and anomeric protons of various nucleotides were measures as a function of increasing concentrations of amino acids, at fixed nucleotide concentrations.

3. RESULTS

3.1 ASSOCIATIONS OF HOMOCODONIC AMINO ACIDS WITH NUCLEOSIDE-5'-PHOSPHATAES

There is an association between amino acids and nucleotides is demonstrated by Fig. 2, wherein is shown a typical plot of the change in the chemical shift of a specific nucleotide resonance vs the concentration of added amino acid (in this case, for the L-PHE/5'-AMP system). Similar assays were performed for all combinations of NMPs and amino acids used in this study. Association constants in replicate experiments agreed to within $0.10m^{-1}$.

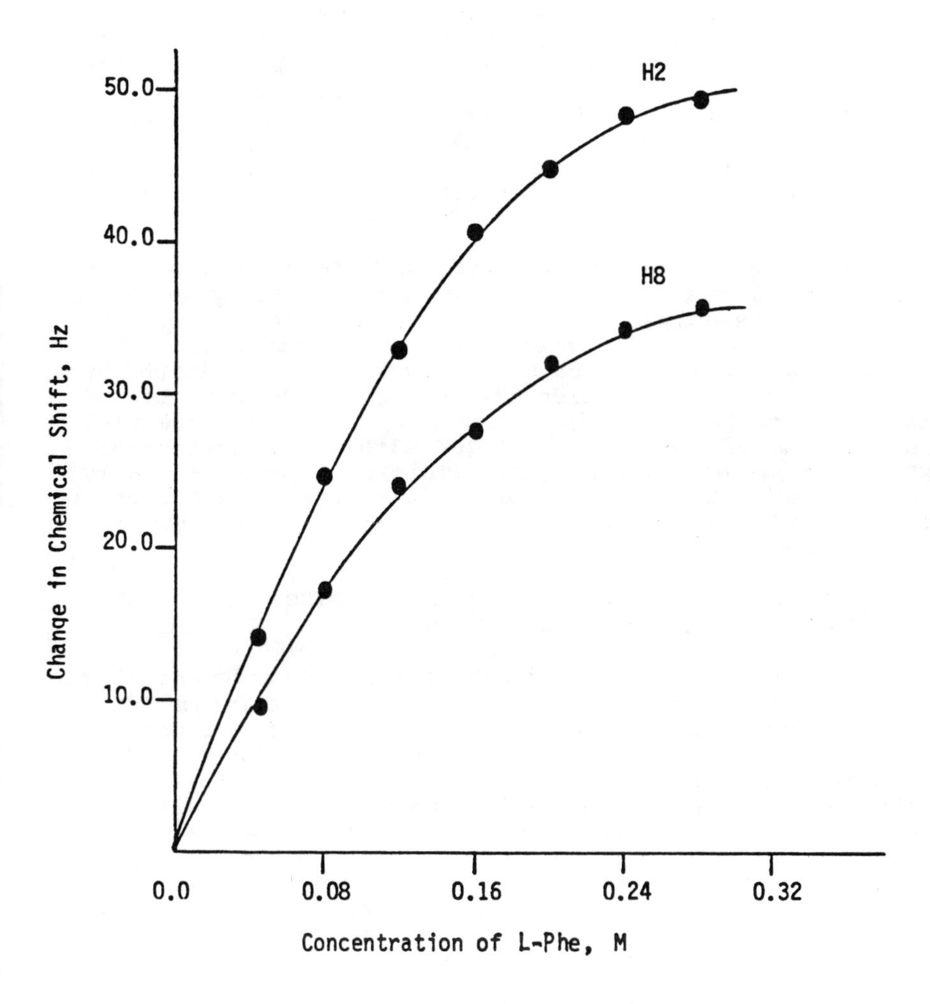

Figure 2. Change in Chemical Shift vs. Concentration of Amino Acid for L-Phe : 5'-AMP at 24°C, pD 7.2.

As may be seen in Fig. 3, the homocodonic amino acids (PHE, LYS, PRO, GLY) associate preferentially with their anticodonic nucleoside-5'-monophosphates. It is noteworthy that PRO and GLY associate only with their anticodonic nucleoside-5'-monophosphates among the four different nucleoside monophosphates investigated, 5'-AMP, 5'-UMP, 5'-CMP, and 5'-GMP. LYS associates most strongly with its anticodonic nucleoside-5'-monophosphate, UMP; a weaker association is seen with CMP. PHE associates with all four nucleoside-5'-monophosphates. At a 95% confidence level, PHE has a stronger affinity for its anticodonic nucleoside-5'-monophosphate, AMP, than for UMP, CMP, or GMP.

The association of PHE with all four NMPs may be attributed to the relatively high hydrophobicity of PHE: the NMPs, themselves being hydrophobic, are likely to associate with it by hydrophobic interactions. The preferential association of PHE with AMP may be explained by the fact that the nucleotide AMP is the most hydrophobic of the four nucleotides investigated. The preferential associations of LYS, PRO, and GLY with their anticodonic NMPs is not as easily explained. This preference is likely due to the combined effect of hydrophobic interactions, electrostatic interactions, hydrogen bonding, and van der Waals forces.

3.2 ASSOCIATIONS OF HOMOCODONIC AMINO ACIDS WITH DINUCLEOSIDE-(3'-5')-MONOPHOSPHATES

All the hocodonic amino acids associated preferentially with their anticodonic DNMPs over the codonic DNMPs (Fig. 4). LYS, PRO, and GLY associate with their anticodonic DNMPs; no association is seen with the codonic DNMPs. Association of PHE with its anticodonic DNMP, (3'-5')ApA, is approximately twenty four-fold stronger than that with the codonic DNMP, (3'-5')UpU. Another important observation is that the associations of homocodonic amino acids with their anticodonic DNMPs are significantly stronger than with their anticodonic NMPs (Fig. 3). Clearly, the anticodonic preference is enhanced when the number of bases in the anticodonic nucleotide is increased from one to two.

3.3 ASSOCIATION OF AMINO ACIDS WITH DIFFERENT PHOSPHATE ESTERS AND PHOSPHODIESTERS OF NUCLEOSIDES

It is apparent that the location of the phosphate group in the nucleotide has a significant influence on the strength of the association with an amino acid. The 5'-phosphate esters of ribonucleosides are the favored structures for stronger associations with amino acids (Fig. 5). The 3'-phosphate esters of ribonucleosides associate

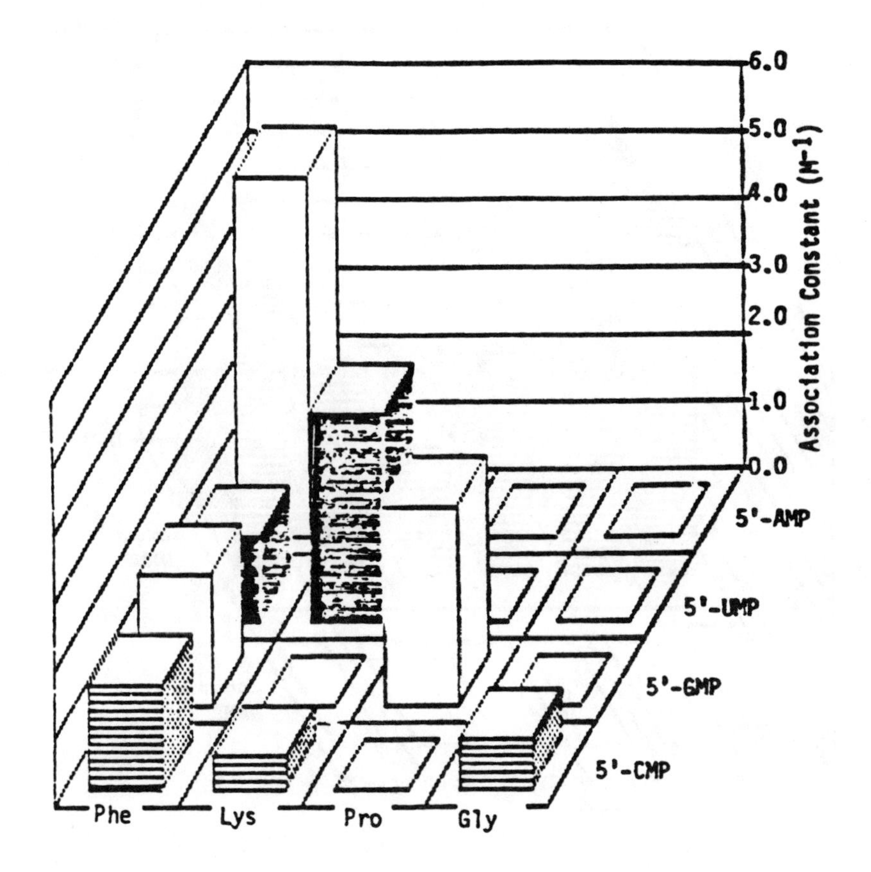

Figure 3. Associations of amino acids with nucleoside 5'-monophosphates.

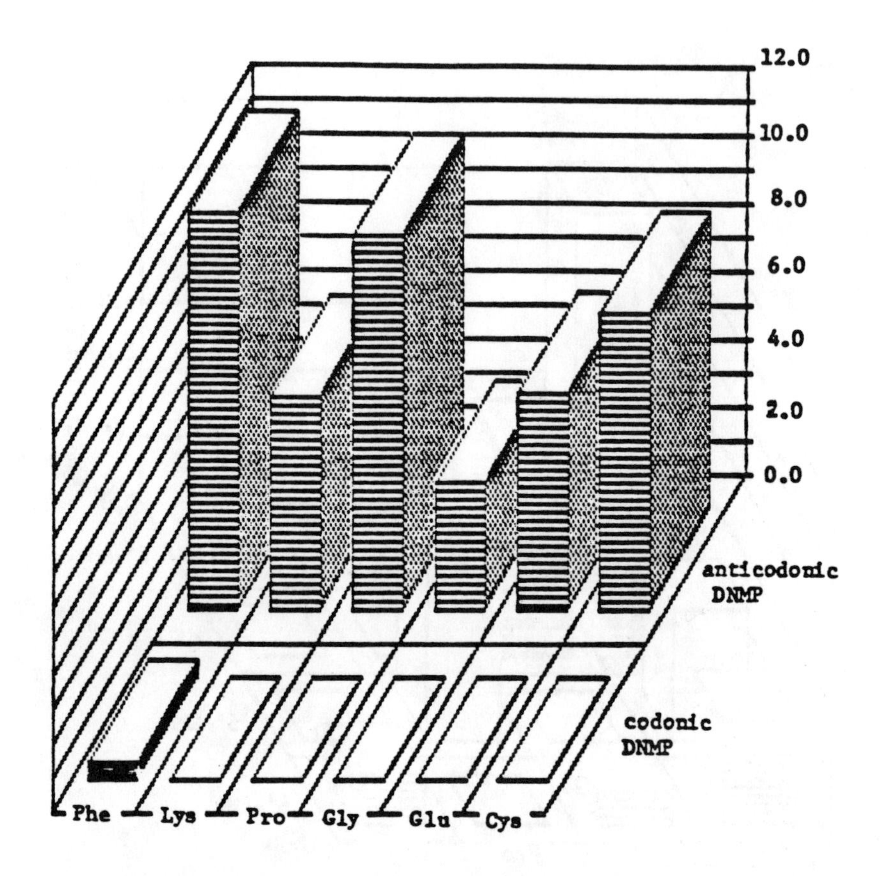

Figure 4.
Associations of amino acids with codonic and anticodonic dinucleoside monophosphates.
The association constants used are the averages of all the association constants at different sites of the nucleotide investigated.
DNMP = Dinucleoside monophohphate

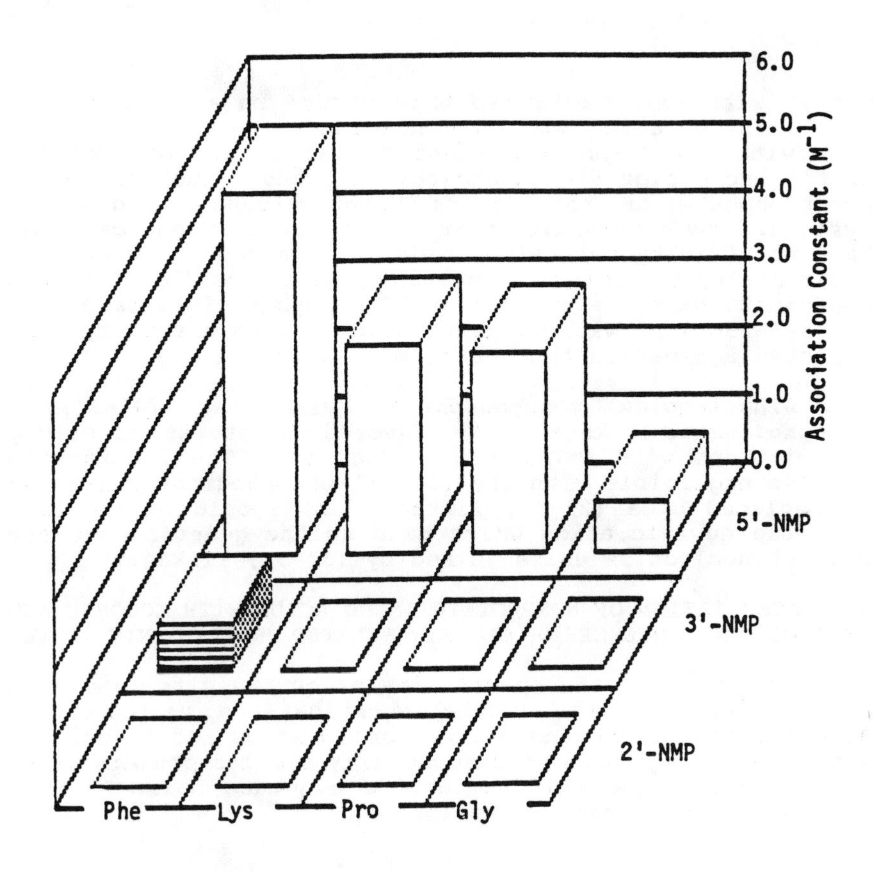

Figure 5.

Associations of amino acids with different isomers of anticodonic nucleoside monophosphates.

The association constants used are the average association constants.

NMP = Nucleoside monophosphate

weakly with amino acids in this study; no association is seen with the amino acids in this study; no association is seen with the 2'-phosphate ester. It should be recalled that nucleoside-5'-monophosphates and nucleoside-3'-monophosphates are the isomers commonly found in biological systems. The observation that nucleoside-2'-monophosphates have no affinity for amino acids in this assay may reflect their rarity in biological systems. It is possible that the conformation of the nucleoside-2'-phosphate is unfavorable for interactions with an amino acid and, as a result, were selected against in the course of evolution.

Dinucleoside monophosphates with the (3'-5') - phosphodiester linkage are the favored structures for strong associations with amino acids (Fig. 6). The interaction becomes negligible with the (3'-5')-phosphodiester linkage is replaced by a (2'-5') linkage. It should be recalled that the nucleic acids which make up the genetic code are made of nucleotide units joined by (3'-5') linkages.

3.4 ASSOCIATION OF NONPROTEIN AMINO ACIDS WITH CODONIC AND ANTICODONIC SEQUENCES OF CLOSELY RELATED PROTEIN AMINO ACIDS

As a test of the specificity of our results, and as a preliminary foray into further investigations, we looked for associations between nonprotein amino acids and codonic or anticodonic sequences of structurally similar protein amino acids. The amino acid/nucleotide sequence pairs studies showed no demonstrable association.

4. DISCUSSION

It is evident that there are reproducibly measurable interactions between the amino acid and nucleotide sequences studied here. Although these interactions are weaker than those typical of binding interactions found in present-day living systems, these data are consistent and reproducible. Furthermore, our results agree well with other data reported in the literature (Lacey et al., 1984).

Also noteworthy is the observation that the associations at different sites of the same nucleotide are markedly different. We surmise that the association constant as determined by a given proton's chemical shift data may be a measure of how tightly the amino acid is held at that position. However, it is also possible that the nucleotide and amino acid are distributed according to different equilibria amongst various bound configurations in which one or the other proton interacts preferentially with the amino acid.

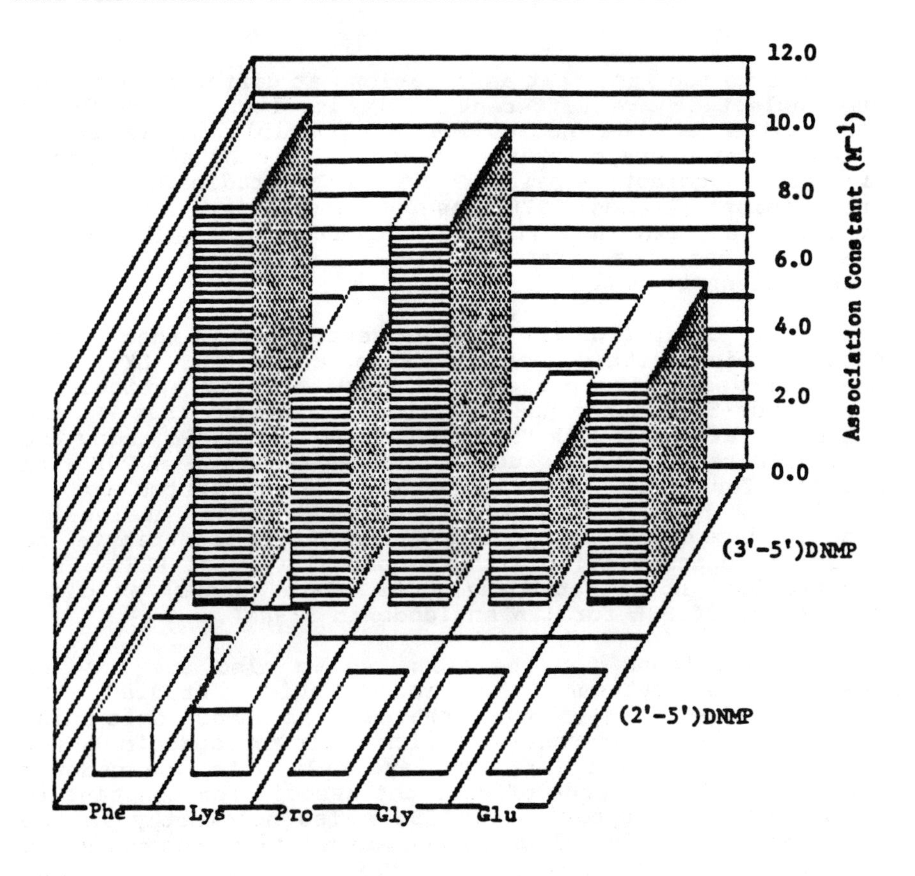

Figure 6.
Associations of amino acids with anticodonic dinucleoside monophosphates and (2'-5') isomers of anticodonic dinucleoside monophosphates.
The association constants used are the averages of all the association constants at different sites of the nucleotide investigated.
DNMP = Dinucleoside Monophosphate

Due to the fact that associations at different sites of the nucleotide are different it is important to study as many sites of the nucleotide as possible in order to calculate an average apparent association constant for each amino acid-nucleotide pair. Previous NMR studies of similar interactions (Reuben, 1978; Dasgupta and Podder, 1979; Lacey et al., 1984) examined only one such resonance. This study has provided the most complete data sets for such interactions to date.

It is clear from the data presented herein that in all cases examined, the homocodonic amino acids (PHE, LYS, PRO, and GLY) associated preferentially with their anticodonic nucleoside-5'-monophosphates. It is noteworthy that PRO and GLY associate only with their anticodonic 5'-NMPs among the four different 5'-NMPs studied. The association of PHE with all the four NMPs investigated may be attributed to the relatively high hydrophobicity of PHE: The nucleoside bases, themselves being hydrophobic, may contribute to nonspecific association with the phenyl ring of PHE. This observation notwithstanding, the data shows a clear preference of PHE for its anticodonic 5'-NMP.

As was found for the NMPs, so we also see for the associations between the amino acids studied and dinucleoside-(3'-5')-monophosphates: the homocodonic amino acids associate preferentially with their anticodonic DNMPs. Again LYS, PRO, and GLY associate only with the codonic sequences. In the case of PHE, the association constant is by a factor of approximately 25. Also noteworthy is the observation that in all cases the association constants are higher than those found for the NMPs. These observations would seem to indicate that the anticodonic preference is real. Indeed, analysis of associations between two heterocodonic amino acids, L-GLU, and L-CYS show the same kind of anticodonic sequence preference, with no observable association with their codonic sequences.

There is also strong evidence for the steroselective nature of amino acid-nucleotide interactions as found in the study of the four heterocodonic amino acids, ALA, ARG, Met, and TRY, for which the anticodonic and codonic dinucleoside monophosphates have the same structure. In these cases selectivity was found in the stronger association with the (3'-5') isomer than with the inverted (5'-3') sequence. In the case of MET, there was no observable binding with the (5'-3') sequence at all. In one case examined with a TNDP, TYR shows a five-fold enhancement of association with the TNDP over the DNMP. Further, studies show that the location of the phosphate group in the nucleotide has a strong influence on the association: The 5'-phosphate esters of ribonucleosides are favored. The 3'-phosphate esters show some binding, but less than the 5'-phosphate esters, and no association is observed with 2'-phosphate esters. Using DNMPs with a (2'-5') linkage results in negligible association. (Fig. 7)

A fascinating observation is that the use of an amino acid in the D enantiomeric configuration shows no evidence of association with its anticodonic sequence CpU(3'-5'). This strongly suggests that a specific enantiomeric structure is required for these amino acid-nucleotide interactions to take place. Further evidence for structural requirements come from the studies of CYS and homoCYS, which differs from CYS only in the addition of a CH_2 group is its side chain: CYS associates well with ApC(3'-5'), but homoCYS shows no such association. Similar results are seen with lysine and ornithine. Clearly there is a definite structural requirement for these specific, in-solution interactions to occur.

5. SUMMARY

The results of this study provide strong support for a physical basis for the origin of the genetic code. The important findings may be summarized as follows:

5.1 The associations between homocodonic amino acids and nucleotides are selective and preferentially anticodonic. With even as simple a system as a mononucleotide and an amino acid there appears to be this preferred association with the anticodonic nucleotide.

5.2 The anticodonic preference over the other nucleotides is enhanced significantly with dinucleoside monophosphates.

5.3 A particular nucleotide structure is required for association with an amino acid. The nucleoside-5'-monophosphates and the (3'-5') linked DNMPs are the preferred geometries for strong associations. These are the geometries most commonly found in biological systems. Nucleoside-2'-phosphates show no association with amino acids, and dinucleoside-(2'-5')-monophosphates show poor associations with amino acids. These findings provide a possible explanation for why the present genetic material consists only of (3'-5') linked nucleotides.

5.4 Preliminary data suggest that a particular enantiomeric structure of the amino acid is required for specific associations with nucleotides.

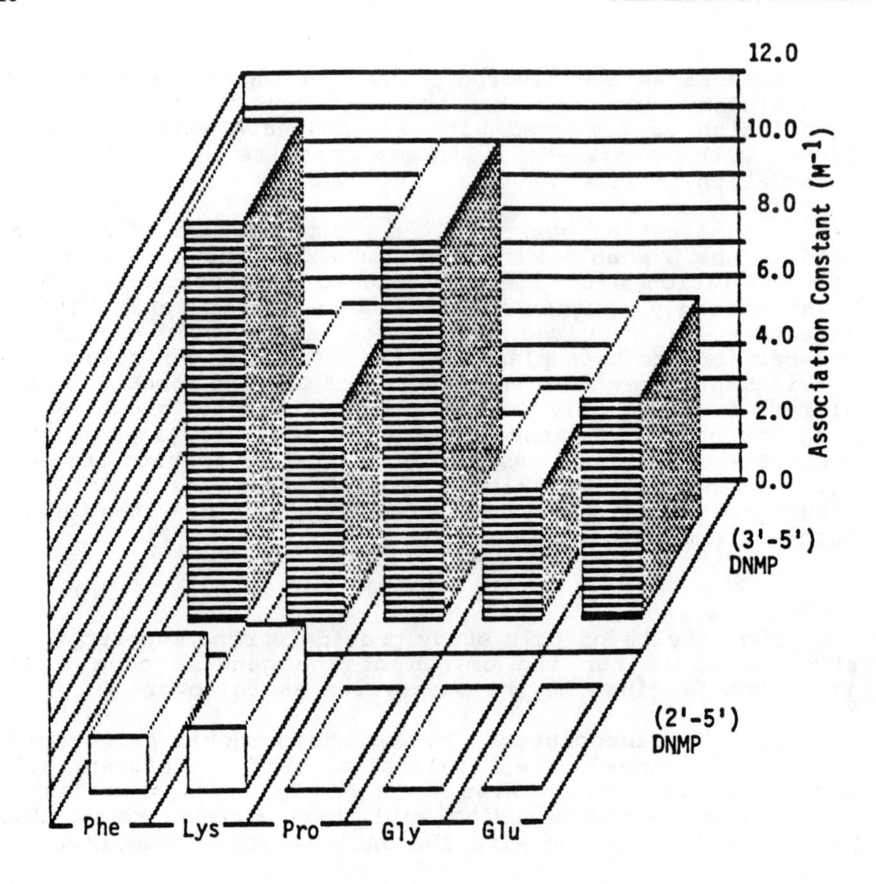

Figure 7. Associations of amino acids with anticodonic dinucleoside monophosphates and (2'-5') isomers of anticodonic dinucleoside monophosphates.

The association constants used are the average association constants.

DNMP = Dinucleoside monophosphate

These data have provided new insight into a possible anticodonic basis for the origin of the genetic code. It is possible that the selective associations of amino acids and anticodonic nucleotides allow selective activation of amino acids in a primitive translation system, establishing a kinetic means of genetic coding.

The exact nature of the interactions between amino acids and their anticodonic nucleotide sequences remains to be determined. Two-dimensional NMR spectroscopic methods may provide the best probe in such a study. Further insight would be gained by experiments designed to detail the contributions of various groups on the various molecules to the thermodynamics of the association, taking into account the differential ionization states of those groups which titrate in the physiological pH range.

The results presented in this paper provide strong evidence for a physical basis for the origin of codon assignments in the genetic code, albeit from the standpoint of the anticodon.

REFERENCES

Crick, F.H.C., J. Mol. Biol. 38 367, 1968

Dasgupta, G., and Podder, S.K., Indian J. Biochem. Biophys. 16 316, 1979

Eigen, M., Naturwissenschaften 58 465, 1971

Gamow, G. Nature 173 318, 1954

Jardetzky, O., and Roberts, G.C.K., in NMR in Molecular Biology, Academic Press, Inc., NY 1981

J ukes, T.H., Nature 262 22, 1973

Jukes, T.H., Comp. Biochem. 24 235 1977

Kowalsky, A., J. Biol. Chem. 237 1087, 1962

Kowalsky, A., and Cohn, M., Ann. Rev. Biochem. 33 481, 1964

Lacey, J.C., Mullins, D.W., and Kahled, M.A., Origins of Life 14 505, 1984

Nirenberg, M.W., and Matthei, J.H., Proc. Nat. Acad. Sci. 47 1588, 1961

Orgel, L.E., J. Mol. Biol. 38 381 1968

Pauling, L., and Delabruck, M., Science 92 77, 1950

Reuben, J., FEBS Letters 94 20, 1978

Wagner, K.G., and Lawaczeck, R., J. Mag. Res 8 164, 1972

Watson, J.D., in Molecular Biology of the Gene, W.A. Benjamin, Inc., 3rd edition, 1975

Woese, C.R., in, The Genetic Code, Harper and Row, NY 1967

ROLE OF NITRILES AND OTHER REACTIVE MOLECULES IN CHEMICAL EVOLUTION

Mohindra S. Chadha
Bhabha Atomic Research Centre
Bombay-400 085, India.

ABSTRACT

Work in the area of chemical evolution during the last 4 decades has revealed the formation of a large number of biologically important molecules produced from simple starting materials under plausible primitive earth conditions. Much of this work has resulted from studies carried out under conditions meant to simulate those prevailing on the early Earth, other planets or their satellites. During the last 25 years, identification of chemical constituents of interstellar medium has revealed that a number of the molecules identified are the same as characterised in laboratory experiments and bear a chemogenic relationship to those discovered in the meteorites. Investigations on the carbon chemistry of some meteorites have also provided extremely valuable information leading to a coherent picture substantiating the tenets of Chemical Evolution as enunciated by some of the early workers in this area.

Eventhough the conditions of the laboratory experiments are vastly different from those of the cool, low density interstellar medium, or those under which the chemical constituents of meteorites were produced, yet the similarities are too obvious to go unnoticed. In the present survey an attempt has been made to highlight the commonness of occurrence or formation of some of the molecules like NH_3, HCN, H_2CO, $HC \equiv C-CN$ etc. Also their reactivities and the nature of their products under different conditions are discussed.

1. INTRODUCTION

Oparin-Haldane's model for a highly reducing early Earth atmosphere received substantial support from Urey's CH_4-NH_3-H_2-H_2O model. This led to Urey-Miller experiments in 1953, which yielded dramatic results. Serious doubts have however, been expressed in recent years about the highly reducing nature of the early

Earth atmosphere. The more acceptable models adopt a CO (CO_2), N_2 and H_2O scenario with only transitory or localised existence of more reducing conditions. Nevertheless, experimental data from a variety of studies using mixtures of simple sources of C and N and also H and O under different conditions of energy inputs are available. Also studies on simulated Jovian atmosphere and Titan atmosphere have brought into focus the key role which some of the molecules like HCN, HC ≡ CH, HC ≡ C-CN would have played in cosmic chemistry. The importance of H_2CO in the early Earth experiments is well documented. An understanding of the formation and reactivities of these and related molecules together with identification of many of these molecules in the interstellar medium leads one to believe that it is the chemical nature of such molecules which dictates the nature of more complex products (like those observed in the primordial Earth experiments or characterised from the meteorites). The chemical nature of the precursor molecules provides a rationale for the formation of amino acids, nucleic acid bases, nucleosides, nucleotides, polypeptides and polynucleotides. After this is established, the next formidable task is to find the transition from products of chemical evolution to the formation of early cell, an area in which serious effort is being made in many laboratories.

A survey of the existence of molecules like H_2, N_2, CO, CH_4 and NH_3; the formation of HCN, H_2CO, HC≡CH, HC ≡ CCN; and reactions of compounds like HCN, H_2CO, HC ≡ CH-CN to give aminonitriles, purines and pyrimidine and the role of water in the formation of amino acids and hydroxy acids is given below.

2. OCCURRENCE AND FORMATION

2.1 HYDROGEN

The ubiquitous existence of hydrogen in the composition of the Sun and the universe at large bestows on it a key role. Urey (1952) had postulated that the atmosphere of the primitive Earth was rich in hydrogen. The two preponderant elements which constitute the Sun are H_2 (87.0 %) and He (12.9%). Even the interstellar clouds are predominently hydrogen and helium. It is safe to state that without abundant H_2, interstellar chemistry would be quiet depleted. More than 80 molecules have so far been characterised in the interstellar medium (Taylor and Williams, 1993) and more are bound to be discovered. Of these, nearly

80% have hydrogen associated with them. It is presumed that the main processes by which the chemistry is initiated are through ion-molecule reactions (Duley and Williams, 1984). Some of the intersteller molecules relevant to our discussion on chemical evolution are given in Fig. 1.

H_2	CO	HC≡C–CN
HCN	H_2O	H_2C=CH–CN
NH_3	CH_2O	H_2C=NH
CH_4	HCO_2H	H_3CCN, H_3CNH_2

Fig. 1. Some intersteller molecules of interest in Chemical Evolution.

2.2 NITROGEN

Nitrogen constitutes about 0.02% of the composition of Sun, 2.7% of Mars, 3.5% of Venus and 82-94% of Titan's atmosphere. The present day composition of the Earth's atmosphere has ≃ 80% N_2, but if its atmosphere were reconstructed (no weathering, no life) it would be about 1.9% according to Owen (1982).

Nitrogen is an important component of more than 30% of all interstallar molecules found so far including HCN and HC ≡ C-CN and it is also richly represented in the cometary material.

2.3 METHANE AND AMMONIA

While the presence of methane and ammonia in the primitive Earth's atmosphere was strongly favoured by Urey (1952), but a number of geologic, thermodynamic and geochemical considerations by Rubey (1955), Holland (1962) and Abelson (1966) have cast serious doubts on this. Nevertheless, there is strong evidence for the existence of NH_3 in the interstellar region and that of CH_4 in Titan's atmosphere (Owen, 1982). According to a recent model, methane could have resulted from the volcanic outgasing on the primitive Earth if the upper mantle was more reduced than the present mantle (Kasting, 1993).

2.4 HYDROGEN CYANIDE

Hydrogen cyanide has been observed in volcanic emissions (Mukhin, 1974). Hydrogen cyanide and its congeners CN, HNC, NH_2CN, C_2CN endow the intersellar medium very richly, also CN, CN^+, HCN and CH_3CN have been characterised in the comets whereas Titan's atmosphere contains HCN, C_2N_2 and HC_3N.

Formation of HCN in simulated Jovian atmosphere experiments was demonstrated by Woeller and Ponnamperuma (1969) and in the simulated Titan atmosphere by Gupta et al. (1981). In the arc discharge experiments through CH_4 and NH_3, HCN is one of the major product, while nitriles are also formed (Ponnamperuma and Woeller, 1967).

Formation of HCN and other nitriles by Fischer Tropsch reaction of CO, NH_3, H_2 at 500°C has been demonstrated by Anders et al. (1974), by photolysis of a mixture of CH_4 + NH_3 by Ferris and Chen (1975a) and photolysis of NH_3 in the presence of acetylene by Ferris and Ishikawa (1988). There are numerous other scenarios which also demonstrate the ease of formation of HCN.

2.5 CARBON MONOXIDE

The presence of carbon monoxide in the early Earth's atmosphere has been attributed to volcanic eruptions. It is a component of volcanic eruptions even today, though carbon dioxide predominates due to the high degree of oxidation of the Earths mantle.

Carbon monoxide has been detected in the interstellar medium not only as $^{12}C^{16}O$ but also in all combinations of ^{12}C and ^{13}C with ^{16}O, ^{17}O and ^{18}O (Turner, 1980).

2.6 FORMALDEHYDE

Garrison et al. (1951) were the first to carry out a simulated early Earth experiment. They irradiated a mixture of CO_2, H_2, Fe^{2+} with He atoms in a cyclotron and detected the formation of traces of CH_2O and formic acid.

Formaldehyde and higher aldehydes are formed by the photolysis of H_2O in the presence of CH_4 (Ferris and Chen 1975b), photolysis of $CO-H_2O$ mixture (Park and Getoff, 1988), the photolysis of methanol (Allamandola et al. 1988). Schlesinger and Miller (1983) showed CH_2O to be a major product resulting from the passage of electric discharge through mixtures of gases aimed at simulating primitive atmospheres.

Formaldehyde was one of the abundant molecules to be detected in the interstellar medium (Mann and Williams 1980). The presence of polyoxymethylene, a CH_2O polymer has also been inferred on the comet Halley by Huebner (1987).

2.7 CYANOACETYLENE

This three carbon compound was identified by Sanchez, Ferris and Orgel (1966), when an electric discharge was passed through a mixture of N_2 and CH_4.

Though the atmosphere of Titan has been shown to have acetylene, the formation of cyanoacetylene in a simulated Titan atmosphere experiment was definitely established by Gupta et al. (1981) wherein electric discharge was passed through CH_4 and N_2. This was in addition to HCN, C_2H_2, CH_3CN and a number of olefins as products in their experiment.

2.8 AMINONITRILES

The presence of NH_2CN, in the interstellar medium has been demonstrated. The formation of $H_2N\ CH_2CN$ on the passage of semi-corona or arc discharge through a mixture of CH_4 and NH_3 was reported by Ponnamperuma and Woeller (1967). The formation of $H_2N\ CH_2\ CN$ and its C-methyl and N-methyl substituented compounds reported by the same authors was confirmed further by GC-MS studies of the hydrolytic products of the resulting mixture (Chadha et al. 1971a and 1971b).

2.9 METHYLAMINE, METHYLCYANIDES AND METHANIMINE

Both methylcyanide and methylamine as well as their homologues have been identified in experiments where electric discharge was passed through mixtues of CH_4 and NH_3 (Ponnamperuma and Woeller 1967, Chadha and Ponnamperuma 1970 - unpublished). Methanimine can be considered as an intermediate in the formation of aminoacetonitrile from CH_2O, NH_3 and HCN (Chadha et al. 1971c). All these compounds have been identified in the interstellar medium.

2.10 CYANOETHYLENE (VINYL CYANIDE)

Cyanoethylene has been identified in the interstellar medium and has been shown as a likely intermediate in the formation of β-alanine and of pyrimidines in the Jovian atmosphere experiments as discussed by Chadha et al. (1975).

3. REACTIONS OF SOME OF THE ABOVE MOLECULES AS SUCH OR IN COMBINATION

3.1 METHANE, AMMONIA, WATER

Passage of electric discharge through a mixture of CH_4, NH_3, H_2 and water vapour gave a number of amino acids (Miller, 1953). Aminonitriles appear to be the intermediates in the formation of some of the amino acids (Ponnamperuma and Woeller 1967).

When a mixture of CH_4 and NH_3 in experiments meant to simulate Jupiter atmosphere was subjected to semicorana or electric discharge, a number of compounds e.g. HCN, $HC \equiv CH$, aminonitriles (Ponnamperuma and Woeller 1967) were formed, some of which were shown to be precursors of amino acids (Chadha et al. 1971a, 1971b).

The synthesis of adenine by the electron bombardment of a mixture of CH_4, NH_3, H_2O and H_2 was reported by Ponnamperuma et al. (1963).

3.2 HYDROGEN CYANIDE

Hydrogen cyanide in aqueous ammonia solution has been shown to be the precursor of adenine by Oro and Kimball (1961). When HCN was exposed to UV, the formation of adenine and guanine was reported by Ponnamperuma (1965). The self condensation of HCN in mildly basic conditions has been shown to give diaminomaleonitrile (DAMN), which is the key intermediate in the formation of purines (Ferris and Orgel, 1966). The formation of adenine and guanine from HCN is rationalised on logical reaction mechanisms.

Several workers (Abelson, 1966, Matthews and Moser, 1966, 1967, Harada, 1967) have discussed the role of HCN oligomers in prebiotic chemistry. In a comprehensive study, HCN oligomer on hydrolysis has been shown to give purines, pyrimidines and a number of amino acids (Ferris et al. 1978). A solution of HCN in water on -irradiation, followed by hydrolysis, was shown to give glycine, alanine, valine, serine, threonine, aspartic acid and glutamic acid (Sweeney et al., 1976; Draganic et al., 1978). HCN, its tetramer and a number of related compounds (cyanamide, dicyanamide, carbodiimide, dicyandiamide and cyanogen) are known to be efficient condensing agents in prebiotic experiments (Hulshof and Ponnamperuma, 1976). HCN is indeed a kingpin in the formation of a number of biologically important molecules of prebiotic origin.

The possible role of HCN in a Strecker type synthesis in Miller's experiments (1953, 1957) or via aminoacetonitrile in the Jovian atmosphere experiments (Chadha et al., 1971) is inescapable. This can be explained as outlined in Fig. 2.

$$RCHO + NH_3 + HCN \rightleftharpoons RCH(NH_2)CN + H_2O$$

$$RCH(NH_2)CN + 2\,H_2O \longrightarrow RCH(NH_2)COOH + NH_3$$

Fig. 2. Possible role of HCN in Strecker type synthesis.

The validity of the above equilibria is demonstrated through the incorporation of alanine if AAN is subjected to hydrolytic conditions as was shown by Chadha et al. (1971), where iminoacetic propionic acid was shown to be one of the products. This is illustrated in Fig. 3.

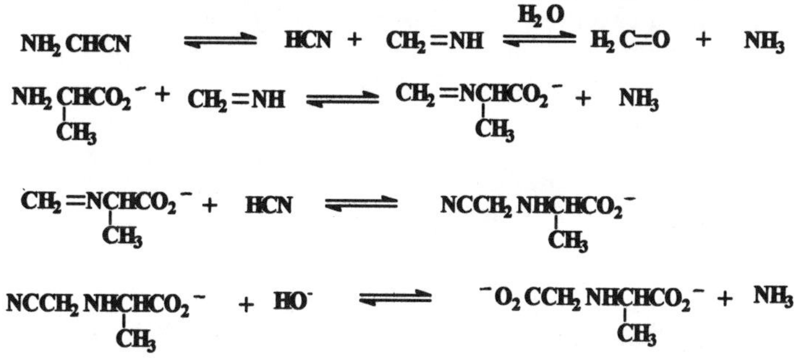

$$NH_2CHCN \rightleftharpoons HCN + CH_2{=}NH \overset{H_2O}{\rightleftharpoons} H_2C{=}O + NH_3$$

$$NH_2\underset{\displaystyle CH_3}{\underset{|}{C}}HCO_2^- + CH_2{=}NH \rightleftharpoons CH_2{=}N\underset{\displaystyle CH_3}{\underset{|}{C}}HCO_2^- + NH_3$$

$$CH_2{=}N\underset{\displaystyle CH_3}{\underset{|}{C}}HCO_2^- + HCN \rightleftharpoons NCCH_2NH\underset{\displaystyle CH_3}{\underset{|}{C}}HCO_2^-$$

$$NCCH_2NH\underset{\displaystyle CH_3}{\underset{|}{C}}HCO_2^- + HO^- \rightleftharpoons {}^-O_2CCH_2NH\underset{\displaystyle CH_3}{\underset{|}{C}}HCO_2^- + NH_3$$

Fig. 3. Possible role of methyleneimine in Chemical Evolution

The above 2 schemes bring out the importance of a number of molecules viz. HCN, NH_3, H_2O, $H_2C=O$, $CH_2=NH$, and NH_2 CH_2CN in the formation of amino acids via the Strecker type synthesis.

3.3 AMINOACETONITRILES

The reactivity of aminoacetonitrile (AAN) or its role as an intermediate in the formation of amino acids has been brought out in the above discussions. The presence of 2 reactive functional groups viz. NH_2 and CN in this simple molecule bestows it with high degree of reactivity. Its ability to polymerise into high molecular weight complex compounds possibly through amidine groups has been speculated (Chadha et al. 1971c), which may have structural features like those given below (Fig. 4).

$$(NH_2CH_2CN)_n \longrightarrow NH_2CH_2(\text{-}\overset{NH}{\overset{\|}{C}}\text{-NHCH}_2\overset{NH}{\overset{\|}{C}}\text{NHCH}_2)_{n\text{-}x}CN$$

Fig. 4. Amidine polymers from AAN

The formation of heterocyclic compounds from AAN cannot, however be ruled out.

The formation of amidinic polymers is demonstrated through the hydrolysis of polymeric materials, which under mild conditions should give peptidic material and then amino acids under more drastic conditions. These could be represented as shown in Fig. 5.

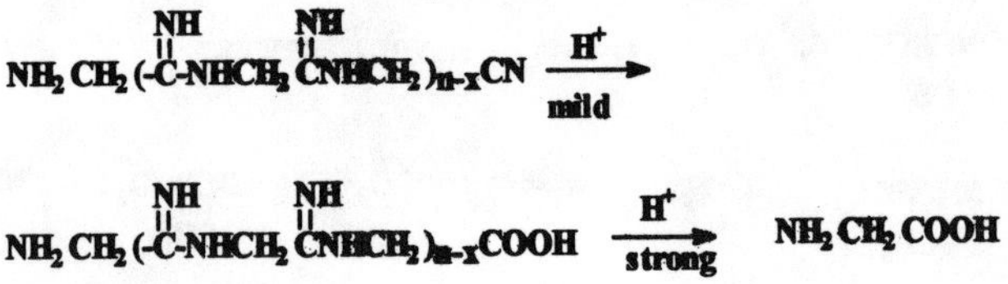

$$NH_2CH_2(\text{-}\overset{NH}{\overset{\|}{C}}\text{-NHCH}_2\overset{NH}{\overset{\|}{C}}\text{NHCH}_2)_{n\text{-}x}CN \xrightarrow[\text{mild}]{H^+}$$

$$NH_2CH_2(\text{-}\overset{NH}{\overset{\|}{C}}\text{-NHCH}_2\overset{NH}{\overset{\|}{C}}\text{NHCH}_2)_{n\text{-}x}COOH \xrightarrow[\text{strong}]{H^+} NH_2CH_2COOH$$

Fig. 5. Peptides and amino acids from amidine polymer

It has been shown (Woeller and Ponnamperuma, 1969) that in the unpolymerised products of CH_4-NH_3 discharge experiments, compounds like propionitrile, C-methyl and N-methylacetonitrile are formed. Their incorporation into polymeric material and the formation of alanine and sarcosine in the products of hydrolysis of the polymer can be easily explained (Fig. 6).

NH₂CHCN
R
+
R'NHCH₂CN
NH R NH
NH₂CH(-C-N-CHCNHCH₂)CN
R R'
H⁺
NH₂CHCOOH
R
+
NH-CHCOOH
R' R
etc.
R = H or CH₃
R' = H or CH₃

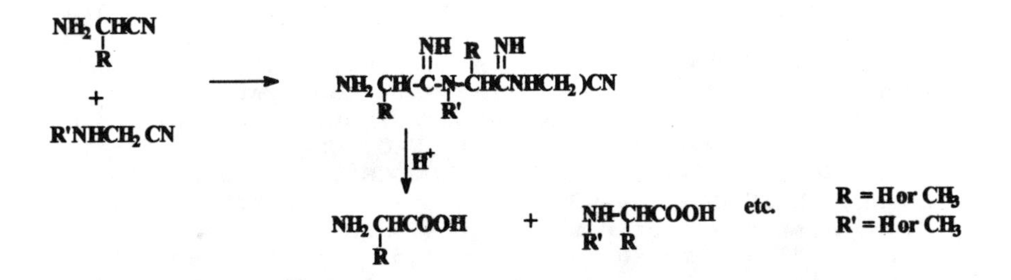

Fig. 6. Formation of substituted amino acids from amidine polymers.

The precursor to the aspartic acid could be $H_2N-\overset{CH_2-CN}{CH}-CN$, which although not detected in the CH_4-NH_3 reaction, can be easily visualised on simple chemical considerations, it is logical to infer that the species like H_2N, CH_2, CN resulting from the high energy reactions could interact to give the dicyano precursor.

3.4 CYANOACETYLENE

Cyanoacetylene has been shown by Ferris et al. (1968) to be a precursor of aspartic acid and cytosine (Fig. 7).

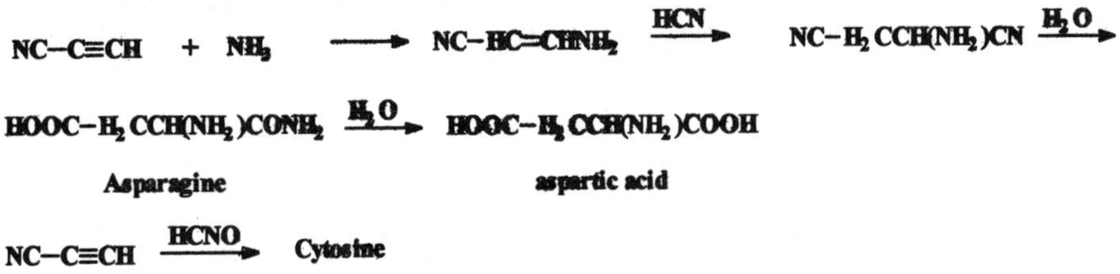

NC−C≡CH + NH₃
NC−HC=CHNH₂
HCN
NC−H₂CCH(NH₂)CN
H₂O
HOOC−H₂CCH(NH₂)CONH₂
H₂O
HOOC−H₂CCH(NH₂)COOH
Asparagine
aspartic acid
NC−C≡CH
HCNO
Cytosine

Fig. 7. Cyanoacetylene in aspartic acid and cytosine synthesis.

3.5 CYANOETHYLENE

Cyanoethylene is the likely intermediate in the formation of -alanine in the Jovian atmosphere experiments as discussed by Chadha, Molton and Ponnamperuma (1975).

3.6 CARBON MONOXIDE

According to Anders (1977), carbon monoxide is the stable form of carbon at high temperature, whereas if equilibria were attained on cooling, then CO would transform to CH_4, by reacting with the very abundant hydrogen. Anders has shown that most of the interstellar molecules, namely H_2O, $CH_3C \equiv CH$, CH_3OH, CH_2O, CH_3CHO, CH_3OCH_3, HCOOH, $HCOOCH_3$, NH_3, HCN, HNCO, CH_3NH_2, $CH{\equiv}C\text{-}CN$, $CH_2{=}CH\text{-}CN$ are seen in Fischer-Tropsch Type syntheses or in meteorites. In fact, Anders et al. (1974) listed a number of small molecules identified in their syntheses which were subsequently discovered in interstellar space : C_2H_5OH, CH_3CHO, CH_3OCH_3, HCO_2CH_3, CH_3NH_2 and $CH_2{=}CH\text{-}CN$. According to Anders et al. (1973), organic compounds in meteorites seem to have been formed by catalytic reactions of CO, H_2 and NH_3 in the Solar nebula at 360-400°K at (4 to 10) x 10^{-6} atm. They conclude that these reactions may be a source of prebiotic carbon compounds on the inner planets and interstellar medium.

3.7 FORMALDEHYDE

The formation of formaldehyde from CO and H_2 and from CH_4 and H_2O as well as in Miller Type experiments is well known. Its polymerisation to a variety of sugars (trioses, tetroses, pentoses and hexoses has been established by Gabel and Ponnamperuma (1967). More recently the role of the dimeric compound glycolaldehyde to yield specific pentameric and hexameric sugars has been emphasised by Eschenmoser's group (Muller, D. et al. 1990).

The important role of CH_2O in the Strecker synthesis has already been discussed above. Also the possible role of CH_2O in the formation of serine from glycine and of hydroxymethyluracil has been demonstrated (Choughuley et al. 1972). Formic acid, a product of prebiotic experiments and also a compound already identified in the interstellar medium can be a useful reducing agent as shown by the hydride type reduction of hydroxymethyluracil (HMU) to thymine (Choughuley et al. 1977).

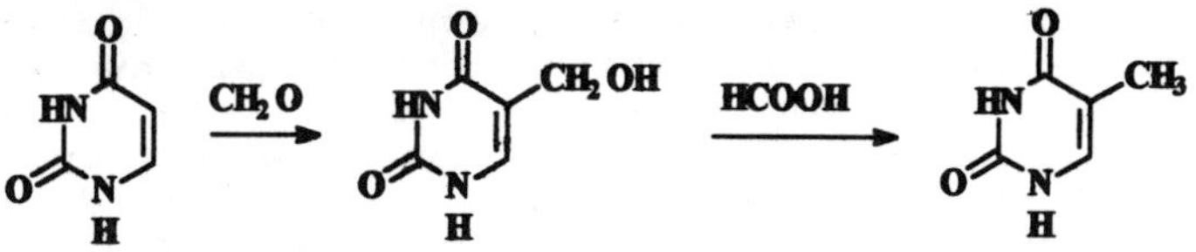

Uracil $\xrightarrow{CH_2O}$ HMU $\xrightarrow{HCOOH}$ Thymine

Fig. 8. Formaldehyde and formic acid in alkylation reactions

4. SIMILARITIES IN THE NATURE OF PRODUCTS ON THE INTERSTELLAR MEDIUM, METEORITES AND CHEMICAL EVOLUTION EXPERIMENTS

In the discussion above frequent references have been made to the nature of chemical components characterised in the interstellar medium and those resulting from the chemical evolution experiments in the laboratory. For this to happen, inspite of extreme differences between the conditions is quite remarkable. According to Townes (1971), a typical dark cloud with 10 H_2 molecules cm^{-3} might have an abundance of CO $\sim 10^{-1}$ to $10^{-2} cm^{-3}$, HCN $\sim 10^{-4}$ cm^{-3}, $NH_3 \sim 10^{-3}$ cm^{-3}; HCHO $\sim 10^{-4}$ to 10^{-5} cm^{-3}. Also the conditions of interstellar clouds are vastly different (temp., radiation flux etc) as compared to those of the terrestrial planets and in the laboratory.

Notwithstanding the vast differences discussed above, the discovery of such a large number of organic molecules in the interstellar space reveals the universality of organic compounds. Chemical evolution has advanced much beyond that expected in the cold ($\simeq$ 20 k), high vacuum conditions ($\lesssim 10^{-5}$ torr) that prevail in the interstellar clouds. Already fairly complex organic compounds have been discovered in the interstellar medium, and it is not unlikely that aminoacetonitrile (8 atoms), glycine (10 atoms) and related compounds may be found in the interstellar medium.

The universality and inevitability of the formation of some organic compounds having direct bearing on the formation of precursors of polymeric compounds like proteins, nucleotides, sugars and lipids which are essential for life today has become emphatically clear from studies on the carbonaceous chondrites during recent years. The most dramatic results were obtained by Ponnamperuma's group from studies on Murchison meteorite (Kvenvolden et al. 1970) followed by those on Murray. Their findings have been confirmed and extended by many other scientists. The presence of amino acids and their precursors, mono and dicarboxylic acids, N-heterocyclics and hydrocarbons as well as many other compounds demonstrates the range of chemistry which has taken place in the outer space. Studies of the Murchison and other meteorites have revealed the presence of at least 40 amino acids with similar abundances of D and L isomers. The population consists of both protein and non-protein amino acids. A comparison of some of the amino acids resulting from the primitive Earth experiments with those from meteorites like Murchison is indeed very revealing (Wolman et al. 1972). These results as well as those discussed earlier in this survey, provide persuasive support for the theory of Chemical Evolution.

ACKNOWLEDGEMENT

Grateful thanks are extended to the Indian National Science Academy, New Delhi for award of INSA Senior Scientist position to the author for carrying out studies on Chemical Evolution and Origin of Life.

REFERENCES

Abelson, P.H. 1966. Chemical events on the primitive Earth. Proc. Natl. Acad. Sci. U.S.A. 1365-1372.

Allamandola, L.J., Sandford, S.A. and Valero, G.J. 1988. Photochemical and thermal evolution of interstallar/precometary ice analogs. Icarus, 76, 225-252.

Anders, E., Hayatsu, R. and Studier, M.H. 1974. Interstellar Molecules : Origin by catalytic reactions on grain surfaces. Astrophysics. J. 192, L101-L105.

Anders, E. 1977. Proceedings of Robert A. Welch Foundation Conference on Chemical Research XXI, Cosmochemistry, p-723.

Chadha, M.S. and Chaughuley, A.S.U. 1984. Synthesis of prebiotic molecules - role of some carbonyl compounds in prebiotic chemistry, Origins of Life, 14, 469-476.

Chadha, M.S. and Ponnamperuma, C. 1970. Formation of methylamine and homologs in experiments in Jovian Atmosphere (unpublished).

Chadha, M.S., Molton, P.M. and Ponnamperuma, C. 1975. Aminonitriles : Possible role in chemical evolution, Origins of Life 6, 127-136.

Chadha, M.S., Lawless, J.G., Flores, J.J. and Ponnamperuma, C. 1971a. Experiments in Jovian Atmosphere in chemical evolution and the origin of life. Eds., R. Buvet and C. Ponnamperuma. North Holland Pub. Co., Amsterdam, 143-151.

Chadha, M.S., Flores, J.J., Lawless, J.G. and Ponnamperuma, C. 1971b. Organic Synthesis in a Simulated Jovian Atmosphere-II. Icarus 15, 39-44.

Chadha, M.S., Replogle, L., Flores, J.J. and Ponnamperuma, C. 1971c. Possible role of Aminoacetonitrile in Chemical Evolution. Bioorganic Chemistry. 1, 269-274.

Chadha, M.S., Subbaraman, A.S., Kazi, Z.A. and Choughuley, A.S.U. 1978. A Possible Prebiotic Synthesis of Thymine in H. Noda. Origins of Life. Centre for Academic Publications, Japan. 197-204.

Choughuley, A.S.U., Subbaraman, A.S. and Kazi, Z.A. 1972. Methyleneaminoacetonitrile : Possible role in Chemical Evolution. Indian J. Biochem. & Biophys. 9, 144-147.

Choughuley, A.S.U., Subbaraman, A.S. and Kazi, Z.A. 1975. Methyleneaminoacetonitrile : Possible role in Chemical Evolution-II. Origins of Life 6, 537-539.

Choughuley, A.S.U., Subbaraman, A.S., Kazi, Z.A. and Chadha, M.S. 1977. A possible synthesis of Thymine : Uracil-formaldehyde-formic acid reaction. Biosystems. 9, 73-76.

Duley, W.W. and Williams, D.A. 1984. Interstellar Chemistry, Academic Press, London.

Ferris, J.P. and Chen, C.T. 1975a. Phytosynthesis of organic compounds in the atmosphere of Jupiter, Nature (London) 258, 587-588.

Ferris, J.P. and Chen, C.T. 1975b. Chemical evolution XXVI. Photochemistry of methane, nitrogen and water mixtures as a model for the atmosphere of the primitive earth. J. Am. Chem. Soc., 2962-2967.

Ferris, J.P. and Ishikawa, Y. 1988. The formation of HCN and Acetylene Oligomers by photolysis of Ammonia in the presence of Acetylene : Application to the Atmosphere Chemistry of Jupiter. J. Am. Chem. Soc., 110, 4306-4312.

Ferris, J.P. and Orgel, L.E. 1966.. An unusual photochemical rearrangement in the synthesis of adenine. J. Am. Chem. Soc., 88, 1074.

Ferris, J.P., Sanchez, R.A. and Orgel, L.E. 1968. Studies in prebiotic synthesis III. Synthesis of pyrimidines from cyanoacetylene and cyanate, J. Molec. Biol. 33, 693-704.

Ferris, J.P., Joshi, P.C., Edelson, E.H. and Lawless, J.G. 1978.. HCN: A plausible source of purines, pyrimidines and amino acids on the primitive earth, J. Molec. Evol. 11, 293-311.

Gabel, N.W. and Ponnamperuma, C. 1967. Model for origin of monosacharides, Nature 216, 453.

Garrison, W.M., Morrison, D.C., Hamilton, J.G., Benson, A.A. and Calvin, M. 1951. Reduction of carbon dioxide in aqueous solutions by ionizing radiation. Science 114, 416-418.

Gupta, S., Ochiai, E. and Ponnamperuma, C. 1981. Organic synthesis in the atmosphere of Titan, Nature 293, 725-727.

Harada, K. 1967. Formation of amino acids by thermal decomposition of formamide-oligomerization of hydrogen cyanide, Nature 214, 479-480.

Hayatsu, R., Studier, M.H., Matsuoka, S. and Anders, E. 1972. Origin of organic matter in the solar system VI. Catalytic synthesis of nitriles, nitrogen bases and porphyrin-like pigments. Geochim. Cosmochim. Acta 36, 555-571.

Holland, H.D. 1962. Model for the evolution of the earth's atmosphere in A.E.J. Engel, H.L. James and B.F. Leonard eds., Petrologic studies: A Volume to Honour A.F. Buddington, New York, Geol. Soc. Am., 44-477.

Hulshof, I. and Ponnamperuma, C. 1975. Prebiotic condensation reactions in an aqueous solution: A review of condensing agents. Origins of Life. 7, 197-224.

Huebner, W.F. 1987. First polymer in space identified in Comet Halley, Science, 237, 628-630.

Kastings, J.F. 1993. Earth's early atmosphere: How reducing was it? Abstracts 10th international conference on the origin of life, Barcelona (Catolonia), Spain. p. 89.

Kastings, J.F. 1990. Bolide impacts and the oxidation state of carbon in the earth's early atmosphere, Origins Life Evol. Biosphere. 20, 199-231.

Kvenvolden, K.A., Lawless, J., Pering, K., Peterson, E., Flores, J., Ponnamperuma, C., Kaplan, I.R. and Moore, C. 1970. Evidence for Extraterrestrial Amino Acids and Hydrocarbons in the Murchison Meteorite, Nature 288, 923-926.

Mann, A.P.C. and Williams, D.A. 1980. A list of interstellar molecules, Nature, 283, 721-725.

Matthews, C.N. and Moser, R.E. 1966. Prebiological protein synthesis. Proc. Natl. Acad. Sci. U.S. 56, 1087-1094.

Matthews, C.N. and Moser, R.E. 1967. Peptide synthesis from hydrogen cyanide and water. Nature 215, 1230-1234.

Miller, S.L. 1953. A production of amino acids under premitive earth conditions, Science 117, 528-529.

Miller, S.L. 1955. Production of some organic compounds under possible primitive earth conditions. J. Am. Chem. Soc. 77, 2351-2361.

Miller, S.L. 1957a. Mechanism of synthesis of amino acids by electric discharges, Biochem. & Biophys. Acta. 23, 480-489.

Mukhin, L.M. 1974. Evolution of organic compounds in volcanic regions. Nature, 251, 50-51.

Muller, D., Pitsch, S., Kittaka, A., Wagner, E., Wintner, C.E. and Eschenmoser, A. 1990. Chemistry of aminonitriles. Aldomerization of glycolaldehyde phosphate to rac-hexose. 2,4,6-triphosphates and (In presence of formaldehyde), rac-pentose 2,4-diphosphates; rac-allose 2,4,6-triphosphate and rac-ribose 2,4'-diphosphate are the main reaction products. Helv. Chim. Acta., 73, 1410-1468.

Oro, J. and Kimball, A.P. 1961. Synthesis of purines under possible primitive earth conditions 1. Adenine from hydrogen cyanide, Arch. Biochem. & biophys. 94, 217-227.

Owen, T. 1982. Planetary atmospheres and the search of life. The Physics Teacher, 90-96.

Park, H.R. and Getoff, N. 1988. Photoinducted transformation of carbon monoxide in aqueous solution, Z. Naturforsch. 43a, 430-434.

Ponnamperuma, C. 1965. In Origins of Prebiological systems, Academic Press, New York. 221.

Ponnamperuma, C. and Woeller, F. 1967. o-Aminonitriles formed by an electric discharge through a mixture of anhydrous methane and ammonia. Currents Modern Biol., 1, 156-158.

Ponnamperuma, C., Lemmon, R.M., Mariner, R. and Calvin, M. 1963. Formation of adenine by electron irradiation of methane, ammonia and water. Proc. natl. Acad. Sci. U.S. 49, 737-740.

Rubey, W.W. 1955. Development of the hydrosphere and atmosphere, with special reference to probable composition of the earth atmosphere. In A. Poldervaat. ed., Crust of the earth. New York, Geol. Soc. Amer. 631-650.

Sanchez, R.A., Ferris, J.P. and Orgel, L.E. 1966. Cyanoacetylene in prebiotic synthesis, Science 154, 784-785.

Schlesinger, G. and Miller, S.L. 1983. Prebiotic synthesis in atmospheres containing methane, carbon monoxide and carbon dioxide. J. Molec. Evol. 19, 376-390.

Sweeney, R.E., Toste, A.P. and Ponnamperuma, C. 1976. Formation of amino acids by cobalt-60 irradiation of hydrogen cyanide solutions. Origins of Life. 7, 187-189.

Taylor, S. and Williams, D. 1993. Star-studded chemistry, Chemistry in Britain. 29(8), 680-683.

Townes, C. 1971. In Highlights of Astronomy 2, Trans. Intern. Astron. Union., D. Reidel, Dodrecht, 359.

Turner, B. 1980. Interstellar molecules. J. Mol. Evol. 15, 79-101.

Urey, H.C. 1952a. The Planets: Their Origin and Development, Yale University Press. Conn. New Haven, pp 245.

Urey, H.C. 1952b. On the early chemical history of the Earth and the origin of life. Proc. Natl. Acad. Sci. U.S.A. 38, 351-363.

Woeller, F. and Ponnamperuma, C. 1969. Organic synthesis in a simulated Jovian atmosphere. Icarus 10, 386.

Wolman, Y., Haverland, W.J. and Miller, S.L. 1972. 'Non-protein Amino Acids from Spark Discharges and Their Comparison with Murchison Meteorite Amino Acids'. Proc. Nat. Acad. Sci. U.S. 69(4), 809-811.

FORMATION OF AMINO ACID PRECURSORS BY COSMIC RADIATION IN PRIMITIVE TERRESTRIAL AND EXTRATERRESTRIAL ENVIRONMENTS[1]

Kensei Kobayashi
Department of Physical Chemistry, Yokohama National University
Hodogaya-ku, Yokohama 240, Japan

Takeshi Saito
Institute for Cosmic Ray Research, University of Tokyo
Tanashi, Tokyo 188, Japan

Tairo Oshima
Department of Life Science, Tokyo Institute of Technology
Midori-ku, Yokohama 227, Japan

ABSTRACT

There has been some controversy whether bioorganic compounds were formed on primitive earth or formed in space and supplied to the earth before the origin of life on the earth. In order to verify the possibility of the formation of amino acids under both primitive terrestrial and extraterrestrial conditions, several kinds of gas mixtures and ice mixtures ("simulated planetary atmospheres and cometary ices") were irradiated with high energy particles. We found wide variety of amino acids in the hydrolysates of the products. The present results strongly suggest that amino acid precursors could be formed by cosmic ray radiation at any place where carbon and nitrogen compounds exist such as in primitive earth and in cometary nuclei.

1. INTRODUCTION

Bioorganic compounds such as amino acids are believed to be supplied to the primitive earth before the origin of life on the earth. Were the bioorganic compounds formed on the primitive earth, or formed in extraterrestrial environments and delivered to the earth? A large number of experiments have be performed to obtain evidences of amino acid formation on the earth or in space.

It has been reported that, if the primitive earth had the strongly reduced atmosphere like a mixture of methane, ammonia and water, amino acids could be easily

[1] Supported in part by a Grant for Basic Experiments Oriented to Space Station Utilization, Institute of Space and Astronautical Science, and a Grant-In-Aid (No. 05833007) from the Ministry of Education, Science and Culture, Japan.

formed by such energy sources as spark discharge (Miller, 1953) or ultraviolet lights (Sagan and Khare, 1971). On the other hand, it is known that a wide variety of organic compounds are present in extraterrestrial environments like interstellar molecular clouds and comets. It has been estimated, however, that the primitive earth atmosphere was not strongly reduced but was only slightly reduced, where major carbon sources were carbon dioxide and carbon monoxide (Kasting, 1990). It is difficult to synthesize amino acids from a slightly reduced gas mixture by conventional energy sources like spark discharges (Schlesinger and Miller, 1983).

Among various types of energies possibly obtained on the primitive earth and in space, we considered cosmic ray as a possible energy sources for abiotic synthesis. Here we experimentally verified the possible formation of amino acids or their precursors from simulated primitive earth (Kobayashi et al., 1990), other planetary atmospheres (Kobayashi et al., in press), and cometary ice materials.

2. EXPERIMENTAL

A gas mixture of carbon monoxide, carbon dioxide, and nitrogen at various mixing ratios, simulating the primitive earth atmosphere, was enclosed in a glass tube with and without liquid water. The gas mixture was irradiated with protons of 2.5—4.0 MeV generated from a Van de Graaff accelerator (Tokyo Institute of Technology), protons of 40 MeV and helium nuclei of 65 MeV from an SF cyclotron (INS, University of Tokyo), and electrons of 400 MeV and 1 GeV from an electron synchrotron (INS, University of Tokyo). The temperature of the gases was kept at a fixed temperature by circulating water during irradiation.

Various kinds of gas mixtures simulating planetary atmospheres (e.g., a mixture of methane, ammonia and water; see Table 1) was also irradiated with high energy protons from a Van de Graaff accelerator to simulate extraterrestrial planetary atmospheres.

An aqueous solution in the irradiation tube was recovered from the tube after irradiation: When no water was contained in the starting material, water was admitted into the tube to dissolve the product. A part of the aqueous product was acid-hydrolyzed with 6M hydrochloric acid at 110°C for 24 h. Amino acids in the hydrolyzed and unhydrolyzed products were analyzed by ion-exchange chromatography (IEC), gas chromatography (GC), mass spectrometry (MS) and/or GC/MS.

3. RESULTS AND DISCUSSION

3.1 AMINO ACID FORMATION ON THE PRIMITIVE EARTH

A wide variety of amino acids was found in all the hydrolysates of irradiation products, when carbon monoxide and nitrogen were used in the starting materials. Glycine was predominant, followed by aspartic acid, serine, alanine and ß-alanine. The D/L ratios were measured as ca. 1 by GC.

The amount of products was in proportion to the total energy deposited, and is independent of the kinds of irradiated particles (protons, helium nuclei or electrons) as well as of initial energy. This indicates that not only primary cosmic rays but also secondary particles, nuclear and electro-magnetic cascades generated in the atmosphere,

Table 1. G-Values of amino acids by proton irradiation of gas mixtures.

Type of	Partial pressure of starting gases / Torr						G-Value × 100**		
atmosphere	CO_2	CO	CH_4	N_2	NH_3	H_2O	Gly	Ala	Asp
Primitive Earth	0	280	0	280	0	20	2.12	0.326	0.053
	0	280	0	280	0	0	2.38	0.310	0.256
	280	140	0	140	0	0	0.64	0.040	0.056
Comet	0	350	0	0	350*	20	23.1	2.62	0.654
Titan	0	0	350	350	0	0	1.25	0.859	0.003
Jupiter	0	0	350	0	350	0	4.55	2.08	0.002

* Partial pressure measured before liquid water was admitted.

**All the products were acid-hydrolyzed before analysis.

would contribute the abiotic synthesis of amino acid precursors. Cosmic rays would contribute the formation of organic compounds until they lose their kinetic energies

The G-values (number of formed molecules per 100 eV) of glycine were obtained as about 0.02 when a 1:1 mixture of carbon monoxide and nitrogen was irradiated with protons. It should be noted that the G-value of glycine by spark discharges is about 10^{-6} from the same gas mixture (Kobayashi, unpublished), and no amino acids were detected when the same gas mixture was irradiated with a mercury lamp (Bar-Nun and Chang, 1983). The present work shows that high energy charged particles as cosmic rays are a more effective energy source for prebiotic synthesis of amino acids than conventionally considered energy sources like spark discharges and ultraviolet light. The energy flux of cosmic rays at 0.062 cal/cm^2 yr 2π which is given by integrating from 10^9 eV to 10^{12} eV including heavy cosmic ray nuclei (calculated after Meyer et al., 1974). This energy deposit corresponds to the production of glycine of 2.3 μmol/m^2 yr in the case of a 1:1 mixture of carbon monoxide and nitrogen. Energy deposit by electric discharges to the earth atmosphere, estimated as 4 cal/cm^2 yr (Miller and Urey, 1959), gives only 10 nmol glycine/m^2 yr in the same gas mixture. Ultraviolet lights with shorter wavelengths might be effective for amino acid synthesis from the gas mixture. The energy fluxes from the Sun lights 4 billion years ago are given as 0.16 cal/cm^2 yr at λ <110 nm corresponding to photolysis energy of nitrogen (estimated after Heroux and Hinteregger, 1978). It should be noted that this flux value is comparable with the cosmic ray energy discussed above. Dependence of G-values on the irradiated wavelengths in the amino acid formation is required to compare the formation rates between cosmic rays and the Sun light.

3.2 AMINO ACID FORMATION IN SPACE

Table 1 shows G-values of amino acids when various kinds of gas mixtures were irradiated with protons of 2.8—4.0 MeV. Amino acids were detected in hydrolysates of every type of gas mixtures, if proper carbon compounds (carbon monoxide or methane) and nitrogen compounds (nitrogen or ammonia) were in the starting gas mixtures. Even if there is no water in the starting materials, amino acid precursors, which give free amino acids after acid hydrolysis, were produced. When a ice mixture of methane (or propane), ammonia and water was irradiated with high energy protons, several kinds of amino acids were detected in the hydrolysates of the resulting non-volatile residues recovered on the metal block (Kasamatsu et al., 1994). The present results show that amino acid precursors can be formed by irradiation of not only gas mixtures but also ice mixtures with high energy particles, if proper carbon and nitrogen sources are contained in them. It is suggested that there are amino acid precursors in cometary environments, though free amino acids have not been detected there (Kissel and Krueger, 1987).

4. CONCLUSION

It is concluded that the amino acids or precursors of amino acids could be formed by cosmic ray radiation at any place where carbon and nitrogen compounds exist. Carbon monoxide is a major carbon compound in space, and energy sources of cosmic rays are distributed uniformly in the disc of our Galaxy. It may safely be said that the formation of bioorganic compounds is a universal and inevitable event in our space. Thus it is not relevant to discuss whether bioorganic compounds were created on the primitive earth or in the extraterrestrial environments, they might be produced in both. It will be of great interest to estimate and compare the amounts of bioorganic compounds that were formed on the earth and that brought by the extraterrestrial objects, by taking into consideration time of formation and molar fraction of gases.

The authors wish to express their thanks to Dr. M. Tsuchiya, Mr. T.Kaneko and Mr. T. Kasamatsu, Yokohama National University, and Dr. J.Koike, Tokyo Institute of Technology, for their kind help and useful discussion. They are indebted to Mr. K. Kawasaki, Tokyo Institute of Technology, and the staff of the SF cyclotron and the Electron synchrotron, INS, University of Tokyo, for their kind assistance in the operation of the accelerators.

REFERENCES

Bar-Nun A. and Chang S., 1983: *J. Geophys. Res.*, **88**, 6662-6672.
Heroux L. and Hinteregger H.E., 1978: *J. Geophys. Res.*, **83**, 5305-5308.
Kasamatsu T., Kaneko T., Koike J., Oshima T. and Kobayashi K., 1994: *Viva Origino*, **22**, in press.
Kasting J.M., 1990: *Origins of Life*, **20**, 199-231.
Kissel J. and Krueger F.R., 1987: *Nature*, **326**, 755-760.
Kobayashi K., Tsuchiya M., Oshima T. and Yanagawa H., 1990: *Origins of Life*, **20**, 99-109.
Kobayashi K., Saito T., Koike J., Oshima T., Yamamoto T., Kaneko T. and Tsuchiya M., *Adv. Space Res.*, in press.
Kobayashi K., 1992: unpublished.
Meyer P., Ramaty R. and Webber W.R., 1974 (October): Physics Today, 23-30.
Miller S.L., 1953: *Science*, **117**, 528-529.
Miller S.L. and Urey H.C., 1959: *Science*, **130**, 245-251.
Sagan C. and Khare B.N., 1971: *Science*, **173**, 417-420.
Schlesinger G. and Miller S.L., 1983: *J. Mol. Evol.*, **19**, 376-382.

HYDROGEN CYANIDE POLYMERS: PREBIOTIC AGENTS FOR THE SIMULTANEOUS ORIGIN OF PROTEINS AND NUCLEIC ACIDS

Clifford N. Matthews
Department of Chemistry, University of Illinois at Chicago,
Chicago, Illinois 60680, USA.

ABSTRACT

Hydrogen cyanide polymers—heterogeneous solids varying in color from yellow to orange to red to black—may be among the organic macromolecules most readily formed within the solar system. Current studies of these ubiquitous compounds point to the presence of polyamidine structures readily converted by water to polypeptides. Implications for prebiotic chemistry are profound. Primitive Earth may have been covered by HCN polymers through bolide bombardment or terrestrial synthesis, producing a proteinaceous matrix able to promote the molecular interactions leading to the emergence of life. Most significant would have been the parallel synthesis of polypeptides and polynucleotides arising from the powerful dehydrating action of polyamidines on available sugars, phosphates and nitrogen bases. On our dynamic planet, this polypeptide-polynucleotide symbiosis mediated by polyamidines may have set the pattern for the evolution of protein-nucleic acid systems controlled by enzymes, the mode characteristic of all life today.

1. INTRODUCTION

Which came first, amino acids or their polymers? Our continuing investigations support the hypothesis that the direct synthesis of heteropolypeptides from hydrogen cyanide and water is a preferred pathway in prebiotic and extraterrestrial chemistry, proceeding by way of HCN polymers with polyamidine structures. In this paper we summarize the evidence, old and new, for this controversial view and point to the significance of HCN polymers for the origin of proteins, nucleic acids and life.

2. HYDROGEN CYANIDE POLYMERS: SYNTHESIS AND STRUCTURE

Liquid HCN (bp 25°C) polymerizes spontaneously to a dark brown or black solid at low temperatures in the presence of a base such as an amine or ammonia (Matthews and Moser, 1967). Polymerization also occurs readily in non-aqueous solvents or in water (Völker, 1960; Matthews and Moser, 1966, 1967). Two types of structural units appear to be present in these solid materials. Most stable are the ladder polymers (Völker, 1960) shown in Figure 1, formally derived from the olefinic tetramer of HCN, diaminomaleonitrile (A), which is usually found among the products. It seems probable, however, that polymerization to the substituted polymethylene B proceeds by way of an HCN dimer (Völker, 1960; Kliss and Matthews, 1962) for which several structures have been proposed (see Evans et al., 1991). Cyclization to C and D then leads to ladder structures possessing conjugated >C=N- bonds, as proposed by Völker (1960) on the basis of extensive physical and chemical investigations further supplemented by the work of Umemoto et al. (1987).

More controversial is the existence of the polyamidine structures shown in Figure 2. Polyaminomalononitrile (F) can be considered an addition polymer of the reactive trimer aminomalononitrile (E), though again it is possible that polymerization occurs through an HCN dimer ((Kliss and Matthews, 1962; Matthews and Moser, 1967; Yang et al., 1976). Cumulative reactions of HCN on the highly activated nitrile groups of F then yield the heteropolyamidines G which are readily converted by water to heteropolypeptides (Matthews and Moser, 1966, 1967) with release of ammonia and CO_2. Overall, this series of reactions constitutes a route for the direct synthesis of polypeptides without the intervening formation of a α-amino acids (Matthews and Moser, 1966, 1967; Matthews, 1992) (Figure 3).

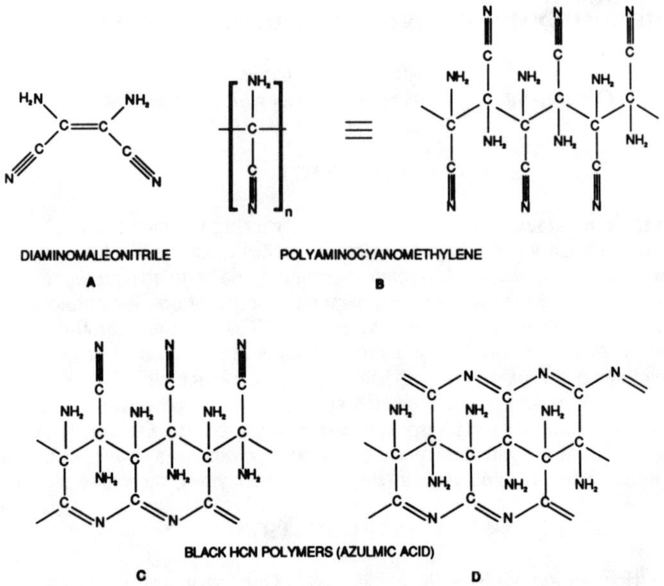

Fig. 1. HCN polymers formally derived from HCN tetramer (diaminomaleonitrile).

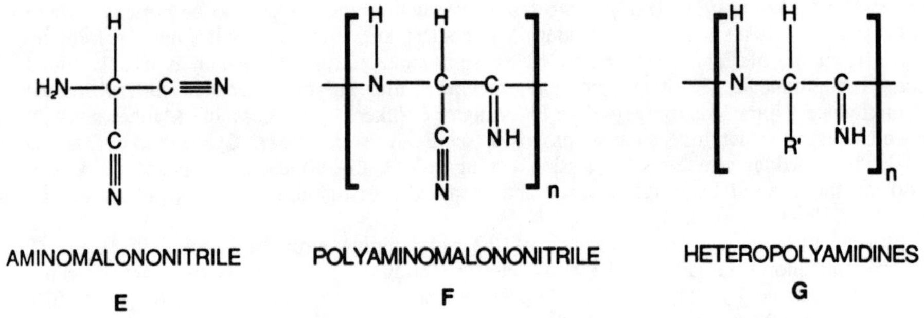

Fig. 2. HCN polymers formally derived from HCN trimer (aminomalononitrile)

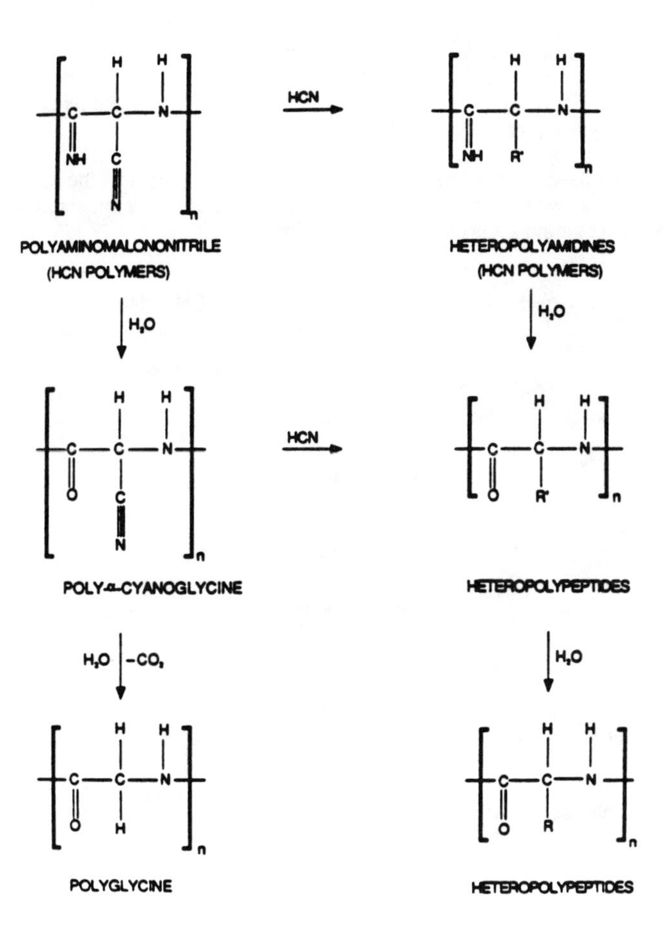

Fig. 3. Polypeptides from polyamidines. Cumulative reactions of HCN on polyaminomalononitrile yield heteropolyamidines (with side chains R′) which are converted stepwise by water to heteropolypeptides (with side chains R)

Several kinds of experiments (see Figure 4) have provided results consistent with this polyamidine model. In general, water-soluble, yellow-brown solids can be extracted from the products of each of the following types of reactions:

1. base-catalyzed polymerization of liquid HCN, alone, in water or in non-aqueous solvents (Matthews and Moser, 1967);
2. electric discharge experiments producing HCN from methane-ammonia mixtures (Matthews and Moser, 1966);
3. alkaline hydrolysis of aminoacetonitrile (Moser and Matthews, 1968), aminomalononitrile (E, HCN trimer) (Moser et al., 1968), and diaminomaleonitrile (A, HCN tetramer) (Moser et al., 1968), all of which are ready sources of HCN at high pH;
4. HCN modification of the reactive nitrile side chains of poly-α-cyanoglycine, (see Figure 3), a polyamide analog of polyaminomalononitrile (F) synthesized from the N-carboxyanhydride related to α-cyanoglycine (Warren et al., 1974; Minard et al., 1975).

Acid hydrolysis of these yellow-brown polymers yields not just glycine, the major product, but other α-amino acids as well, such as alanine, aspartic acid, glutamic acid, serine and threonine, together with some α-amino acids not found in proteins.

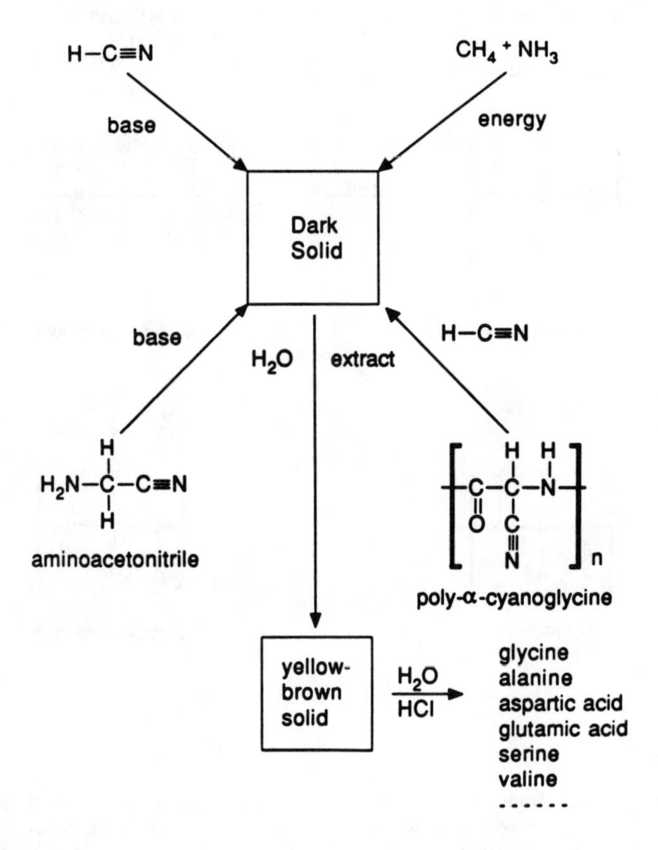

Fig. 4. Water-soluble yellow-brown solids obtained from four kinds of experiments yield similar kinds of α-amino acids after acid hydrolysis.

Non-destructive analysis of the total solid product obtained from HCN, as well as of separate components, became possible with the advent of cross-polarization magic-angle spinning solid-state NMR spectroscopy (^{13}C and ^{15}N). In particular, the unambiguous presence of secondary amide groups, as in peptides, has been established by double-cross-polarization studies on polymers synthesized from equimolar amounts of $H^{13}CN$ and $HC^{15}N$ (Matthews et al., 1984; McKay et al., 1984; Garbow et al., 1987, Matthews and Ludicky, 1991).

An integrated analytical approach now being used for the separation and identification of these intriguing materials, involving pyrolysis mass spectrometry, Fourier transform infrared photoacoustic spectroscopy, and supercritical fluid extraction chromatography, has revealed fragmentation patterns and chemical functionalities consistent with the presence of polymeric peptide precursors—polyamidines — both in HCN polymers and in the Murchison meteorite (Liebman et al., 1993). Taken together, these various analytical studies point to the presence of both peptide and ladder structures in HCN polymers exposed to water. (see Table 1). Hybrid polymers with peptides attached to ladders through their free amino groups — multimers — may be major components.

TABLE 1. Summary of evidence reported by Matthews et al. for the presence of peptide-like segments in hydrogen cyanide polymers and the Murchison meteorite.

	Analytical mode	Previous Studies (see Matthews, 1992)	Current Studies (see Liebman et al., 1993)
		Solid state (^{13}C, ^{15}N) NMR spectroscopy:	Photoacoustic Fourier transform infrared spectroscopy (FTIR-PAS):
1	Non-destructive identification	HCN polymer extracts (H_2O) and insoluble residues contain polypeptide segments	aqueous extracts of HCN polymer and the Murchison meteorite possess polypeptide segments
		Liquid phase (Sephadex) column chromatography:	Supercritical fluid extraction chromatography (SFE-Transcap FTIR-Mic):
2	Separation by fractionation	HCN polymer extracts (H_2O) contain heteropolypeptide segments	Aqueous extracts of HCN polymer and the Murchison meteorite possess polypeptide segments
		Hydrolysis (GC-MS):	Analytical Pyrolysis (DIP-MS):
3	Analysis by fragmentation	HCN polymers and the Murchison meteorite yield α-amino acids (also α-hydroxy acids)	Hydrocarbons, fatty acids, and polypeptides are present in Murchison meteorite. HCN polymers yield polypeptides.

3. HYDROGEN CYANIDE POLYMERS IN THE SOLAR SYSTEM

Hydrogen cyanide polymers — heterogeneous solids ranging in color from yellow to orange to red to black — may be among the organic macromolecules most readily formed within the solar system. The non-volatile black crust of comet Halley, for example, might consist largely of such polymers. (Matthews and Ludicky, 1992; Matthews, 1992). It seems likely, too, that HCN polymers are a major constituent of the dark, C≡N bearing solids identified spectroscopically in the dust of some other comets, on the surfaces of several asteroids, within the rings of Uranus and covering the dark hemisphere of Saturn's satellite Iapetus (Cruikshank et al., 1991). HCN polymerization could account also for the yellow-orange-red coloration of Jupiter and Saturn, as well as for the orange haze high in Titan's atmosphere (Matthews, 1982).

On primitive Earth which came first, amino acids or their polymers? The ubiquitous presence of HCN in reducing environments invites the reexamination and possible reinterpretation of almost all previous research concerned with the origin of α-amino acids, including simulations of the chemistry of primitive atmospheres and studies of aqueous cyanide chemistry (see reviews by Miller and Orgel (1974); Oro' and Lazcano-Araujo (1980); Ponnamperuma (1983); Ferris and Hagan (1984)).

These investigations of reactions ostensibly yielding α-amino acids actually supply evidence for the abundant prebiotic and extraterrestrial existence of polymeric protein ancestors —heteropolypeptides synthesized directly from hydrogen cyanide and water (Matthews and Moser, 1966, 1967; Matthews, 1992, 1994.) The detection of amino acid polymers in some of these experiments (Lowe et al, 1963; Matthews and Moser, 1966, 1967; Woeller and Ponnamperuma, 1969; Su et al., 1989; Khare et al., 1989; McDonald et al., 1991) adds to the plausibility of this conclusion. That all these reactions may proceed through a common preferred pathway is in accord with the well-recognized principle of maximum simplicity known as Occam's Razor i.e. Entitities are not to be multiplied beyond necessity.

Implications for prebiotic chemistry are profound. Primitive Earth may have been covered by HCN polymers, either through cometary deposition or by photochemical reactions in a reducing atmosphere. As polyamidines settled onto land and sea together with other organic products, a proteinaceous matrix developed able to take part in and promote the interactions leading to the emergence of life. These polyamidines could have been the original condensing agents directing the synthesis of nucleosides and nucleotides from available sugars, phosphates, and nitrogen bases. Most significant would have been the parallel synthesis of polypeptides and polynucleotides arising from the dehydrating action of polyamidines on nucleotides:

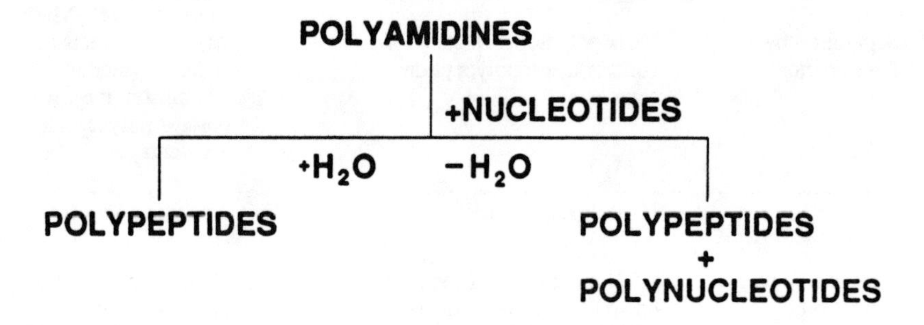

On our dynamic planet this polypeptide-polynucleotide symbiosis mediated by polyamidines may have set the pattern for the evolution of protein-nucleic acid systems controlled by enzymes, the mode characteristic of life today (Matthews 1986, 1988, 1990, 1992).

REFERENCES

Cruikshank, D. P., W. K. Hartmann, D. J. Tholen, L. J. Allamandola, R. H. Brown, C. N. Matthews, and J. F. Bell, 1991: Solid C≡N Bearing Material on Outer Solar System Bodies, *Icarus*, **94**, 345-353.

Evans R. A., P. Lorencak, T. K. Ha, and C. Wentrup, 1991: HCN Dimers: Iminoacetonitrile and N-Cyanomethanimine, *J. Am. Chem. Soc.*, **113**, 7261-7276.

Ferris, J. P., and W. J. Hagan, 1984: HCN and Chemical Evolution: The Possible Role of Cyano Compounds in Prebiotic Synthesis, *Tetrahedron*, **40**, 1093-1120.

Garbow, J. R., J. Schaefer, R. A. Ludicky, and C. N. Matthews, 1987: Detection of Secondary Amides in HCN Polymers by Dipolar Rotational Spin-Echo ^{15}N NMR, *Macromolecules*, **20**, 305-309.

Khare, B. N., C. Sagan, W. R. Thompson, L. Flynn, and M. A. Morrison, 1989: Amino Acids and Their Polymers in the Lower Clouds of Jupiter? Preliminary Findings, *Origins of Life*, **19**, 495-496 .

Kliss, R. M. and C. N. Matthews, 1962: Hydrogen Cyanide Dimer and Chemical Evolution, *Proc. Natl. Acad. Sci.* U.S., **48**, 1300-1306 .

Liebman, S. A., R. A. Pesce-Rodriquez, and C. N., Matthews, 1993: Organic Analysis of Hydrogen Cyanide Polymers: Prebiotic and Extraterrestrial chemistry, *Advances in Space Research*, in press .

Lowe, C. U., M. W. Rees, and R. Markham, 1963: Synthesis of Complex Organic Compounds from Simple Precursors: Formation of Amino-Acids, Amino-Acid Polymers, Fatty Acids and Purines from Ammonium Cyanide, *Nature*, **199**, 219-222.

Matthews, C. N., and R. E. Moser, 1966: Prebiological Protein Synthesis, *Proc. Natl. Acad. Sci. U.S.* **56**, 1087-1094.

Matthews, C. N., and R. E. Moser, 1967: Peptide Synthesis from Hydrogen Cyanide and Water, *Nature* **215**, 1230-1234.

Matthews, C. N., 1982: Heteropolypeptides on Titan? *Origins of Life*, **12**, 281.

Matthews, C. N., R. A. Ludicky, J. Schaefer, E. O. Stejskal, and R. A. McKay, 1984: Heteropolypeptides from Hydrogen Cyanide and Water: Solid-State ^{15}N NMR Investigations *Origins of Life*, **14**, 243-249.

Matthews, C. N., 1986: Simultaneous Synthesis of Polypeptides and Polynucleotides? Hydrogen Cyanide Polymers as Prebiotic Condensing Agents, *Origins of Life*, **16**, 500.

Matthews, C. N., 1988: Cosmic Metabolism: The Origin of Macromolecules. In *Bioastronomy—The Next Steps*, G. Marx, (Ed.), Kluwer, Dordrecht, 167-178.

Matthews, C. N., 1990: Simultaneous Origin of Polypetides and Polynucleotides: Hydrogen Cyanide Polymers as Prebiotic Condensing Agents. In *Prebiological Self Organization of Matter*, C. Ponnameruma and F. R. Eirich, (Eds.) Deepak, Hampton VA; 171-182.

Matthews, C. N. and R. A. Ludicky, 1991: Hydrogen Cyanide Polymers: Solid State NMR (^{15}N, ^{13}C) Investigations. In *Solid State NMR of Polymers*, L. Mathias (Ed.) Plenum Press, New York, N.Y. 331-342.

Matthews, C. N. and R. Ludicky, 1992: Hydrogen Cyanide Polymers on Comets, *Adv. Space Research*, **12**, 4 (21) -4(32).

Matthews, C. N. 1992: Dark Matter in the Solar System: Hydrogen Cyanide Polymers, *Origins of Life*, **21**, 421-434.

Matthews, C. N. 1992: Origin of Life: Polymers Before Monomers? In *Environmental Evolution*, L. Margulis and L. Olendzenski, (Eds.) MIT Press, Cambridge MA: 29-38.

Matthews, C. N. 1994: Cyanide Polymers in the Solar System: Miller-Urey Revisited, *Origins of Life*, in press.

McDonald, G. D., B. N. Khare, W. R. Thompson, and C. Sagan, 1991: $CH_4/NH_3/H_2O$ Spark Tholin: Chemical Analysis and Interaction with Jovian Aqueous Clouds, *Icarus*, **94**, 354-367.

McKay, R. A., J. Schaefer, E. O. Stejskal, R. A. Ludicky, and C. N. Matthews, 1984: Double-cross-polarization Detection of Labelled Chemical Bonds in HCN Polymerization, *Macromolecules*, **17**, 1124-1130.

Miller, S. L. and L. E. Orgel, 1974: *The Origins of Life on the Earth*, Prentice-Hall, Englewood Cliffs, N. J.

Minard, R. D., W. Yang, P. Varma, J. Nelson, and C. N. Matthews, 1975: Heteropolypeptides from Poly-α-Cyanoglycine and Hydrogen Cyanide: A Model for the Origin of Proteins, *Science*, **190**, 387-389.

Moser, R. E., and C. N. Matthews 1968: Hydrolysis of Aminoacetonitrile: Peptide Formation, *Experientia*, **24**, 658-659.

Moser, R. E., A. R. Claggett, and C. N. Matthews, 1968: Peptide Formation from Aminomalononitrile. Peptide Formation from Diaminomaleonitrile. *Tetrahedron Lett.*, 1599-1608.

Oró, J. and A. Lazcano-Araujo, 1980: The Role of HCN and its Derivatives in Prebiotic Evolution. In *Cyanide in Biology*, B. Vennesland, E. E. Conn, C. J. Knowles, J.Westley, and F. Wissing, (Eds.), Academic Press New York, N. Y., 577.

Ponnamperuma, C. (Ed.), 1983: Cosmochemistry and the Origin of Life, In *Cosmochemistry and the Origin of Life*, Reidel, Dordrecht, 1-34.

Su, Y-L., Y. Honda, P. E. Hare, and C. Ponnamperuma, 1989: Search of Peptide-Like Materials in Electric Discharge Experiments, *Origins of Life*, **19**, 237-238.

Umemoto, K., M. Takahashi, and K. Yokota, 1987: Studies on the Structure of HCN Oligomers, *Origins of Life*, **17**, 283-293.

Völker, T., 1960: Polymeric Hydrogen Cyanide, *Angew. Chem.*, **72**, 379-384.

Warren, C. B., R. D. Minard, and C. N. Matthews, 1974: Synthesis of α-Cyanoglycine N-Carboxyanhydride and α-Cyanoglycine, *J. Org. Chem.*, **39**, 3375-3378.

Woeller, F., and C. Ponnamperuma, 1969: Organic Synthesis in a Simulated Jovian Atmosphere, *Icarus*, **10**, 386-390.

Yang, W., R. D. Minard, and C. N., Matthews, 1976: Azacyclopropenylidenimine: A Low-Energy Dimer of Hydrogen Cyanide, *J. Theor. Biol.*, **56**, 111-123.

A PLAUSIBLE ROUTE FOR THE SYNTHESIS OF BIO-ORGANIC COMPOUNDS FROM INORGANIC CARBON ON THE PRIMITIVE EARTH. AN OVERVIEW.

G. Albarrán, A. Negrón-Mendoza, S. Ramos-Bernal
Instituto de Ciencias Nucleares, UNAM
México, D.F. 04510 México.

C.H. Collins
Instituto de Química, UNICAMP
Campinas, S.P. 13083-970 Brasil

ABSTRACT

The composition of the early atmosphere is now considered as a mixture of CO, CO_2, N_2 and H_2O. Yet, it has been pointed out that the yields of organic compounds of biological significance in such gas mixtures are extremely low. This paper describes the synthesis of reduced organic compounds through the radiolysis of oxidized inorganic carbon-containing compounds in the atmosphere, ocean and sea floor of the primitive Earth. Studies of several radiation-induced reactions in aqueous solutions of carbonates/bicarbonates or in solid carbonates suggest that these pathways may have played an important role in primordial organic chemistry.

1. INTRODUCTION

Chemical evolution of organic matter is a part of an integral evolution of the planet and the universe. Thus, the chemistry involved on the prebiotic Earth must have been constrained by the global physical-chemistry of the planet; which determined the temperature, pressure and chemical composition of various environments. These considerations need to be taken into account for possible pathways for prebiotic synthesis of organic compounds. In this context, there is no longer a consensus on the prevalence of reducing conditions in the early Earth atmosphere. The primitive Earth's atmosphere is now considered as a mixture of CO_2, N_2, H_2O and CO. It has been pointed out that the yield and product diversity of organic compounds of biological significance in such gas mixtures is low. So it is important to re-examine the prebiotic synthesis of organic compounds, and investigate those pathways that can yield organic compounds in a neutral or weakly reduced atmosphere and from oxidized carbon in the ocean and sediments.

1.1 PRIMITIVE SCENARIO

During the accretion, hydrogen, helium and other primordial gases escaped into space. After the Earth reached its present size and structure, and most of the primary atmosphere had been lost, a new atmosphere was formed by outgassing (volcanism, hydrothermal phenomena and related processes). There are suggestions that the outgassing was a gradual process (MacDonald, 1959). Nevertheless, modern dynamical calculations suggest that accretion was rapid and the burial of accretional energy yielded hot, convecting planetary interiors (Wetherill, 1980). Convective motion of the Earth´s interior would have brought all portions of the mantle close to the surface in a short period of geological time.

It is assumed, firstly, that the origin of the early Earth's atmosphere was released from the planet's interior rapidly and very early in the history of the solar system. Secondly, that the early Earth lacked extensive, stable continental platforms, since in the Archean Earth the principal land areas were volcanic islands and small microcontinental block (Lowe, 1994). Also, it is probable that the early Earth's atmosphere contained a very large partial pressure of carbon dioxide (Walker 1985). Thus, the carbon reservoirs were CO_2 in the atmosphere. Carbon would have been transferred from the atmosphere to the fluid reservoir of the ocean as carbonates and bicarbonates. From the ocean the carbon was transferred to seafloor sediments and much latter to the continents as solid mineral carbonates.

The purpose of this work is to present some considerations about the behavior of the systems carbon dioxide/bicarbonates/carbonates (in solution) and mineral carbonates (in solid state) in prebiotic environments and, in particular, to study the these compounds as starting materials for the synthesis of reduced carbon compounds using localized sources of energy. Also, it is important to analyze if solid state radiation-chemical processes could have provided pathways for the formation of organic compounds in hydrothermal vents.

To reach these objectives we need to analyze: The sources and reservoirs of carbon in the early Earth, particularly as carbon dioxide/bicarbonate/carbonate in the system atmosphere-ocean-sediments. Some sources of energy, like ionizing radiation, to carry out these reactions. The use of this source implies that we need to review the radiolytic behavior of these compounds in different conditions and physical states.

2. CARBON RESERVOIRS

2.1 IN THE ATMOSPHERE

The key element of life, carbon, was widespread on the early Earth. The carbon reservoir in the atmosphere was CO_2. It appears that the partial pressure of CO_2 could be as high as 10 bars. Although the geochemical atmospheric cycles of carbon dioxide are well known, there is still debate on the processes that control the carbon dioxide partial pressure (Walker, 1985). Some fraction of the carbon dioxide presumably is stored in the mantle. The rest is returned to the atmosphere and ocean as a component of volcanic and metamorphic emanations. The interest in carbon dioxide is due to its close association with the precipitation of carbonate minerals. Also, according to the pH of sea water, carbon existed as oceanic bicarbonate or as CO_2 at the interface. Fluctuations in the partial pressure of carbon dioxide imply changes in these ratios.

2.2 IN THE OCEANS AND ON THE SEAFLOOR

Carbonates and bicarbonates are encountered in the oceans with mineral carbonates in seafloor sediments. However, the lifetime of the seafloor is ephemeral. In a somewhat short period, approximately 60 million years on the average today, the seafloor returns to the mantle by way of subduction (Sprague and Pollack, 1980; Sclater et al., 1981). Its sediments are subjected to high temperatures, and carbonate minerals are largely decarbonated.

2.3 IN THE CONTINENTS

The relative abundance nowadays of carbon on the continental platforms is, of course, a consequence of their stability. They provide a storage place for carbon with a long life in geological terms.

Early continent formation was about 3.3-3.1 Ga and the formation of enormous continental blocks were 2.7-2.5 Ga. The later produced large-scale CO_2 depletion and the precipitation of carbonates (Lowe, 1994). The amount of this sedimentary carbonate is huge even when compared with other forms of carbon in the biosphere. Sedimentary organic carbon reservoir is between $1.2\text{-}1.4 \times 10^{22}$ g and carbonate carbon as the other sedimentary carbon species is about four times more abundant (Schidlowski, 1991). Indeed, carbonate formation might have been extensive much earlier than this. Such carbonated rocks do seem prevalent in the Archean period. Carbonate carbon and organic carbon species can be traced back in sedimentary rocks to 3.8 Ga ago (Schidlowski, 1988).

3. SOURCES OF ENERGY

Naturally occurring abiotic processes could have caused the synthesis of organic molecules that were the precursors of terrestrial life. To carry out these syntheses, raw materials and an input of energy were required to induce chemical reactions. Considering the energy impinging on the Earth now, there are several sources: sunlight, lightning, radioactivity, volcanism and sound waves. These sources probably were important on the early Earth and they have been proposed as prebiotic sources of energy. A comparison of the relative importance for the production of organic molecules, on the early Earth was first assessed by Miller and Urey (1959) and reviewed by Chyba and Sagan (1991). The evaluation of these energy sources is difficult. Ultraviolet light was probably the largest source of energy on the primitive Earth. It effectiveness could be diminished by the possible decomposition of the products that may occur in the upper atmosphere, before they reach the protection of the oceans. Electrical storms were probably important for synthesis of organic compound on the primitive Earth. Other methods of energy input such as radioactivity, sonic energy and vulcanism have been considered as minor partners. Still, considering the characteristics of the primitive Earth, the later types of localized sources may have been important in primitive environments. They may have been an alternative sources to induce synthesis of reduced carbon compounds starting in a neutral atmosphere, in the ocean or from solid sediments such as carbonates.

3.1 IONIZING RADIATION

Radioactivity is the property of some nuclei to undergo a spontaneous disintegration accompanied by the emission of radiation (Draganic et al., 1990). Most of the high energy radiation originates from the interior of the Earth.

The interaction of high energy radiation with matter leads to physical and chemical changes. The net result is excitation and ionization, producing short lived chemical intermediates (Spinks and Woods, 1990).

In the early time after accretion far more radioactive matter was present in the Earth than is left now. The basis of using ionizing radiation as energy source stems from the presence of radionuclides in the crust. The principal radionuclides are ^{40}K, ^{235}U, ^{238}U and ^{232}Th (Albarrán et al., 1988). These isotopes have half-lives as long as 10^9 years. Based on recent determination of the concentrations of these radioisotopes (Klement, 1982, Kathren, 1984), it is possible, by backward extrapolation, to calculate the associated radiation doses on the primeval Earth, 4.0 Ga ago (Table 1).

As is shown in Table 1 the highest values originate from ^{40}K. This isotope is widely distributed in the hydrosphere, in rocks and sediments. The ^{40}K content of the Earth's crust 4 Ga ago was about eight times what it is now. In certain regions, particularly in solid-liquid and solid-gas interphases, conditions for formation and conversion of organic molecules by high energy radiation should have been favorable.

TABLE 1. ESTIMATED ANNUAL DOSE (Gy/year) AT THE EARTH'S CRUST 4.0 Ga AGO

MATERIAL	^{40}K	^{238}U	^{232}Th
Sedimentary rock carbonates	2.1×10^{-3}	1.07×10^{-3}	1.69×10^{-4}
Deep sea sediments carbonates	2.2×10^{-3}		
Seawater	2.8×10^{-4}	1.46×10^{-6}	1.99×10^{-9}

3.2 VULCANISM

It has been suggested that heat from vulcanism might have been an abundant and localized energy source in the primitive Earth. Evaluation of its efficiency and abundance is difficult, although many geologists assume that 4 Ga years ago there was considerably more volcanic activity than at the present time, so the scenario of high volcanic activity might have been frequent in the Archean. Another source of heat is released at steam spouts, fumaroles and similar locations.

An important phenomenon, first discovered in 1977 at the Galapagos is venting of hydrothermal system on the seafloor (Corliss et al., 1979). It is possible that at prebiological times organic materials were synthesized in the regions in which high pressures, temperatures and mineral redox buffers were obtained. Besides this, reactions resulting from ionizing radiation may occur in these hydrothermal systems (Ferris, 1992). The hydrothermal vents are considered as dynamic systems with potential capacity for abiotic organic geochemical processes (Holm, 1992, Holm and Hennet, 1992).

4. SOME CHEMICAL REACTIONS OF THE SYSTEM CARBON DIOXIDE/BICARBONATE/CARBONATE RELEVANT TO CHEMICAL EVOLUTION

4.1 RADIATION EFFECTS ON CO_2/ HCO_3^-/ CO_3^{2-}

As was pointed out in Section 1 most of Earth's carbon was in an oxidized inorganic form at the surface, either dissolved in the ocean, in the atmosphere or as solid carbonate minerals in sediments. These carbon forms could suffer either photolysis or radiolysis and could be reduced to produce organic compounds containing one or two carbon atoms.

In this section, we review briefly literature data on the effects of ionizing radiation on the system carbon dioxide/bicarbonate/carbonate. The effect of the irradiation depends on several variables. Some of these variables are physical state, concentration of the target molecules, pH, presence of other compounds, radiation dose, type of radiation, etc. In this survey we focus our attention on the general aspects of radiolysis in different physical states. For atmospheric carbon dioxide we also analyze its photolytic behavior.

4.1.1 CO_2 at the Atmosphere

From photochemical calculations it is known that if any reduced carbon species in the early atmosphere existed they would have been rapidly oxidized by photochemical reactions involving water vapor and its photolysis products (Kuhn and Atreya, 1979). Rapid early outgassing combined with photochemical reactions in the atmosphere should have yielded a situation in which most of Earth´s carbon was in oxidized form.

Carbon dioxide in the atmosphere could be photolyzed in presence of H_2O yielding H_2CO as principal product. Nevertheless, due to its high reactivity only 1 % of this formaldehyde survived. This percentage could be enough to reach a significant amount delivered into the oceans through its incorporation into rain droplets (Pinto, et al., 1980). Formaldehyde dissolved in the ocean could have had a key function in the formation of more complex organic molecules.

4.1.2 CO_2 in the Oceans

Several well documented radiation induced-reactions for the system $CO_2/HCO_3^-/CO_3^{2-}$ are known. Some of these experiments focused on the problem of chemical evolution, while others were made with different objectives.

Nevertheless, we will apply this knowledge to the synthesis of reduced carbon and then go into the question of evaluating the extent to which these pathways were probable on the primeval Earth.

In the radiolysis of aqueous solutions, radiation energy is deposited mainly in water molecules, producing the following radicals: H ([1]G = 0.5), e_{aq} (G = 2.7); ·OH (G = 2.8). Hydrogen peroxide and molecular hydrogen are also formed with yields of G = 0.71 and G = 0.45, respectively. These radicals attack the solute and form transient species that give rise to the observed products (Draganic, and Draganic, 1973). This effect is known as the indirect action of ionizing radiation and is the main pathway for the radiolysis of compounds in dilute aqueous solutions. Irradiation of oxygen-free CO_2-aqueous solution at pH 3.5, in the presence of ferrous ions was carried out by Garrison et al. (1951). CO_2 was reduced to formic acid. The same system was studied by Getoff et al.,1960 and Getoff, 1962, under various conditions of irradiation. They detected aldehydes, glycol, oxalic acid and formic acid. The products distribution were strongly dependent on the pH of the irradiated solution. Note that the pH controls that of the chemical forms CO_2, HCO_3^-, CO_3^{2-} predominate. It was observed that the number of molecules of formic acid produced decreases in going from acidic to a basic medium.

The synthesis of organic compounds using oxygen-free diluted aqueous solution of bicarbonate irradiated by gamma rays was studied by Draganic et al. (1991). The reaction mechanism established that the main pathway of the radiolysis was due to the reaction of hydrated electrons with CO_2 present in equilibrium with HCO_3^-. This reaction forms a carbonyl radical-ion that reacts giving rise to stable products. The main radiolytic products were formate (G = 2.2) and oxalate (G = 0.05). Trace amounts of formaldehyde and an unidentified polymer were also present. The dosage curve suggests that the formaldehyde is a secondary product due to the radiolytic decomposition of formate. The study was carried out in neutral and alkaline pH. At pH >10 the radiolytic products decreased.

Kolomnikov, et al.(1982) irradiated aqueous solutions or suspensions of different mineral carbonates. Their results showed that all radiolytic products were dependent on radiation dose, pH, and the nature of the metal ion present in the solution. Nevertheless, in all the

---

[1]G is defined as a radiochemical yield which is equal to the number of molecules formed, or destroyed, in a system when 100 eV are discharged in it.

experiments performed using Mg^{2+}, Ca^{2+}, Tl^{+}, Cu^{2+}, Hg^{+}, Hg^{2+} carbonates, oxalic acid was always present (7.5×10^{-2} - 1.3×10^{-2}M). In some of these mineral carbonates formic acid was formed. Also traces of glycolic acid and formaldehyde were observed in some systems.

4.1.3 Irradiation of Metal Carbonates in Solid State.

The effects produced by the irradiation of metal carbonates in the solid state are similar to those produced by irradiation of aqueous solutions, but the mechanism is completely different. Ionizing radiation in solids produces defects in the lattice structure. These defects are responsible for the chemical reactions. The complexity of such reactions is considerable and many transient and stable radicals can be formed and trapped in the lattice.

Irradiation of solid carbonates by gamma or X rays produce molecule-ions such a CO_3^{3-}, CO_3^{-} and CO_2^{-}. These species were identified by ESR methods (Serway and Marshall, 1967); Marshall et al. 1964). Other studies have been made with barium and calcium carbonates in the solid state (Albarrán et al., 1987a and 1987b). The detection of radiolytic products was made by dissolving the solid in water. This solution was analyzed by conventional methods. For example, the self-radiolysis of $Ca^{14}CO_3$ (doses 5.15 - 6.63 MGy) produces precursors that give upon dissolution formic (> 80 %) and oxalic (< 12 %) acids. Also glycolic, glyoxylic and acetic acids and formaldehyde were detected in low yields (Albarrán et al., 1987a and 1987b). It appears very likely that there are several interactions with the water and the radicals trapped in the lattice. Water acts as H-donor.

Primary carbonate sediments like $CaCO_3$ and other mineral carbonates have been widely distributed on the Earth. Further, they may be exposed to environmental radiation for very long periods, producing radiolytic compounds or trapped molecule-ions. But these carbonates are deposited in water. They can only be dissolved by change to acid pH or by exposure to hydrostatic pressure. The latter will only occur if shallow water calcium carbonates are transported below "calcite compensation depth", where they dissolve. The "calcite compensation depth" is a region in the ocean that represents the depth where $CaCO^3$ is dissolved as fast as it is deposited (Berner, 1971).

4.2 POTENTIAL ORGANIC SYNTHESIS FROM CO_2 IN SUBMARINE HYDROTHERMAL SYSTEMS

The proposal that abiotic synthesis can occur in hydrothermal systems was presented by Ingmanson and Dowler (1977). Since then many developments have been achieved (Shock, 1990 and the references therein). As was mentioned in section 3, for abiotic synthesis on the primeval Earth, it is important to identify the possible sources of energy and the potential raw material for those synthesis. These factors are essential to evaluate critically the potential role for the proposed pathways. Shock (1990) has proposed that organic synthesis in hydrothermal systems is driven mainly through oxidation/reduction reactions. The role of temperature is less important than the kinetic barriers to these reactions. The reactions occur only if kinetic barriers block the establishment of a stable equilibrium (Shock 1990, 1992). With respect to the raw materials in hydrothermal systems, based on present day hydrothermal vent fluids, the predominant form of volatile carbon is as CO_2 (Shock, 1992). Methane and short chain hydrocarbons are lower than CO_2 (Welhan and Lupton, 1987). Non volatile form of carbon is mainly as carbonates (Veizer et al., 1989). Abiotic synthesis of organic compounds in hydrothermal systems may occur through the reduction of the CO_2 and N_2. From these sources, Shock (1992) estimated that the potential organic productivity of submarine hydrothermal synthesis would be ~ $2 \times 10^7 - 2 \times 10^8$ kg/year now and could have been considerably higher on the primeval Earth.

It has been hypothesized that synthetic processes driven by ionizing radiation may occurs in hydrothermal systems (Ferris, 1992). However, data on actual radiolyis in those conditions are lacking.

It should be kept in mind that there are ample spectra of probable environments in these hydrothermal systems. Organic synthesis in submarine hydrothermal is more likely in fluids circulating through the flanks rather than in those high-temperature mid-ocean ridges or venting at black smokers (Shock, 1990, 1992).

From experimental results, it has been suggested that abiotic synthesis from inorganic starting materials also polymerization reactions, may be possible in hydrothermal systems. However, well constrained experiments at properly controlled hydrothermal conditions and more theoretical calculations are needed to evaluate the role of these environments to abiotic synthesis. These achievements would represent a significant break through the study of chemical evolution.

5. CONCLUDING REMARKS

Carbon as CO_2 in the atmosphere, by its fixation into the oceans (HCO_3^-) or into sediments as CO_3^{2-} could have played an important role, as a raw material, for prebiotic synthesis of aldehydes and carboxylic acids such as formaldehyde, glyoxal, formic acid, oxalic acid, and other compounds containing one or two carbon atoms. By irradiation of oxidated inorganic compounds it is possible to obtain reduced organic compounds. On the primitive Earth the main source of ionizing radiation came from the radioactive isotope ^{40}K, was 8 times more abundant than it is now. The radiation dose estimated from ^{40}K 4.0 Ga ago was around 2.2×10^{-3} Gy/year. The possible amount of organic material that might have been formed by the radiolytic effect of ^{40}K in mineral carbonates can be estimated. For example, the yield of formic acid produced by the β or γ radiolysis of solid $CaCO_3$, after dissolution is $G = 1.8 \times 10^{-2}$ molecules/100 eV. The formic acid that might have formed from ^{40}K radiolysis of carbonates may be $\approx 10^{-10}$ g/year/kg of carbonate. This amount is not neglible and supports the hypothesis that primary carbonates in the ocean or as deposits in the seafloor, exposed to natural ionizing radiation should be considered as a meaningful pathway for primordial organic synthesis.

It is also important to remark that ionizing radiation may have a role in the formation of organic compounds in hydrothermal vents (Ferris, 1992). However, data on actual radiolysis in conditions of hydrothermal environments are lacking. The importance of ionizing radiation in hydrothermal vents remain to be confirmed by experimental evidence. Still, from the results of the radiolysis of solutions of carbonates and bicarbonates it is expected that ionizing radiation may drive some organic synthesis in these environments.

REFERENCES

Albarrán, G., K.E. Collins, C.H. Collins, WY,DLY: Formation of Organic Products in Self-Radiolyzed Calcium Carbonate, J. Mol. Evol., **25**, 12-14.

Albarrán, G. and C.H. Collins, 1987b: Ion-moderated Partition Chromatographic Determination of $Ca^{14}CO_3$ and $Ba^{14}CO_3$ Self-radiolysis Products, J. Chromatogr. **395**, 623-629.

Albarrán, G., A. Negrón-Mendoza, C. Treviño, J.L. Torres, 1988: Role of Ionizing Radiation in Chemical Evolution Studies, Radiat. Phys. Chem, **31**, 821-823.

Berner, R.A., 1971, Principles of Chemical Sedimentology, McGraw Hill Book, New York.

Chyba, C., and C. Sagan, 1991: Electrical Energy Sources for Organic Synthesis on the Early Earth, Origins Life Evol. Biosphere, **21**, 3-17.

Corliss, J.B., J.Dymond, L.I. Gordon J.M. Edmond, R.P. Vond Herzen, R.D. Ballard, K. Green, D. Willians, A. Bainbridge, K.Crane, and T.H. Van Andel, 1979: Submarine Thermal Springs on the Galapagos Rift, Science, **203**, 1073-1083

Draganic, I.G., and Z.D. Draganic, 1973: The Radiation Chemistry of Water, Academic Press, New York.

Draganic, I.G., Z.D. Draganic, and J.P. Adloff, 1990: Radiation and Radioactivity on Earth and Beyond, CRC Press, Inc. Boca Raton, Florida.

Draganic, Z.D., A. Negrón-Mendoza, K. Schested, S.I. Vujosevic, R. Navarro-González, M.G. Albarrán, I.G. Draganic, 1991: Radiolysis of Aqueous Solutions of Ammonium Bicarbonate Over a Large Dose Range, Radiat. Phys. Chem., **38**, 317-321.

Ferris, J.P. 1992: Chemical Markers of Prebiotic Chemistry in Hydrothermal Systems, Origins Life Evol. Biosphere, **22** 109-134.

Garrison, W.M., D.C. Morrison, J.G. Hamilton, A.A. Benson, M. Calvin, 1951: Reduction of Carbon Dioxide in Aqueous Solutions by Ionizing Radiation, Science, **114**, 416-418.

Getoff, N., G. Scholes, J. Weiss, 1960: Reduction of Carbon Dioxide in Aqueous Solutions Under the Influence of Radiation, Tetrahedron Letters, **18**, 17-23.

Getoff, N., 1962: Synthese Organischer Stoffe aus Kohlensaure in Wasseriger Losung unter Einwirkung von Co^{60} gamma Strahlung, Int. J. Appl. Radiat. Isot., **13**, 205- 213.

Holm, N.G., 1992: Why are Hydrothermal Systems Proposed as Plausible Enviroments for the Origin of Life?, Origins Life Evol. Biosphere, **22**, 5-14.

Holm, N.G., and R.J.-C. Hennet, 1992: Hydrothermal Systems: Their Varieties, Dynamics, and Suitabi-lity for Prebiotic Chemistry, Origins Life Evol. Biosphere, **22**, 5-31.

Ingmanson, D.E., and M.J. Dowler, 1977: Chemical Evolution and Evolution of the Earth's crust, Origins Life Evol.Biosphere, **8**, 221-224.

Kasting, J.F., K.L. Zahnle, J.C.G. Walker, 1983: Photochemistry of Methane in the Eart's Early Atmosphere, Precambrian Res. **20**, 121-148.

Kathren, R.L., 1984: Radioactivity in the Environment: Sources, Distribution, and Surveillance, Harwood Academic Publishers, New York.

Klement, A.W. 1982: Handbook of Environmental Radiation, A. Brodsky (Ed.) CRC Press, Inc., Boca Raton, Florida.

Kolomnikov, I.S., T.V. Lysyak, E.A. Konash, E.P. Kaliazin, A.V. Rudnev, Yu.Ya. Kharitonov, 1982: Formation of Organic Products from Metal Carbonates and Water by the Action of Ionizing Radiation, Dokl. Phys. Chem., **25**, 596-597.

Kuhn, W.R. and S.K. Atreya, 1979: Ammonia Photolysis and the Greenhouse Effect in the Primordial Atmosphere of the Earth, Icarus, **37**, 207-213.

Lowe, D.R., 1980: Archean Sedimentation, Ann. Rev. Earth Planet. Sci., **8**, 145-67.

Lowe, D.R., 1994: Early Environments: Constraints and Opportunities for Early Evolution, Nobel Symposium, in press.

Marshall, S.A., A.r. Reinberg, R.A. Serway, J.A. Hodges, 1964: Electron Spin Resonance Absorption Spectrum of CO_2^- molecule-ions in Single Crystal Calcite, Mol. Phys., **8**, 225-231.

MacDonald, G.J.F., 1959: Calculations on the Thermal History of the Earth, J. Geophys. Res. **64**, 1967.

Miller, S. L., and H. C. Urey, 1959: Organic compounds Synthesis on the Primitive Earth, Science, 130, 245.

Pinto, J.P., G.R. Gladstone, Y.L. Yung, 1980: Photo-chemical Production of Formaldehyde in Earth's Primitive Atmosphere, Science, **210**, 183-185.

Sclater, J.G., B. Parsons, and C. Jaupart, 1981: Oceans and Continents: Similarities and Differences in the Mechanims of heat loss, J. Geophys. Res. **86**, 11535.

Serway, R.A., and S.A. Marshall, 1967: Electron Spin Resonance Absorption Spectra of CO_3^- and CO_3^{3-} Molecule-Ions in Irradiated Single-Crystal Calcite, J. Chem. Phys. **46**, 1949-1952.

Schidlowski, M., 1988: A 3,800-million-year isotopic record of life from carbon in sedimentary rocks, Nature, **333**, 313-318.

Schidlowski, M., 1991: Organic Carbon Isotope Record: Index Line of Autotrophic Carbon Fixation over 3.8 Gys of Earth History, Southeast Asian Earth Sci., **5**,1-4.

Shock, E.L., 1990: Geochemical Constraints on the Origin of Organic Compounds in Hydrothermal Systems, Origins Life Evol. Biosphere, **20**, 331-367.

Shock, E.L., 1992 a: Chemical Environment of Submarine Hydrothermal Systems, Origins Life Evol. Biosphere, **22**, 67-107.

Shock, E.L., 1992 b: Hydrothermal Organic Synthesis Experiments, Origins Life Evol. Biosphere, **22**, 135-146.

Spinks, J.W.T., and R.J. Woods, 1990: An Introduction to Radiation Chemistry, Wiley & Sons, N.Y. 3th edition.

Sprague, D.S., H.N. Pollack, 1980: Heat flow in the Mesozoic and Cenozoic, Nature, **285**, 393-395.

Veizer, J.,J. Hoefs, R.H. Ridler, L.S. jensen and D.R. Lowe, 1989, Geochim. Cosmochim. Acta, **53**, 845-857.

Walker, J.C.G., 1985: Carbon Dioxide on the Early Earth, Origins of Life Evol. Biosphere, **16**, 117-127.

Welhan, J.A. and J.E. Lupton, 1987: Light hydrocarbon gases in Guaymas Basin Hydrothermal fluids: Thermogenic versus abiogenic origin, Amer. Assoc. Petr. Geol. Bull., **71**, 215-223.

Wetherill, G.W., 1980: Formation of the Terrestrial Planets, Ann. Rev. Astron. Astrophys. **18**, 77-113.

PART II
GEOPHYSICAL ASPECTS OF SELF-ORGANIZATION

EARLY TERRESTRIAL LIFE: PROBLEMS OF THE OLDEST RECORD

Manfred Schidlowski
Max-Planck-Institut für Chemie
(Otto-Hahn-Institut)
D-55020 Mainz (Germany)

ABSTRACT

A review is presented of the currently available empirical (paleontological and biogeochemical) evidence encoded in, and retrievable from, the sedimentary record which spans 3.8 Gyr (or ~85%) of the Earth's history. The balance of this evidence is overwhelmingly in favour of the existence of bacterial (prokaryotic) ecosystems as early as 3.8 Gyr ago, becoming virtually unassailable as from 3.5 Gyr ago. Current problems related to the documentation of early life processes in the terrestrial rock record principally center around (1) the metamorphic impairment and partial obliteration of the oldest (Isua) segment of the paleobiological record, (2) the virtual absence (or non-documentation) of an archaebacterial lineage that should have been conspicuous, if not dominant, in the oldest terrestrial ecosystems, and (3) the potential role of impact interference (if any) with the lifestyle, proliferation and preservation of the oldest microbial communities.

1. INTRODUCTION

The origin and establishment of life on Earth continue to stand out as major challenges to the human mind. Although there is little doubt that life must have emerged at a certain stage of either cosmic or planetary evolution as an intrinsically new property of matter (cf. Sagan, 1974; McKay, 1991; Ponnamperuma, 1993; and others), crucial details of this process still await elucidation. However, the last decades have witnessed concerted efforts to resolve the residual questions that have involved all major scientific disciplines and embarked on an impressive array of exploratory concepts [see Oró et al. (1990) and Greenberg et al. (1993) for recent overviews].

In principle, the problem of the origin and early evolution of life can be approached from two sides, with the shift of the respective frontiers apt to progressively diminish the realm of uncertainty between the explored domains. One of these frontiers is ultimately driven by the remarkable brand of prebiological organic chemistry proceeding in specific astrophysical environments such as interstellar molecular clouds. Radio astronomical molecular spectroscopy has, of late, confirmed the presence there of an impressive inventory of organic molecules (cf. Irvine and Knacke, 1989) that could serve as intermediates in the prebiotic synthesis of nucleic acids, proteins and sugars. Besides, it is well known that the submicron-sized frozen dust particles that make up the bulk of matter in the interstellar medium are preferential sites of condensation reactions that consume about half of the oxygen-, carbon- and nitrogen-containing volatile species present in these environments. Impinging cosmic UV radiation has been shown to act as main promoter of a vigorous photochemistry in the outer mantles of these grains, giving rise to high-molecular-weight organic polymers of prebiotic significance (Greenberg and Mendoza-Gómez, 1993). Forming an important constituent of the protosolar nebula, the organic component of this interstellar dust is ultimately consumed by star formation, a major part of it ending up in the chondritic fraction of newly formed planets. Residual material of the nebula preserved in the marginal outer reaches of the primary dust cloud constitutes a reservoir of cometary matter whose organic component potentially provides seeding material for initiating a prebiological organic chemistry in suitable (specifically water-bearing) planetary environments (cf. Kissel and Krüger, 1987).

While these findings testify to the proceedings of abundant organic reactions of apparently prebiological significance in outer space, a second frontier can be opened by tracing the fossil evidence of terrestrial life back into the geological past, attempting to constrain the time of transition from a prebiotic to a biologically dominated world (with the concomitant onset of a paleontological and biogeochemical record). Earth-based scenarios for the origin of life (Oparin, 1936; Miller, 1955; Miller et al., 1976; Chang, 1993) have mostly centered around the abiotic synthesis of organic compounds from the Earth's primary degassing products such as $H_2O, CO_2, CO, CH_4, NH_3$, and others (with solar UV radiation, electric discharges, or telluric heat as possible energy sources) and the subsequent establishment of a primordial heterotrophic biosphere which, after exhaustion of the "primeval broth", had subsequently switched over to (photo)autotrophy. Of late, this "classical" scenario has been challenged by models postulating a chemoautotrophic origin of life, envisioning the chemical energy of inorganic reactions (such as pyrite formation) as driving force for the buildup of organic substances (Wächtershäuser, 1988). However intellectually fascinating and internally stringent in their own rights, these models are nevertheless fraught with various uncertainties, providing at best approximations to the intricate sequence of events that consequently led to the establishment of biology on our planet. In contrast, the paleontological and biogeochemical information encoded in the sedimentary record constitutes factual (empirical) evidence which, on the other hand, does not always lend itself to a quick and unambiguous interpretation.

The following discussion is supposed to critically address the principal uncertainties that currently beset the interpretation of the paleobiological evidence preserved in the oldest (Archaean) geological record. The ultimate aim of this discourse is to arrive at a more refined demarcation of the boundary line that sets off the major segment of the Earth's history characterized by the presence of life processes from those remote parts of the geological past that withhold relevant evidence.

2. BEGINNINGS OF BIOLOGY ON THE EARLY EARTH: EMPIRICAL APPROACHES TO A FRONTIER

A multifaceted problem such as the origin of life lends itself to a variety of interdisciplinary approaches spanning the range from biological to geological sciences. The following attempt of a status report constitutes an update of a recent summary (Schidlowski, 1993a) that had focused on the currently available empirical evidence, i.e., paleontological documents and biogeochemical data that are encoded in, and may be retrieved from, the hitherto known geological record. Paleobiological documents of such kind may be justly categorized as *prima facie* evidence of ancient life, irrespective of the fact that their detailed interpretation may be occasionally beset with difficulties. It should be stated at the outset that geological evidence pertaining to the antiquity of life on Earth cannot predate the start of the presently known sedimentary record (3.8 Gyr ago) as the sole carrier of relevant information (see Fig.1).

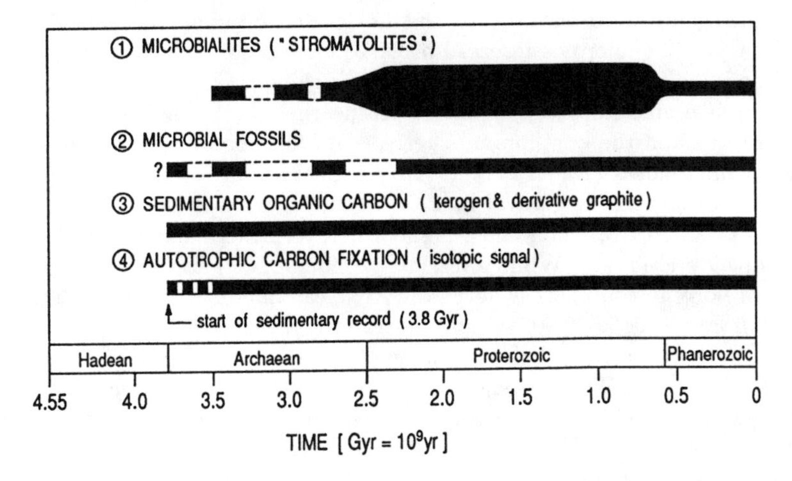

Figure 1. Principal categories of paleontological (1,2) and biogeochemical (3,4) evidence over 3.8 Gyr of geological history. The bulging stromatolite diagram (1) reflects the proliferation of microbial ecosystems after the formation of extensive marine shelves during the Proterozoic and their decline after the advent of heterotrophic (animal) grazers 0.6-0.7 Gyr ago. The extension to 3.8 Gyr of the microfossil record (2) is contingent on the biogenicity of the principal cell-like morphotype (*Isuasphaera isua*) from the Isua Suite. The isotopic signature of biological (autotrophic) carbon fixation as preserved in sedimentary organic carbon bears a metamorphic overprint as from t> 3.5 Gyr (broken line), whose direction and magnitude is, however, well understood (for further information see text).

2.1. THE EARLY PALEONTOLOGICAL RECORD

Having been demonstrably dominated by prokaryotic and eukaryotic microbial ecosystems prior to the emergence of the first Ediacaran-type metazoan faunas at the dawn of the Phanerozoic, the Precambrian Eon (~0.6 to 3.8 Gyr) has furnished an impressive, though initially patchy, record of cellular microfossils and biosedimentary structures ("stromatolites") that reflect the activities of mat-forming prokaryotic microbenthos (Fig.1). Stand-outs among the hitherto described Late Precambrian (Proterozoic) microfloras are the ~0.9 Gyr-old Bitter Springs assemblage of Australia (Schopf, 1968) and the ~2.0 Gyr-old microbial community of the Gunflint iron-formation of Canada (Tyler and Barghoorn, 1954; Barghoorn and Tyler, 1965; Cloud, 1965). Recent findings by Han and Runnegar (1992) apparently confirm previous inferences (Kazmierczak, 1979; Pflug and Reitz, 1985) that the evolution of the early microbial world had culminated in the appearance of the eukaryotic (nucleated) cell already some 2.1 Gyr ago, i.e., considerably earlier than previously assumed (cf. Cloud, 1976; Schopf and Oehler, 1976).

When going back in time to the Early Precambrian or "Archaean" Era (2.5–3.8 Gyr), the microfossil record becomes increasingly scant (Awramik, 1982; Schopf and Walter, 1983). Naturally, this patchiness (Fig.1) is due to gaps in the occurrence and discovery of microfossil-bearing chert lithologies rather than to interruptions in the continuum of terrestrial life; moreover, new discoveries of Archaean microfloras (Naqvi et al., 1987; Venkatachala et al., 1990) and more detailed investigations (e.g., Walsh, 1992) of the previously described ones (e.g., Muir and Grant, 1976; Knoll and Barghoorn, 1976; Schopf and Walter, 1982) are coming up at a rapid rate. While several of the older reports appear to be dominated by evidence on the "dubiofossil" level, the Archaean microfloras described from the Australian Fortescue (~ 2.7 Gyr) and Warrawoona Groups (~3.5 Gyr) have come to be generally accepted as being made up of *bona fide* microfossils.

Specifically, the Warrawoona community has been accorded benchmark status as the oldest unassailable evidence of a cellularly preserved microflora. With the Warrawoona Group bracketed between 3.3 and 3.5 Gyr, and the fossiliferous chert lithology having been assigned an age of ~3.465 Gyr, this ancient microbial assemblage appears astoundingly modern and diverse. Following the pioneering studies and attendant discussions by Awramik et al. (1983) and Buick (1984), subsequent work by Schopf and Packer (1987) and Schopf (1993) has shown that the preserved prokaryotic community is dominated by various taxa of filamentous (trichomic) and colonial coccoidal (usually sheath-enclosed) microorganisms of preferentially cyanobacterial and flexibacterial affinities. It should be noted that the Warrawoona rocks also harbour the oldest stromatolites (Fig.1) hitherto encountered in the geological record (Walter et al., 1980) which gives supportive biosedimentary evidence for the presence of benthic microbial ecosystems within the primary depositional environment. Altogether, both the morphological diversity and the excellent preservation of the constituent microbiota have endowed the Warrawoona microflora with an air of authenticity, thereby defining a piece of firm ground within a terrain often swamped by evidence considered contentious or equivocal. Given the remarkable degree of morphological and possibly also physiological diversity of this assemblage, there is little doubt that the ancestral lines of bacterial (prokaryotic) life as well as the evolution of any preceding lineages of chemosynthetic (archaebacterial) precursor forms must considerably predate Warrawoona times.

With the existence of bacterial precursor floras to the Warrawoona community thus highly probable (if not almost certain), assemblage of cell-like morphologies reported from $\sim$ 3.8 Gyr-old metasediments within the Isua Supracrustal Belt of West Greenland (Pflug 1978, 1987; Pflug and Jaeschke-Boyer, 1979; Pflug and Reitz, 1992) were bound to stir up considerable interest. This holds particularly for a globular, supposedly sheath-enclosed cell-like morphotype described as *Isuasphaera isua* Pflug that had been recovered from a metachert lithology collected during the 1976 Isua field campaign of the Max-Planck-Institute for Chemistry, Mainz (cf. Schidlowski et al., 1979). The biogenicity of this and related cell-resembling morphologies had been subsequently disputed on variable grounds, notably the improbability of survival of microfossils during the amphibolite-grade metamorphism to which the Isua rocks had been subjected (Bridgwater et al., 1981), or the invocation of convergences with purely mineralogical features such as dissolution cavities (Roedder, 1981).

Meanwhile, however, there is ample evidence that both macro- and microfossils may preserve at least their basic identities through low- to medium-grade metamorphism (notably of greenschist and lower amphibolite facies), and this to a degree permitting even a palynostratigraphic approach to the study of metamorphosed sediment series (Pflug and Reitz, 1992). This would imply that objections to the proposed biogenicity of the Isua micromorphs can be based solely on a critique of the objects as such and not on petrological criteria related to the metamorphic reconstitution of their host rocks.

Since the structurally more differentiated septate filaments and coccoidal colonial unicells characterizing the younger record (and notably the Warrawoona microflora) are absent from the Isua assemblage, the cellular evidence is markedly empoverished and decidedly less unequivocal as compared to younger microfossil assemblages. Part of the depauperation of the Isua record can be attributed to the amphibolite-grade metamorphic overprint of the host rock that was bound to erode a wealth of informative morphological detail. However, in spite of the obvious reduction of the morphotype inventory and the conspicuous absence (or secondary obliteration) of structural detail, there is a striking resemblance of *Isuasphaera* as the preponderant Isua morphotype to a possible counterpart in the foregoing Archaean-Proterozoic record known as *Huroniospora* sp. whose biogenic interpretation appears to be well founded and generally accepted (cf. Barghoorn and Tyler, 1965). It has been shown also that the near-perfect morphological analogy between the two forms is matched by selected microchemical characteristics of their residual carbon coatings as revealed by Laser Raman spectroscopy (Pflug, 1987). Hence, even with due application of critical standards, selected microstructures from the Isua metasedimentary sequence seem to qualify for at least dubiofossil status as they clearly stand in the continuity of the previous record [see also Weber (1988) for a more recent re-investigation of *Isuasphaera isua*]. Moreover, finds in the Isua iron-formation of possible aggregate-forming, *Siderocapsa*-type iron bacteria displaying intricate external cell wall structures as lately reported by Robbins (1987) would lend additional credence to the probable existence of a fairly advanced early microbial world during Isua times.

Summing up the available evidence, the statement seems warranted that at least part of the morphological evidence reported from the 3.8 Gyr-old Isua Metasedimentary Suite may suggest microbial affinities, irrespective of the extraordinary degree of uncertainty surrounding individual morphotypes and of occasional convergences with purely mineralogical features. While there may be differences of opinion whether or not the degree of structural differentiation, and the quality of preservation, of the observed morphotypes justify a taxonomic description as genera and species, it certainly cannot be excluded at this stage that the inventory as such contains at least some elements of a structurally degenerated "sticks and balls" assemblage that would result from an intense metamorphic alteration of filamentous and coccoidal members of a Warrawoona-type Archaean microfossil community.

2.2 PALEOBIOLOGICAL INFERENCES FROM THE EARLY CARBON RECORD

Apart from the paleontological record, there also exists a biogeochemical record of life, primarily expressed in the form of sedimentary organic carbon. Carbon is present in sedimentary rocks in two modifications, namely, (1) as oxidized or carbonate carbon (C_{carb}) and (2) as reduced or organic carbon (C_{org}) representing the fossil residuum of primary biogenic substances. For the most part, the organic moiety is made up of kerogen and related substances, i.e., acid-insoluble, high-molecular-weight (polycondensed) organic substances representing the end-product of the diagenetic alteration of organic debris that had come to be buried in sediments. On average, sedimentary rocks contain between 0.5 and 0.6 % organic carbon (Ronov, 1980). Students of the carbon record had been intrigued since long by observations that the C_{org} content of the average sediment seems to holds through to the very beginning of the record, with the C_{org} burden of Precambrian sediments hardly smaller than that of geological younger rocks (Schidlowski, 1982). For instance, the (largely graphitized) C_{org}-content of the 3.8 Gyr-old Isua metasediments has been found to amount to >0.6 % in carbon-rich members of the sequence.

Moreover, observed $^{13}C/^{12}C$ fractionations between sedimentary carbonate and organic carbon have been shown to closely resemble respective fractionations in the contemporary world (cf. Schidlowski, 1988). The characteristic difference of -20 to -30 permil in terms of the conventional $\delta^{13}C$ notation between C_{org} and C_{carb} is known to stem from the fact that all common pathways of autotrophic carbon fixation (notably the photosynthetic ones) discriminate against the heavy carbon isotope (^{13}C), the resulting bias in favour of ^{12}C principally deriving from a kinetic isotope effect inherent in the first irreversible enzymatic carboxylation reaction. When carbonate and organic carbon enter newly-formed sediments, they tend to preserve, with just little diagenetic change, the isotopic signatures acquired in their primary surficial environments, thus propagating the kinetic isotope effect inherent in photosynthetic carbon fixation from the biosphere into the rock section of the carbon cycle. Accordingly, the C_{org} (kerogen) fraction of sedimentary rocks preserves the transcript of the isotope-discriminating properties primarily of ribulose-1,5-bisphosphate carboxylase, the key enzyme of the Calvin cycle that channels most of the carbon transfer from the inorganic to the organic world (Hayes et al., 1983; Schidlowski et al. 1983; Schidlowski, 1988).

In this way, the sedimentary carbon isotope record conveys a coherent signal of photosynthetic carbon fixation over the geologic past. Neglecting contentious detail regarding possible changes in the magnitude of this fractionation over geologic time (cf. DesMarais et al., 1992; Schidlowski, 1993b), the isotopic index line of photosynthetic carbon fixation has been demonstrated to persist over the whole of the unmetamorphosed rock record back to $\sim$3.5 Gyr ago, a result fully consistent with the reported existence of prolific microbial ecosystems at Warrawoona times.

Further, when making allowance for the metamorphic overprint suffered by currently known pre-Warrawoona rocks, biologically mediated carbon isotope fractionations can be traced back, with fair confidence, to the very beginning of the sedimentary record. This holds particularly for the $\sim$3.8 Gyr-old Isua Metasedimentary Suite that exemplarily displays the discontinuity in $\delta^{13}C$ between the unmetamorphosed and the metamorphosed Archaean record. While the long-term averages for sedimentary $\delta^{13}C_{carb}$ and $\delta^{13}C_{org}$ have been given as $+0.5 \pm 2.5$ °/oo [PDB] and -26.0 ± 7.0°/oo [PDB], respectively (Schidlowski et al., 1975, 1983; Veizer et al., 1980), the corresponding means for the Isua Suite are -2.3 ± 2.2°/oo and -13.0 ± 4.9°/oo (Schidlowski et al., 1983). The observed isotope shifts are, however, fully consistent with the predictable effects of a high-temperature isotopic reequilibration between coexisting carbonate and organic carbon in response to the amphibolite-grade metamorphism experienced by the Isua rocks. Both currently available thermodynamic data on $^{13}C/^{12}C$ exchange between C_{carb} and C_{org} as a function of increasing metamorphic temperature (Bottinga, 1969), and observational evidence from geologically younger metamorphic terrains (Valley and O'Neil, 1981; Wada and Suzuki, 1983; Arneth et al., 1985) make it virtually certain that the "normal" sedimentary $\delta^{13}C_{carb}$ and $\delta^{13}C_{org}$ records had originally gone back to 3.8 Gyr ago, and that the anomalous means encountered in the Isua rocks are clearly due to a metamorphic overprint (for details see Schidlowski et al., 1979, 1983; Schidlowski 1987, 1988). This would argue persuasively for the operation as early as 3.8 Gyr ago of autotrophic carbon fixation, and specifically photosynthesis, as the quantitatively most important carbon fixing process. As a consequence, the impact of biology on the terrestrial carbon cycle dates back very early in the geologic past, being largely synchronous with the formation of the oldest terrestrial sediments.

2.3. EARLY ARCHAEAN SEDIMENTARY ENVIRONMENTS: WERE THEY COMPATIBEL WITH THE EXISTENCE OF LIFE?

With both the paleontological and biogeochemical records indicating that the Archaean Earth had hosted a patchy veneer of preferentially prokaryotic life since at least 3.5, if not 3.8 Gyr ago, the question might arise whether the paleoenvironmental, lithological and mineralogical evidence retrieved from the oldest sedimentary sequences give complementary support to the inferred or documented presence of biological activity in those times. This should be particularly relevant for the Isua supracrustals where both strands of paleobiological information have been gravely impaired or modulated by high-temperature metamorphism. In this context, the occurrence of conglomerate horizons in the lowermost "Quartzitic Unit" and notably in the middle "Carbonate-Bearing Siliceous Schist" member of the Isua Suite (cf. Bridgwater and McGregor, 1974; Allaart, 1976; Schidlowski et al., 1979) appears to be of crucial significance as it gives testimony to the presence of running water and thus of a hydrosphere on the terrestrial surface as from 3.8 Gyr ago (Fig. 2). Being characterized by a unique set of chemical properties (such as high polarity, a high dielectric constant, the capability to form hydrogen bonds, and others), liquid water is so intimately involved in the chemistry of life processes as to make it an absolute prerequisite for the existence of life as such (cf. Brack, 1993).

Figure 2. Loosely packed conglomerate (with prominent major boulder) from the middle part of the Isua Suite, attesting to the presence of running water on the terrestrial surface 3.8 Gyr ago (i.e., ~0.7 Gyr after the Earth's formation).

In view of this relationship, it has been even suggested that water be viewed as a surrogate indicator of life (McKay, 1991). Applying such criteria, the existence of life in Isua times would be a very reasonable, if not cogent postulate.

Also, the intercalation in the Isua Suite of banded iron-formation (Appel, 1980) is likely to be of paleobiological and paleoenvironmental significance. Banded iron-formation (BIF) and related lithologies have been found to figure as characteristic members of the Archaean-Proterozoic record until their factual disappearance during the time interval 1.8 - 2.0 Gyr ago. Current models of their genesis include concepts that envision links between the iron-bound oxygen constituent of these sedimentary iron ores and the O_2-release by contemporaneous microbial ecosystems principally made up of photosynthetic prokaryotes (Cloud, 1973; Schidlowski, 1990). As already pointed out by Cloud (1976), the occurrence in Isua times of banded iron-formation might indicate a corresponding persistence of the biological agent believed to be involved in BIF generation.

Decisive support for the operation of life processes during Isua times has also come lately from improved knowledge of the low-temperature aqueous phosphate system, assigning to sedimentary apatite (calcium-fluoro-carbonate-hydroxyphosphate) the role of a potential biomarker mineral (Arrhenius et al., 1993). While it could be shown that phosphate precipitation from sterile solutions with the large Mg/Ca ratio (> 5) and the pH value of sea water yields exclusively Mg-Ca and Ca hydrogen("protonated")phosphates, the formation of apatite as a Mg- and H-free phosphate phase appears to depend on biological mediation. Generalizing from currently available experimental data, the statement seems warranted that the formation of apatite at sea water Mg/Ca ratios is conditional on an effective removal, or containment in the liquid phase, of the quantitatively dominant magnesium component. The exclusion of Mg from processes of marine phosphate formation is apparently due to microbial activity that proceeds within selected semi-closed (cellular or interstitial) microenvironments conducive to a near quantitative complexation of Mg^{2+} with metabolically derived ammonium ion, thereby preventing magnesium involvement in phosphate precipitation (see, inter alia, Lucas and Prévôt, 1985). As specifically pointed out by Arrhenius (1993), such findings would accord biomarker quality to sedimentary apatite if formed in an ocean characterized by quasi-modern Mg/Ca and pH values, and with NH_4^+ concentrations distinctly lower than isomolar with Mg^{2+} (at higher ammonia levels, apatite formation may also proceed inorganically in the pH-range <9 that approaches modern sea water). As we have little reason to expect dramatic changes in the chemistry of the early oceans that might have permitted apatite formation without the help of biological mediation, the occurrence notably in the Isua banded iron-formation of authigenic apatite (in part intergrown with the graphitic derivatives of sedimentary organic matter) is an important corollary in the set of evidence that crucially bears on the problem of the existence of life on Earth 3.8 Gyr ago.

3. OUTLOOK: OPEN QUESTIONS AND PERSPECTIVES

Considering the evidence set out above, there is little doubt that the ancient Earth had hosted a microbial biosphere as from about the onset of the sedimentary record 3.8 Gyr ago. This is likely to place the very beginnings of terrestrial life far back into the "Hadean" Era virtually devoid of geological documents. There is reason to believe that, after its emergence, all subsequent diversification of life had proceeded in an evolutionary continuum, with Darwinian evolution operating on a genetic program based on DNA. Specifically, molecular phylogeny grounded on the evaluation of nucleic acid sequences has given increasing credence to the notion that all terrestrial life is ultimately rooted in a single common ancestor or "progenote" (cf. Woese, 1981).

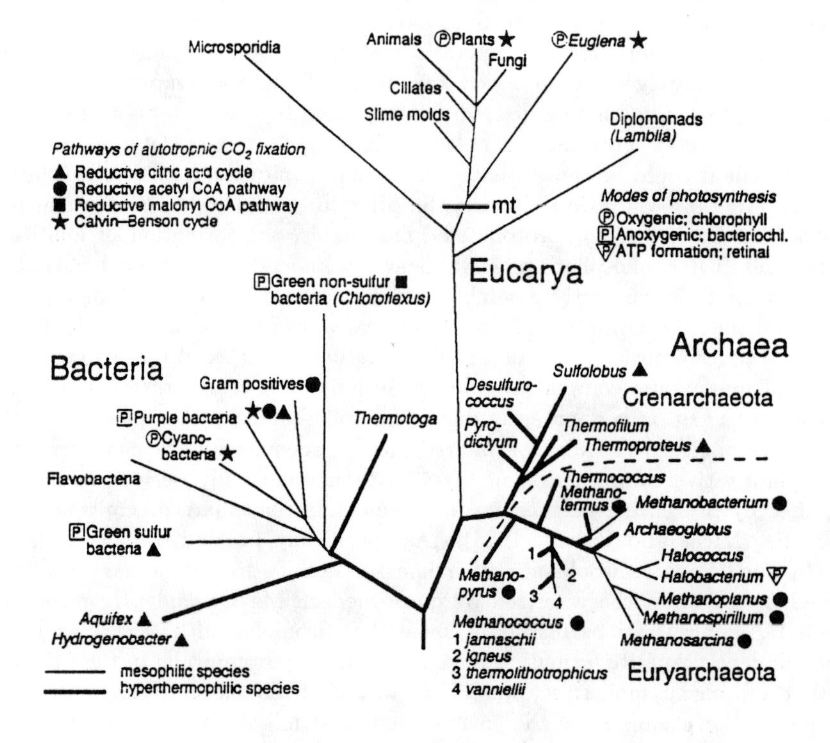

Figure 3. Universal phylogenetic tree based on sequence comparisons of selected semantophoretic molecules, notably 16S/18S rRNA (modified and updated from Kandler, 1994; Kandler, pers. comm., 1995). Note tripartition into the domains Bacteria, Archaea and Eukarya as well as pronounced clustering of hyperthermophilic taxa (strong-lined branches) at the very base of the tree.

A currently constructed phylogenetic tree primarily based on 16S/18S rRNA sequence comparisons (Fig. 3) defines three major domains of organisms (Bacteria, Archaea, Eukarya), with hyperthermophilic chemoautotrophs largely monopolizing the lowermost branches. This is particularly true for most of the archaebacteria and should hold, by inference, also for the ancestral progenote at the very base of the system. It seems, therefore, reasonable conjecture to envision a high-temperature and chemoautotrophic origin of life (Woese, 1987; Wächtershäuser, 1988), with a probable persistence of the hyperthermophilic lifestyle through the extended period of impact frustration that had possibly harrassed early microbial ecosystems at times > 3.9 Gyr ago (cf. Sleep et al., 1989). In accordance with such scenarios, it could be reasonably expected that chemoautotrophic hyperthermophiles of mostly archaebacterial affinity had held dominion over the Earth well until Archaean times.

In contrast to such inferences, the early paleontological record appears to be dominated by evidence on the cyanobacterial level, even in lithologies found in close proximity to high-temperature environments (such as the cherts within the 3.465 Gyr-old Apex Basalts of the Warrawoona Group, cf. Schopf, 1993). Subsequent work will have to reconcile the empirical findings from the geological record with the apparently stringent inferences derived from molecular phylogeny. A promising field for relevant efforts is the paleobiology of Archaean volcano-sedimentary and hydrothermal systems whose potential store of information is as yet virtually untapped.

REFERENCES

Allaart, J.H., 1976: The pre-3760 Myr old supracrustal rocks of the Isua area, central West Greenland, and the associated occurrence of quartz-banded ironstone. In *The Early History of the Earth*, B.F. Windley (Ed.), Wiley, London, 177-189.

Appel, P.W.U., 1980: On the early Archaean Isua iron-formation, West Greenland, *Precambrian Res.* **11**, 73-87

Arneth, J.D., M. Schidlowski, B. Sarbas, U. Goerg, and G.C. Amstutz, 1985: Graphite content and isotopic fractionation between calcite-graphite pairs in metasediments from the Mgama Hills, Southern Kenya, *Geochim. Cosmochim. Acta*, **49**, 1553-1560.

Arrhenius, G., B. Gedulin, and S. Mojzsis, 1993: Phosphate in models for chemical evolution. In *Chemical Evolution: Origin of Life*, C. Ponnamperuma and J. Chela-Flores (Eds.), Deepak Publ., Hampton, Virginia, 25-49.

Awramik, S.M., 1982: The pre-Phanerozoic fossil record. In *Mineral Deposits and the Evolution of the Biosphere*, H.D. Holland and M. Schidlowski (Eds.), Springer, Berlin, 67-81.

Awramik, S.M., J.W. Schopf, and M.R. Walter, 1983: Filamentous fossil bacteria from the Archaean of Western Australia. In *Developments and Interactions of the Precambrian Atmosphere, Lithosphere and Biosphere (Developments in Precambrian Geology*, **7**), B. Nagy, R. Weber, J.C. Guerrero and M. Schidlowski (Eds.), Elsevier, Amsterdam, 249-266.

Barghoorn, E.S. and S.A. Tyler, 1965: Microorganisms from the Guntflint chert, *Science*, **147**, 563-577.

Bottinga, Y., 1969: Calculated fractionation factors for carbon and hydrogen isotope exchange in the system calcite–carbon dioxide– graphite–methane–hydrogen–water vapor, *Geochim. Cosmochim. Acta*, **33**, 49-64.

Brack, A., 1993: Liquid water and the origin of life, *Origins Life Evol. Biosph.*, **23**, 3-10.

Bridgwater, D., and V.R. McGregor, 1974: Field work on the very early Precambrian rocks of the Isua area, southern West Greenland, *Rapp. Gronlands Geol. Unders.*, **65**, 49-54.

Bridgwater, D., J.H. Allaart, J.W. Schopf, C. Klein, M.R. Walter, E.S. Barghoorn, P. Strother, A.H. Knoll, and B.E. Gorman, 1981: Microfossil-like objects from the Archaean of Greenland: A cautionary note, *Nature*, **289**, 51-53.

Buick, R., 1984: Carbonaceous filaments from North Pole, Western Australia: Are they fossil bacteria in Archaean stromatolites?, *Precambrian Res.*, **24**, 157-172.

Chang, S., 1993: Prebiotic synthesis in planetary environments. In *The Chemistry of Life's Origins*, J.M. Greenberg, C.X. Mendoza-Gómez and V. Pirronello (Eds.), Kluwer Acad. Publ., Dordrecht, 259-299.

Cloud, P.E., 1965: Significance of the Guntflint (Precambrian) microflora, *Science*, **148**, 27-45.

Cloud, P.E., 1973: Paleoecological significances of the banded iron formation, *Econ. Geol.*, **68**, 1135-1143.

Cloud, P.E., 1976: Beginnings of biospheric evolution and their biogeochemical consequences, *Paleobiology*, **2**, 351-387.

DesMarais, D.J., H. Strauss, R.E. Summons, and J.M. Hayes, 1992: Carbon isotope evidence for the stepwise oxidation of the Proterozoic environment, *Nature*, **359**, 605-609.

Greenberg, J.M., and C.X. Mendoza-Gómez, 1993: Interstellar dust evolution: A reservoir of prebiotic molecules. In *The Chemistry of Life's Origins*, J.M. Greenberg, C.X. Mendoza-Gómez and V. Pirronello (Eds.), Kluwer Acad. Publ., Dordrecht, 1-32.

Greenberg, J.M., C.X Mendoza-Gómez, and V. Pirronello (Eds.), 1993: *The Chemistry of Life's Origins*, Kluwer Acad. Publ., Dordrecht, 423 pp.

Han, T.M., and B. Runnegar, 1992: Megascopic eukaryotic algae from the 2.1-billion-year-old Negaunee Iron-Formation, Michigan, *Science*, **257**, 232-235.

Hayes, J.M., I.R. Kaplan, and K.W. Wedeking, 1983: Precambrian organic geochemistry: Preservation of the record. In *Earth's Earliest Biosphere: Its Origin and Evolution*, J.W. Schopf (Ed.), Princeton University Press, Princeton, New Jersey, 93-134.

Irvine, W.M., and R.F. Knacke, 1989: The chemistry of interstellar gas and grains. In *Origin and Evolution of Planetary and Satellite Atmospheres*, S.K. Atreya, J.B. Pollack and M.S. Matthews (Eds.), University of Arizona Press, Tucson, Arizona, 3-34.

Kandler, O., 1994: The early diversification of life. In *Early Life on Earth*, S. Bengtson (Ed.), Columbia Univ. Press, New York, 152-160

Kazmierczak, J., 1979: The eukaryotic nature of Eosphaera-like ferriferous structures from the Precambrian Gunflint Iron Formation, Canada: A comparative study, *Precambrian Res.*, **9**, 1-22.

Kissel, J., and F.R. Krueger, 1987: The organic component in dust from comet Halley as measured by the PUMA mass spectrometer on board Vega 1, *Nature*, **326**, 755-760.

Knoll, A.H., and E.S. Barghoorn, 1977: Archean microfossils showing cell division from the Swaziland System of South Africa, *Science*, **198**, 396-398.

Lucas, J., and L. Prévôt, 1985: The synthesis of apatite by bacterial activity: Mechanism, *Sci. Géol. Mém.*, **77**, 83-92.

McKay, C.P., 1991: Planetary evolution and the origin of life, *Icarus*, **91**, 93-100.

Miller, S.L., 1955: Production of some organic compounds under possible primitive Earth conditions, *J. Am. Chem. Soc.*, **77**, 2351-2361.

Miller, S.L. H.C. Urey, and J. Oró, 1976: Origin of organic compounds on the primitive Earth and in meteorites, *J. Mol. Evol.*, **9**, 59-72.

Muir, M.D., and P.R. Grant, 1976: Micropaleontological evidence from the Onverwacht Group, South Africa. In *The Early History of the Earth*, B.F. Windley (Ed.), Wiley, London, 595-604.

Naqvi, S.M., B.S. Venkatachala, M. Shukla, B. Kumar, R. Natarajan, and M. Sharma, 1987: Silicified cyanobacteria from the cherts of Archaean Sandur Schist Belt, Karnataka, India, *J. Geol. Soc. India*, **29**, 535-539.

Oparin, A.I., 1936: *The Origin of Life*. Dover, New York, 270 pp.

Oró, J., S.L. Miller, and A. Lazcano, 1990: The origin and early evolution of life on Earth, *Annu. Rev. Earth Planet. Sci.*, **18**, 317-356.

Pflug, H.D., 1978: Yeast-like microfossils detected in the oldest sediments of the Earth, *Naturwissenschaften*, **65**, 611-615.

Pflug, H.D., 1987: Chemical fossils in early minerals, *Topics in Current Chemistry*, **139**, 1-55.

Pflug, H.D., and H. Jaeschke-Boyer, 1979: Combined structural and chemical analysis of 3.800-Myr-old microfossils, *Nature*, **280**, 483-486.

Pflug, H.D., and E. Reitz, 1985: Earliest phytoplankton of eukaryotic affinity, *Naturwissenschaften*, **72**, 656-657.

Pflug, H.D., and E. Reitz, 1992: Palynostratigraphy in Phanerozoic and Precambrian metamorphic rocks. In *Early Organic Evolution: Implications for Mineral and Energy Resources*, M. Schidlowski, S. Golubic, M.M. Kimberley, D.M. McKirdy and P.A. Trudinger (Eds.), Springer, Berlin, 508-518.

Ponnamperuma, C., 1993: The origin, evolution and distribution of life in the Universe. In *Chemical Evolution: Origin of Life*, C. Ponnamperuma and J. Chela-Flores (Eds.), Deepak, Hampton, Virginia, 1-11.

Robbins, E.I., 1987: *Appelella ferrifera*, a possible new iron-coated microfossil in the Isua iron-formation, southwestern Greenland. In *Precambrian Iron-Formations*, P.W.U. Appel and G.L.LaBerge (Eds.), Theophrastus Publications, Athens, 141-154.

Roedder, E., 1981: Are the 3.800-Myr-old Isua objects microfossils, limonite-stained fluid inclusions, or neither?, *Nature*, **293**, 459-462.

Ronov, A.B., 1980: *Osadochnaya Obolochka Zemli* (*Earth's Sedimentary Shell*). 20th Vernadsky Lecture, Izdatel'stvo Nauka, Moscow, 97 pp.

Sagan, C., 1974: The origin of life in a cosmic context, *Origins Life*, **5**, 497-505.

Schidlowski, M., 1982: Content and isotopic composition of reduced carbon in sediments. In *Mineral Deposits and the Evolution of the Biosphere*, H.D. Holland and M. Schidlowski (Eds.), Springer, Berlin, 103-122.

Schidlowski, M., 1987: Application of stable carbon isotopes to early biochemical evolution on Earth, *Annu. Rev. Earth Planet. Sci.* **15**, 47-72.

Schidlowski, M., 1988: A 3.800-million-year isotopic record of life from carbon in sedimentary rocks, *Nature*, **333**, 313-318.

Schidlowski, M., 1990: Early evolution of life and economic mineral and hydrocarbon resources. In *Precambrian Continental Crust and its Economic Resources* (Developments in Precambrian Geology **8**), S.M. Naqvi (Ed.), Elsevier, Amsterdam, 605-630.

Schidlowski, M., 1993a: The initiation of biological processes on Earth: Summary of empirical evidence. In *Organic Geochemistry*, M.H. Engel and S.A. Macko (Eds.), Plenum Press, New York, 639-655.

Schidlowski, M., 1993b: Proterozoic carbon cycle, *Nature*, **362**, 117-118.

Schidlowski, M., P.W.U. Appel, R. Eichmann, and C.E. Junge, 1979: Carbon isotope geochemistry of the 3.7 x 10^9 yr old Isua sediments, West Greenland: Implications for the Archaean carbon and oxygen cycles, *Geochim. Cosmochim. Acta*, **43**, 189-199.

Schidlowski, M., R. Eichmann, and C.E. Junge, 1975: Precambrian sedimentary carbonates: Carbon and oxygen isotope geochemistry and implications for the terrestrial oxygen budget, *Precambrian Res.*, **2**, 1-69.

Schidlowski, M., J.M. Hayes, and I.R. Kaplan, 1983: Isotopic inferences of ancient biochemistries: Carbon, sulfur, hydrogen and nitrogen. In *Earth's Earliest Biosphere: Its Origin and Evolution*, J.W. Schopf (Ed.), Princeton University Press, Princeton, New Jersey, 149-186.

Schopf, J.W., 1968: Microflora of the Bitter Springs Formation, Late Precambrian, Central Australia, *J. Paleont.*, **42**, 651-688.

Schopf, J.W., 1993: Microfossils of the Early Archean Apex Chert: New evidence of the antiquity of life, *Science*, **260**, 640-646.

Schopf, J.W., and D.Z. Oehler, 1976: How old are the eukaryotes?, *Science*, **193**, 47-49.

Schopf, J.W., and B.M. Packer, 1987: Early Archaean (3.3-billion to 3.5-billion-year-old) microfossils from Warrawoona Group, Australia, *Science*, **237**, 70-73.

Schopf, J.W., and M.R. Walter, 1982: Origin and early evolution of cyanobacteria: The geologicial evidence. In *The Biology of Cyanobacteria*, N.G. Carr and B.A. Whitton (Eds.), Blackwell, London, 543-564.

Schopf, J.W., and M.R. Walter, 1983: Archean microfossils: New evidence of ancient microbes. In *Earth's Earliest Biosphere: Its Origin and Evolution*, J.W. Schopf (Ed.), Princeton University Press, Princeton, New Jersey, 214-239.

Sleep, N.H., K.J. Zahnle, and J.F. Kasting, 1989: Annihilation of ecosystems by large asteroid impacts on the early Earth, *Nature*, **342**, 139-142.

Tyler, S.A., and E.S. Barghoorn, 1954: Occurrence of structurally preserved plants in pre-Cambrian rocks of the Canadian Shield, *Science*, **119**, 606-608.

Valley, J.W., and J.R. O'Neil, 1981: $^{13}C/^{12}C$ exchange between calcite and graphite: A possible thermometer in Grenville marbles, *Geochim. Cosmochim. Acta*, **45**, 411-419.

Veizer, J., W.T. Holser, and C.K. Wilgus, 1980: Correlation of $^{13}C/^{12}C$ and $^{34}S/^{32}S$ secular variations, *Geochim. Cosmochim. Acta*, **44**, 579-587.

Venkatachala, B.S., M. Shukla, M. Sharma, S.M. Naqvi, R. Srinivasan, and B. Udairaj, 1990: Archaean microbiota from the Donimalai Formation, Dharwar Supergroup, India, *Precambrian Res.*, **47**, 27-34.

Wächtershäuser, G., 1988: Pyrite formation, the first energy source for life: A hypothesis, *System. Appl. Microbiol.*, **10**, 207-210.

Wada, H., and K. Suzuki, 1983: Carbon isotopic thermometry calibrated by dolomite-calcite solvus temperatures, *Geochim. Cosmochim. Acta*, **47**, 697-706.

Walsh, M.M., 1992: Microfossils and possible microfossils from the Early Archean Onverwacht Group, Barberton Mountain Land, South Africa, *Precambrian Res.*, **54**, 271-293.

Walter, M.R., R. Buick, and J.S.R. Dunlop, 1980: Stromatolites 3.400 - 3.500 Myr old from the North Pole area, Western Australia, *Nature*, **284**, 443-445.

Weber, B., 1988: *Isuasphaera isua* Pflug 1978: Zur Kenntnis einer problematischen Mikrostruktur aus archaischen Metasedimenten Grönlands, *Z. geol. Wiss.*, **16**, 253-266.

Woese, C.R., 1981: Archaebacteria, *Sci. Am.*, **244**, 98-122.

Woese, C.R., 1987: Bacterial evolution, *Microbiol. Rev.*, **51**, 221-271.

PART III
BIOCHEMICAL ASPECTS OF SELF-ORGANIZATION

ENERGY, MATTER AND SELF-ORGANIZATION IN THE EARLY MOLECULAR EVOLUTION OF BIOENERGETIC SYSTEMS

Herrick Baltscheffsky and Margareta Baltscheffsky
Department of Biochemistry, Arrhenius Laboratories for Natural Sciences, Stockholm University, S-106 91 Stockholm, Sweden

ABSTRACT

Light, heat, energy-rich phosphates and thioesters, as well as compounds capable of undergoing oxidation-reduction reactions may have been significant among energy sources for the self-organization of matter on the early Earth, in the molecular evolutionary process, which gave rise to the first life from which all known organisms appear to have evolved.

In living cells phosphate compounds play a well-known central role in the fundamental reactions of both bioenergetics and genetic information transfer. They may be assumed to have been of paramount importance also in connection with the origins of life and its first emergence. This presentation will focus attention on the possibility that inorganic pyrophosphate (PPi) was a more or less direct predecessor of adenosine triphosphate (ATP) as a central carrier of biologically useful chemical energy. We have earlier shown that PPi can be formed in bacterial photophosphorylation at the expense of light energy as an alternative to ATP and as such can act as energy donor for several energy requiring reactions in this bacterial system. Furthermore, it is now known that PPi can occur in mineral form and can be produced from hot volcanic magma. In addition, a possible evolutionary pathway from a "PPi world" to an "ATP world" will be outlined, with support from earlier and new sequence information. Finally, the support for the idea that the "anastrophe" concept will be useful not only in biological but also in chemical evolution will be discussed, with reference to the above mentioned, possible evolutionary pathway.

1. INTRODUCTION

1.1 ENERGY

The most abundant and continuously flowing form of energy, which may be assumed to have embraced the Earth since its very beginning, is the radiant energy from our own star, the sun. As is well known, in addition to the light visible to the human eye, radiation of both lower wave length and higher energy, in the ultraviolet, and higher wave length and lower energy, in the infrared, may well have been of importance in connection with the origin and early evolution of life on Earth.

In present-day photosynthesis the pigments involved absorb all the way from the UV to the IR, indicating that the amounts of energy liberated within this broad spectrum are useful for living cells. Indeed, the energies liberated from light quanta at wave lengths where bacterial and higher plant chlorophylls absorb, are sufficient to allow, per light quantum absorbed, the formation of a few molecules of cellular "energy currencies" such as ATP and PPi. So, arguments obviously exist in support of an early emergence of light induced energy conversion during the origin and early evolution of life.

1.2 ENERGY AND MATTER

On the other hand, detailed arguments have also been presented for early emergence of chemical energy conversion, involving energy rich phosphate compounds (H. Baltscheffsky, 1967; Ponnamperuma and Chang, 1971; M. Baltscheffsky and H. Baltscheffsky, 1992) and thioesters (de Duve, 1991), or primitive oxidation-reduction reactions (Wächtershäuser, 1992). So, not only light energy, but also chemical energy, in matter, would seem to have been of major significance when life on Earth emerged. The continuous production of energy rich inorganic pyro- and polyphosphates from volcanic magma is a very strong recent additional argument in this direction (Yamagata et al., 1991).

1.3 MATTER AND SELF-ORGANIZATION

The capability of matter to organize itself into increasingly complex structures may be inherent in self-organizing systems or be driven by coupled reactions, where an energy liberating reaction drives an energy requiring one.

The inherently self-organizing systems and their possible roles in the origin and evolution of life, which have been described in greatest detail, are dissipative structures (Nicolis and Prigogine, 1977) and hypercycles (Eigen and Schuster, 1979). Non-equilibrium flows and autocatalytic processes are of fundamental importance in the prebiological self-organization of matter. The conditions for the spontaneous formation of dissipative structures include openness with respect to exchange of energy and matter with the environment. and far from equilibrium conditions. Related to the theory of dissipative structures is the catastrophe theory (Thom, 1972), with its discontinuous effects of continuous causes.

Energy coupling is of basic biological significance and may be assumed to be so also in the process of chemical evolution leading to the origin of life. In every living cell, energy requiring reactions are driven by energy initially obtained from light, as in the case of photosynthesis, or just by being coupled to energy liberating chemical reactions. It is reasonable to assume, that such energy coupling played a central part also during the chemical evolutionary processes giving rise to the first living entities, to the self-organization of matter, which was a prerequisite for life.

In the following part of this presentation, attention will be focussed on phosphate compounds involved in the coupling and conversion of energy in living cells, with particular emphasis on energy liberating inorganic phosphate compounds and on their possible basic function in a pre-nucleotide world preceding and "RNA world" (Gilbert, 1986) and an "ATP world" (Baltscheffsky, H., 1993). Earlier and new indications for an evolutionary pathway from proteins involved in the metabolism of inorganic phosphate compounds to proteins metabolizing nucleotides will be summarized. The case for considering the "anastrophe" concept to be useful also in connection with the chemical, prebiological part of the molecular origin and evolution of life will be discussed.

An evolutionary pathway of the kind indicated above may be visualized as possibly involving two essential steps:

$$\text{PPi metabolizing protein} \xrightarrow{1} \text{PPi and ATP metabolizing protein} \xrightarrow{2} \text{ATP metabolizing protein}$$

In step 1, a protein involved in the energy metabolism of PPi is assumed to have acquired, over gene duplication and mutation in the genome, a broader substrate specificity, so that also ATP is metabolized. Step 2, in an analogous fashion, causes the protein with broad substrate specificity to change to one which is more or less specific for ATP. Whereas the gene duplication event sets the stage for subsequent mutation, it is the major mutational events themselves themselves and their consequences, for example the specificity changes at the enzyme level, which we consider to be the primary "anastrophes" (Baltscheffsky, H., 1993).

2. ENERGY-RICH PHOSPHATES

2.1 BASIC QUESTIONS

Three questions of primary importance in connection with energy-rich phosphate compounds in the origin and early evolution of life have to do with:
(1) their availability on the primitive earth before the origin of life;
(2) inorganic *versus* organic energy-rich phosphate compounds; and
(3) energy-rich phosphate compounds and their reactions in living cells.

Clearly, one task for the investigator is to search for a pattern of continuity between the possible answers to these three questions. Such a pattern may then provide a framework for more detailed search for possible traces of early evolutionary pathways which may still be discernible among the macromolecules of existing organisms.

2.2 AVAILABILITY

In 1991 a great step forward with respect to identification of a major, plausible source for the continuous production, over long periods of time, of energy-rich phosphates on the primitive earth. It was shown (Yamagata et al., 1991) both in experiments simulating magmatic conditions and by analysis of volatile condensates in volcanic gas, that volcanic activity can produce water-soluble polyphosphates (PPi, tri- and tetrapolyphosphate) through partial hydrolysis of magmatic P_4O_{10}:

$$P_4O_{10} \xrightarrow{+H_2O} H_3PO_4 + \underset{\text{PPi}}{H_4P_2O_7} + \underset{\text{tripoly-}}{H_5P_3O_{10}} + \underset{\text{tetrapolyphosphate}}{H_6P_4O_{13}}$$

As was pointed out, this appears to be the only so far identified route for a continuous production of these condensed species on the primitive Earth.

It would seem from the data discussed above that the old problem of availability of energy-rich phosphate compounds on the primitive Earth has got, finally, a most likely solution. It may also be mentioned, that of the energy-rich phosphate compounds obtained in these experiments, PPi was obtained in the highest amounts.

2.3 INORGANIC *VERSUS* ORGANIC PHOSPHATE COMPOUNDS

Current hypotheses on the origin of life include both inorganic (Wächtershäuser, 1992) and organic (Miller, 1986; de Duve, 1991) variants. Notably, involvement of inorganic phosphate compounds as primitive energy donors is compatible with both alternatives, whereas organic compounds, such as, for example ATP and thioesters presuppose the organic alternative.

However, our main arguments in support for the early involvement of PPi and for the idea that an early, central role for PPi in primitive energy conversion may have been taken over by ATP as a consequence of stepwise evolutionary processes are built upon existing structural knowledge about both the substrates and their enzymes, upon the geological findings discussed above, and upon biological and chemical properties discussed earlier (Baltscheffsky, M. and Baltscheffsky, H., 1992; Baltscheffsky, H. and Baltscheffsky, M., 1994). The complexity of the ATP molecule as compared with PPi easily explains why it took over - it may suffice here to mention the capacity of ATP to feed the adenylation reactions of numerous biosynthetic pathways.

In order to obtain support for the hypothesis that evolutionary links may exist from PPi metabolism to ATP metabolism, in addition to earlier indications (Baltscheffsky, H., 1993), one may examine the situation with respect to the high molecular weight inorganic polyphosphate compounds. A very recent report (Reizer, J. et al., 1993) shows that the enzyme exopolyphosphate phosphatase belongs to a superfamily found to contain mainly ATP binding and metabolizing enzymes with a "typical" ATPase domain. A clearer picture of possible evolutionary relationships between PPi, high molecular weight polyphosphates and ATP metabolizing proteins should result from additional structural information about such proteins.

2.4 ENERGY-RICH PHOSPHATE COMPOUNDS AND THEIR REACTIONS IN LIVING CELLS

Among the energy-rich phosphate compounds in living cells on may distinguish between inorganic, organic non-nucleotide and organic nucleotide compounds. Several of those of organic, non-nucleotide nature are well known to participate in key reactions of anaerobic energy metabolism, and they may well, as some others such as carbamyl phosphate and acetyl phosphate, have been significantly participating in chemical evolution leading to the origin of life, as has often been pointed out. And the high molecular weight inorganic polyphosphates are of great potential interest as carriers of chemical energy under prebiological conditions (Kulaev, 1979). On the other hand, there is no obvious, continuous source known for their prebiological production and neither is there any clear indication of a stepwise path from chemical evolution to the origin of life and further to present-day cells. In contrast, PPi seems to us to have provided such a path, with participation still continuing in the energy conversion of living cells, but in an apparently less pronounced role than in early life when ATP took over, as we have surmised.

Although central, our PPi panorama is thus by no means exclusive, as already was indicated above for alternative phosphate compounds. We have earlier taken up the proposal by de Duve (1991) that thioesters played an early, central energetic role as well as his suggestion that acetyl phosphate may have served as a prebiological link between PPi and thioesters (Baltscheffsky, H. and Baltscheffsky, M.,1994).

3. THE ANASTROPHE CONCEPT[1)]

Anastrophes, as contrasted to catastrophes, have been discussed earlier in connection with the origin and evolution of life (Baltscheffsky, H., 1993). It was then primarily used to describe constructive events in biological evolution, but it may be of value to consider its use also in connection with events of the chemical evolution which is assumed to have led to the origin of life.

We shall exemplify and thus open up for comments and criticism what we mean, by giving a sequence of possible evolutionary steps on the path from chemical to biological energy conversion, where each step is significantly constructive and thus, with our suggested terminology, anastrophic. The steps are given below:

1. Prebiotic production of PPi (from P_4O_{10} in volcanic magma?)
2. Prebiotic production of peptide hydrolyzing PPi, liberating energy.
3. Nucleic acid directed production of PPase (PPi synthase).
4. Nucleic acid directed production of ATPase (ATP synthase).

At least the two first of these four steps would have been part of chemical evolution leading to the origin of life. Mutation at the nucleic acid level would only in subsequent steps be causing the constructive changes. Still, if the events enumerated 1 and 2 were of the evolutionary significance which we have assumed, it seems reasonable to us to use the term anastrophic also here.

Whatever terminology will be found in the future to best describe the constructive events in evolution, their pathways can be expected to become increasingly well illuminated by the rapid progress in the relevant areas of biology, chemistry and geology.

1) This contribution is in a "Festschrift" in honor of Cyril Ponnamperuma. May we, in deep admiration of his lasting constructive contributions to many important aspects of the molecular evolution research field, make the first existing extrapolation of the use of the anastrophe concept also to human, cultural evolution and within this frame confess our opinion that in Cyril we have been fortunate to learn to know and respect a friend, a man, a scientist and a visionary of truly anastrophic dimensions.

REFERENCES

Baltscheffsky, H. (1967): Inorganic pyrophosphate and the evolution of biological energy transformation. Acta Chem. Scand. 21, 1973-1974.

Baltscheffsky, H. (1993): Chemical origin and early evolution of biological energy conversion. In C. Ponnamperuma and J. Chela-Flores (Eds.), Chemical Evolution: Origin of Life, Deepak Publ., Hampton, 13-23.

Baltscheffsky, H. and Baltscheffsky, M. (1994): Molecular origin and evolution of early biological energy conversion. In S. Bengtson (Ed.), Early Life onEarth, Nobel Symp. No 84, Columbia Univ. Press, New York, 81-90.

Baltscheffsky, M. and Baltscheffsky, H. (1992): Inorganic pyrophosphate and inorganic pyrophosphatases. In L. Ernster (Ed.), Molecular Mechanisms in Bioenergetics, Elsevier, Amsterdam, 331-348.

Nicolis, G. and Prigogine, I. (1977): Self-Organization in Nonequilibrium Systems, Wiley, New York.

de Duve, C. (1991): Blueprint for a Cell: The Nature and Origin of Life, Patterson, Burlington.

Eigen, M. and Schuster, P. (1979): The Hypercycle, Springer, Berlin.

Gilbert, W. (1986): The RNA world. Nature 318, 618.

Keller, M., Blöchl, E., Wächtershäuser, G. and Stetter, K.O. (1994): Formation of amide bonds without a condensation agent and implications for origin of life. Nature 368, 836-838.

Kulaev, I.S. (1979): The Biochemistry of Inorganic Polyphosphates, Wiley, New York.

Miller, S.L. (1986): Current status of the prebiotic synthesis of small molecules. Chem. Scripta 26B, 5-11.

Ponnamperuma, C. and Chang, S. (1971): The role of phosphates in chemical evolution. In R. Buvet and C. Ponnamperuma (Eds.) Molecular Evolution 1. Chemical Evolution and the Origin of Life , North Holland, Amsterdam, 216-223.

Reizer, J., Reizer, A. and Saier, Jr, M.H. (1993): Exopolyphosphate phosphatase and guanosine pentaphosphate phosphatase belong to the sugar kinase/actin/hsp70 superfamily. TIBS 18, 247-248.

Thom, R. (1972): Structural Stability and Morphogenesis, Benjamin, Reading, Mass.

Wächtershäuser, G. (1992): Groundworks for an Evolutionary Biochemistry: The Iron-Sulphur World. Progr. Biophys. molec. Biol. 58, 85-201.

Yamagata, Y., Watanabe, H., Saitoh, M. and Namba, T. (1991): Volcanic production of polyphosphates and its relevance to prebiotic evolution. Nature 352, 516-519.

THE EVOLUTION OF ORGANISATION AND THE PERILS OF ERRORS IN THE ORIGIN OF LIFE.

Clas Blomberg
Department of Theoretical Physics
Royal Institute of Technology
S-10044 Stockholm - Sweden

ABSTRACT

Some stages in the origin of life are considered in which experiments may not be of great help but where rather arguments of reason and mathematical modelling provide clues about the course of events. In particular, we consider the possibility to have a first unsystematic creation of a manifold at certain stages and a later occurrence of a feedback effect that can establish a certain mechanism. This is discussed in connection with the selection of chiral molecules together with a mathematical model, and also for the origin of the genetic code. Another emphasis is put upon the loss of accuracy by parasites and error propagation in cooperative systems which also is treated by a mathematical model. Spatial structures that can provide a protection against such vulnerable effects are presented.

1. INTRODUCTION

One has now achieved a good knowledge of possible synthesis processes that could have been relevant for the occurrence of catalytic and self-organizing entities in the origin of life. Still, much remains to be understood in the path from the first functioning biomolecules to the real start of life. It is sometimes suggested that a Darwinian type of selection might have provided a road from early self-organizing molecules to the full organisms. This would mean that the molecular evolution proceeded via selection and natural variation caused by errors (mutations) in the reproduction in the same way as the accepted view of evolution of species. This may work nicely for simple, self-reproducing systems as described by the model by Eigen (1971) (see also Eigen and Schuster, 1979). However, the general path to the first proper organisms can hardly be described by such a picture in a completely satisfactory way. The outcome of a Darwinian evolution of such systems is that there will be one dominating surviving molecule species which has the most favourable relation between growth and decay. It will be accompanied by a natural variation caused by errors in reproduction (the superspecies of Eigen, 1971). Such a model has difficulties to describe the occurrence of a manifoldness and also the development of organisation. Further, a cooperative system of that type will show a great vulnerability to some error effects. (Bresch et al, 1980), (Maynard-Smith, 1979)

The main purpose here is to discuss that kind of problems. The emphasis will be put upon two aspects of evolution: One is a possible, primary occurrence of a non-systematic manifold in which feedback may appear and establish a certain mechanism. The other is about effects of accuracy. If a cooperativity has been established between different kinds of molecules that are depending on each other for reproduction (as catalytic enzymes and the information-carrying nucleic acids), errors may be propagated and destroy the entire system. We will discuss this and together with that remedies against the perils of errors. In these questions, the most important is to be able to make proper conclusions based upon reasonability. We should be able to say: 'This is as way it could have happened' or 'That possibility is highly improbable'. Such questions, for instance, concern the development of organisation. To accomplish this, mathematical modelling is an important tool.

An idea here is that in some stages of the origin of life, there occurred (emerged) spontaneously feedback effects which favoured a certain set of molecules which then could be further reproduced. Such a set can be quite large. This is essentially the view of Kauffman (1986) for the occurrence of an autocatalytic set, and it may be relevant for several stages. We will discuss such possibilities for the selection of chirality and for the occurrence of the genetic mechanism. What here is relevant is that functions and organisation to some degree must have been developed together, a possibility that is difficult to reconcile within a strict Darwinian evolution.

2. SELECTION IN A FIRST, METABOLIC STAGE.

There is a much discussed problem what occurred first, proteins or nucleic acids. The most common view today seems to be a support for the nucleic acid. The RNA-world (Joyce, 1989) could certainly act an independent, self-reproducing world with both catalytic activity and information storage and processing. A protein world, it is said, could not transfer information in the way that was needed to start evolution. On the other hand, to get the RNA-world started, there must have been a catalytic support and there must have been a selection of chirality. Before the RNA world there should have been a stage where molecular building-blocks were synthesized, where metabolism was developed and where some selection took place. This state did not necessarily contain self-reproducing entities, and was not necessarily based upon the transfer of some genetic information. This leads back to the idea by Oparin (1957), later considered among others by Dyson (1985), that the first stage was not genetic but could develop metabolic possibilities. An obvious possibility (not the only one) is an early polypeptide stage. Amino acids were certainly produced during the first stages of the origin of life and it appears to be reasonable that peptides with catalytic activity could be synthesized. They could have been of the protenoid type that has been proposed by Fox (Dose and Fox, 1977). Without genetic transfer, there was no proper evolution but a great manifold of catalytic entities could have been developed leading to many possibilities of catalysis, important at the onset of the origin of life.

Thus, we propose that there was a stage in which catalytic entities were developed before template molecules and self-reproducing units. This was important for the foundation of later stages and for the selection of moleculc units. In such a state, a complex self-sustaining set of catalytic entities can appear, as discussed by Kauffman (1986). Such a set would show some stability and also some possibility to evolve. As it would not transfer any information, there could not be any inheritance of its basic properties besides its obvious achievements. For that reason, there is today no trace of such a stage. It must be strongly emphasized that such early polypeptides were necessarily different from what later was developed by the genetic mechanisms.

It is important that this stage laid the foundations of template molecules with self-reproducing and evolutionary possibilities. A specific way of producing their building blocks should have been developed by some mechanism of catalysis. What is very relevant, there must have been a selection of chirality.

This is dealt with in other articles in this volume, with an emphasis on the possible role of weak interaction. Let us briefly discuss possibilities to get a homochiral selection. One way would be a condensate upon an asymmetric mineral which preferably binds one type of chiral molecules. It is, however, questionable if a mechanism like that which does not influence the production of the monomers could sustain a homochiral manifold. There is a proposal by Salam (1991) that there can be a true phase transition leading to a chiral enrichment solely by the effect of weak interactions. However, there is no evidence that this is relevant for this problem (see the discussion by Figureau in this volume).

Some feedback is probably needed. In most discussions about chiral selection, such mechanisms are considered, often together with effects of weak interaction. The model treated by Kontepudi and Nelson (1983) is probably not realistic, but it is not meant to be so. Its purpose is to be considered together with the weak interaction energies in order to show that these energies could lead to a selection of the isomer with lowest energy. A more obvious feedback mechanism is provided by peptides that produce the amino acids by which they are made. It is known that peptides are most easily formed by amino acids of the same chirality. Even in a racemic mixture, peptides get an excess of one type of chiral monomers. There would be a reasonable chance for producing a homochiral or almost homochiral peptide. *If such peptides could catalyse the production of amino acids of their own chirality, then this yields a mechanism for chiral selection.* We show such a feedback selection formulated as a mathematical model in the next section.

If there was such an enrichment of one kind of chiral monomers, then the world could start to be asymmetric. This, we see as one of the most important conditions for the appearance of more complex molecules and of templates that lead to genetic possibilities.

3. A MATHEMATICAL MODEL OF FEEDBACK CHIRAL SELECTION

The proposal in the previous section can be put in a simple model formulation with the purpose to demonstrate the mechanism in an explicit way. We use a relatively simple model in order to be clear. Assume two kinds of monomers (amino acids), A_1 and A_2 and catalytic peptides E_1 and E_2 which catalyse the production of the respective monomers. These represent mirror forms and then be symmetric: E_1 is the mirror form to E_2 which is related to the monomer A_2 in the same way as enzyme E_1 to A_1. In E_1, a sequence of n monomers A_1 is crucial, and there is a similar sequence of A_2 in E_2. We assume that the monomers can either be made in a non-enzymatic way without any chiral enrichment by a rate α times the concentration of precursors M or by the respective catalytic peptides, with a faster rate β. The kinetic equations for the monomers which include decay terms are:

$$\frac{dA_1}{dt} = \alpha \cdot M + \beta \cdot E_1 \cdot M - \gamma \cdot A_1$$

$$\frac{dA_2}{dt} = \alpha \cdot M + \beta \cdot E_2 \cdot M - \gamma \cdot A_2 \qquad (1)$$

The enzymes E are synthesized of the monomers A. They have a crucial sequence of n monomers of a pure chirality. With production rate K and decay rate g, we get:

$$\frac{dE_1}{dt} = K \cdot A [f(A_1/A)]^n - g \cdot E_1$$

$$\frac{dE_2}{dt} = K \cdot A [f(A_2/A)]^n - g \cdot E_2 \qquad (2)$$

A is the total amount of monomers. The function $f(x)$ is the probability to choose one isomer A_1 or A_2 at a place of the relevant sequence of n monomers. It is a function of the proportion (x) of the respective monomer numbers. If the monomers are attached independently in the polymers E, then $f(x)=x$. If monomers are not attached independently, because an energy coupling favours pairs of the same symmetry (as indeed is the case), then $f(x)$ is larger than x. This can be analyzed in terms of a more elaborate model but we will not go into that here. The monomer precursor, M, obeys a further kinetic equation:

$$\frac{dM}{dt} = a - b \cdot M - (\alpha + \beta \cdot (E_1 + E_2)) \cdot M \qquad (3)$$

a is here a constant production rate, and b a decay rate. The last term stands for what is used for the monomer (A_1, A_2) production.

Now, look for stationary solutions of these equations, and their stability. There are two relevant possibilities. One is a completely symmetric situation when $A_1 = A_2\, (x=1/2)$. Stationary quantities can then be calculated in a straightforward way from the equations. The stability can be investigated by conventional methods by linearizing the equations around the stationary solution. We do not write down any details but note that the crucial situation for stability is when the amino acid concentrations, A_1 and A_2 differ but their sum is constant as is also the precursor concentration, M. This provides a stability criterium:

$$\frac{\beta K a}{\gamma g b} \cdot \left(f\left(\frac{1}{2}\right)\right)^{n-1} f'\left(\frac{1}{2}\right) n < 1 \qquad (4)$$

The first factor is a product of growth rates over decay rates. The f-power function comes from the probability that the relevant enzyme sequence of n monomers consists solely of one chiral form. When (4) is fulfilled, the completely symmetric, racemic situation is stable.

For that the probability and the growth rate shall not be too large. If (4) is not valid, a symmetric situation decays and develops to a chiral selection.

To get (4) as above, the concentration of precursors, M, is put equal to a/b. This can be shown to be a good approximation when enzymatic reproduction is low as is the case for the symmetric situation.

The other relevant stable possibility is a state with one type of component dominating, say A_1. For this, one expects that $\beta >> \alpha$, i.e. the enzyme production rate shall be much larger than the non-catalytic one. Then, the precursor concentration M is much lower than in the symmetric case. There will be virtually no enzyme of the type E_2, and a low concentration of the species A_2 (only what is produced non-enzymatically). A stationary solution of this type, if it exists, is always stable. There is a condition for that kind of solution containing similar rate constants as (4), roughly equal to:

$$\frac{\beta K a}{\gamma g b} > 1 \qquad (5)$$

As for (4), there are some simplifying arguments behind this form. It is a necessary condition for a chiral solution. It can be more refined but would then be more complex and less perspicuous.

(5) contains the similar rate constants as (4), with an opposite inequality sign. When (4) is not fulfilled, (5) is certainly so, and the system turns into the chiral state. For certain values of the quotient, both (4) and (5) can be valid. Then both possibilities are stable, and there is an intermediate threshold state such that states on one side of that turns towards the symmetric situations, those on the other side towards the chiral state. A symmetric situation is then stable against small fluctuations, but large ones may transfer the system to the chiral state. As there may be rather few enzymes present in the symmetric state, these may show large fluctuations and a decay to the chiral state may occur after a rare but not impossible creation of one type of enzyme.

4. DEVELOPMENT OF THE GENETIC APPARATUS

The development of the genetic translation apparatus and the origin of the genetic code are features about which there are very many speculations. Some of their development may be analogous to the selection of chirality, and we will say a few words about this.

There are restrictions about how the genetic code could have evolved. The code could only have evolved by becoming more complex. The information that is contained in existing template molecules should otherwise have got lost in the evolution. There could not have been any change from a two-letter code to a three letter code, and there could not have been any essential exchanges between the meaning of particular codons. What could have evolved is a change from a code that used two bases (first probably C and G) to a code that used four bases. Also codes that at the onset had a less specific meaning (they could have coded for a group of similar amino acids) could have got a more precise interpretation. This is what is behind the idea by Crick (1968) that the actual code was established as a 'frozen accident': when it occurred in a certain shape, it could not be changed in any essential way.

This, however, does not mean that the existing genetic code was the sole attempt in this direction at the origin of life. In a world with self-reproducing, self-catalysing RNA-molecules and certainly also amino acids and catalytic polypeptides, there probably occurred a number of nucleic acids with attached amino acids, and these could be attached to template structures. It is reasonable that an early polypeptide production could have occurred in this way, at the onset with no systematics in it.

What we suggest is that there was a significant polypeptide production by adaptors, complexes of nucleic acids and amino acids. There must have been some kind of enzymatic molecules at that time, there were enzymatic RNA's, and there were probably catalytic peptides. There should have been some catalysts that produced monomers, used energy for activation and also supported polymerization. These could well have produced the adaptors and also supported polypeptide production by nucleic acid-amino acid complexes. At the onset, it is reasonable to believe that many different types of adaptors existed and that polypeptides were formed without any plan of the genetic code. There was no selection at that stage. A rich manifold of polypeptides with a number of possible enzymatic activities could have been synthesized. Probably things were not produced completely at random, probably there were some favoured attachments and some that did not occur.

In such a system, a feedback mechanism similar to what we discussed in the previous sections could arise. Produced polypeptides could catalyse the steps of this process, and then proteins could be produced that could catalyse the synthesis of components responsible for their own production. This would lead to an autocatalytic loop as that discussed for the chiral selection and this is similar to the autocatalytic set of Kauffman (1986). *A genetic code would be established if it provided proteins that catalysed the protein production from that code.* In fact, this is the only way by which a genetic code really could have been established in a stable way. At some stage, there may have existed many other possibilities, which would mean variations of the genetic codes, adaptors and so on. This would eventually have lead to the selection of one possibility that could sustain itself and grow out from the unspecific variety and take over completely.

5. THE OCCURRENCE OF DNA

There is an organisation problem concerning the development of DNA.This provides an interesting example of a variation of the egg-or-hen problem: What was here first, the organisation as DNA or the mechanisms that make DNA useful? A new cell component as DNA could not work properly without proper enzymes that provided a stricter control and also handled the translation of its genes but these were of no use before the DNA. There were probably no chemical difficulties to produce DNA in a RNA world. The importance of DNA is that it is a passive information-carrier without any other functions and in that respect it can be better controlled than RNA. It is reasonable that it never had another function. However, this would not make any immediate advantage in a RNA-world where the RNA's served both as catalysts and genes. Unless an organisation was developed for the processing of DNA, and also for providing enzymes which rapidly broke down the RNA's, the introduction of DNA could not have meant any advantage. If DNA took over the function as storing genetic information, RNA should only have kept its present function of transferring the information, and they should not exist as long-lived genes. Control molecules had no function and no meaningful existence unless DNA was there. RNA-ases that can break down RNA-genes efficiently should have been fatal in the RNA-world. Although needed for a meaningful use of DNA, they could not be developed before DNA. In this case, it seems that one must postulate that DNA occurred before at least some of its proper processing enzymes and that gene information of the RNAs was transformed to DNA at an early stage.

6. ERRORS IN COOPERATIVE SYSTEMS

As said, the traditional Darwinian evolution works well for a system with molecule species that are reproduced independently of each other but compete for the same resources (primarily the monomers). Then, the molecules that have the most favourable value of the ratio between growth and decay rates will win. Such a system always develops towards more efficient reproduction, and unless external influences are drastically changed, it would all the time gain stability. In a cooperative system with different types of components, evolution rules are less clear and there may be bad mutation effects that can lead to instabilities. 'Natural selection' favours polymers that are efficiently reproduced with the support of any catalytic power, not necessarily polymers that themselves are efficient catalysts. Such systems may evolve into a stage which is unstable and dies off. (Maynard-Smith, 1979, Bresch et al, 1981, Blomberg and Cronhjort, 1993).

We can think of a system of polymers which stand for different functions: they act as information carriers (genes) or as catalysts (enzymes). The latter also control the accuracy, and their reproduction is encoded by the information carrier. The functions can be provided by the same polymers (as in the RNA world). There are two kinds of dangerous errors which can be coupled (Blomberg and Cronhjort, 1993):

A dangerous erroneous polymer may be such that its gene is efficiently reproduced by the support of existing catalytic polymers but it does not itself provide any catalytic activity (*parasite effect*). If the erroneous genes are not much less stable than the original ones, the whole system may die out. (Bresch et al, 1981). Erroneous polymers may have a catalytic activity but be less efficient than the 'correct' ones. They may introduce further errors in the reproduction, and successively increasing errors in subsequent generations (*error propagation*). (Orgel, 1970, Kirkwood and Holliday, 1975). The situation can be stable if the erroneous enzymes with a low accuracy also have lower production rates than the correct ones. If the error propagation can proceed over the genes, this is not sufficient and it is also necessary that erroneous polymers have high decay rates.

The fatal effect is that erroneous polymers may grow rapidly and completely take over the system temporarily. The active polymers can then not be efficiently reproduced and will decay. The parasites cannot survive without catalytic support from other components and later, the entire system can die out. The effects are highly relevant in an RNA-world where RNA both represents genes and performs the catalysis for the reproduction.

7. A MATHEMATICAL MODEL FOR ERROR PROPAGATION

We illustrate the effects by a model for a system (such as a RNA-world) where the information carriers and the catalytic ones are the same type of molecules. For these, there is a basic kinetic equation:

$$\frac{dA_i}{dt} = \sum_k \sum_m M \cdot K_{mk} \cdot P_{im,k} \cdot A_m A_k - q_i \cdot A_i \tag{6}$$

A_i = concentration of polymer species (RNA) 'i'.
M = Monomer concentration which relates to the growth rates by a further equation and thus limits the growth.
K_{mk} = Rate for reproduction (not necessarily correct) of polymer A_m with catalytic support by A_k.
$P_{im,k}$= Probability that reproduction of A_m by catalyst A_k leads to polymer A_i.
q_i = Decay rate of species 'i'.

This kind of equation is more fully treated in Blomberg and Cronhjort, (1993). We consider here a simplified model with the same features as the full models.

Assume two relevant kinds of polymers: There is a 'correct one' A_0 which shall provide an efficient catalysis with a good accuracy, and there is a second group of polymers which have some part of their sequences in common with A_0 but differ in a particular sequence of n bases. These will be recognized by the catalysing agents, they may have some catalytic activity, but are as a whole less efficient than A_0. All these polymers that differ from A_0 in the particular sequence are assumed equivalent. The total number of polymers of that group are A_T. This separation of two kinds of polymers is used in other models with similar purposes, for instance by Swetina and Schuster (1982) and Kirkwood and Holliday, (1975).

For these two groups, we have equations:

$$\frac{dA_0}{dt} = M \cdot K_0 \cdot P_{00,0} \cdot A_0^2 + M \cdot K_1 \cdot P_{00,1} \cdot A_0 A_T$$
$$+ M \cdot K_0 \cdot \sum_{m>0} P_{0m,0} \cdot A_m A_0 + M \cdot K_1 \cdot \sum_{m>0} P_{0m,1} \cdot A_m A_T - q_0 \cdot A_0 \tag{7a}$$

$$\frac{dA_T}{dt} = M \cdot K_0 \cdot A_0 A_T + M \cdot K_1 \cdot A_T^2 \tag{7b}$$
$$+ M \cdot K_0 \cdot P_{00,0} \cdot A_0^2 + M \cdot K_1 \cdot P_{00,1} \cdot A_0 A_T - q_1 \cdot A_T$$

The notations are the same as above. Index '1' stands for the big group. Both these equations are exact with the assumptions above. K_0 is the production rate with catalytic support by A_0 , K_1 that by the other polymers.

The most important situation is one with a high specificity, in which the 'correct', efficient A_0 occurs in much higher concentration than any other component. The total number of 'other components' can be as large or even larger than the correct ones, but must not be much larger. If the other, large group would dominate, then the efficiency would be deteriorated. Because of low accuracy, few of the efficient species A_0 would be produced and the situation may deteriorate still more. There is a combination of the parasite and error propagation effects: less efficient components may grow fast and exhaust the resources, and the error may grow as the less efficient polymers become dominant. The question is whether the solution of equation (7) can provide a stabilization of the high specific situation with A_0 dominating, or if the other group may grow and destroy the system.

A condition for the high specificity stabilization can be obtained with a good approximation from a consideration of stationary states of equation (7). For a stationary situation with A_0 dominating, all terms in equation (7a) except the first and the last may be neglected. This means that we neglect the possibilities that the reproduction of an erroneous polymer leads to A_0 and also neglect the production of correct A_0 by the catalytic support of the erroneous polymers. These approximations can be qualitatively well motivated. We get an equation for stationary values of $y = A_T/A_0$ a measure of the error level: ($y = 0$ means no error, $y = 1$ that there are as many erroneous polymers as correct ones.)

$$K_0 q_1 \cdot P_{00,0}\, y = K_0 q_0\, y + K_1 q_0\, y^2 + K_0 q_0 \cdot P_{00,0} \tag{8}$$

This equation gets positive y-solutions provided the following inequality is valid.

$$\frac{q_1}{q_0} \cdot P_{00,0} > 1 + 2 \cdot \sqrt{\frac{K_1}{K_0} \cdot P_{00,0}} \tag{9}$$

If the inequality is not valid, the specific solution is unstable and the system will turn into a non-specific state and may be extinguished. For (9) to be fulfilled, the decay rate of erroneous polymers (q_1) must be significantly larger than that of the correct polymer (q_0). The two different error effects add up in the right hand side of (9): The '1' in the right hand side represents the 'parasite effect'. This is the result when the erroneous polymers are inactive (K_1=0). In that case, the system cannot survive when parasites grow as these can not reproduce without support from active polymers. The square root represents the error propagation, and adds to the parasite effect. Its effect is more prominent the larger K_1 is.

8. REMEDIES TO PARASITES

Parasites and other error propagation effects are very perilous for a primitive cooperative system such as a hypercycle system. The probabilities for parasites to occur are not small and certainly leads to difficulties to attain stability of a primitive system. A **homogeneous system** is likely to be exterminated and has small chances to develop.

It has been suggested that a system which achieves a proper **spatial structure** may function better in this kind of situation. There are several reasons for that. In a living cell, catalytic proteins develop together with the cell, and cells are selected due to their function capacity in contrast to a homogeneous situation with free catalysts that do not 'belong' to a certain information molecule and only the latter are selected. If there is a spatial structure, parasites as mutations may kill off the local structure where they arise but may not be able to penetrate and kill other parts. Some kind of spatial structure has repeatedly been suggested as a remedy against parasites, see for instance (Bresch et al, 1980)

In particular, this has been discussed with a connection to hypercycles. Hypercycles, as introduced by Eigen and Schuster (1979) are essentially what we here mean by cooperative systems, a system with templates and catalysts, where the catalysts are reproduced from the template information of the former. Often, the hypercyclic systems are considered as a cyclic set of reactions where a component '1' catalyses the reproduction of component '2', which reproduces the reproduction of component '3' and so on. In that kind of description, the catalysts and the templates are considered as the same kind of molecules (but the catalysts do not catalyse their own reproduction).

The kinetics of hypercycles is described by differential equations of the form:

$$\frac{dX_i}{dt} = \sum_{i=1}^{N} K_i \cdot X_{i-1} X_i - q_i \cdot X_i \tag{10}$$

The X_i represent the N components of the hypercyclic system, and there is a cyclic behaviour such that X_0 and X_N are the same. The growth rate K (or decay rate q) should include a growth limitation: If the system becomes large, the growth shall be slower (or the decay larger). In the Eigen-Schuster formalism, this is usually introduced by a term which leads to a constant total concentration. In our work, we usually consider an explicit dependence of activated monomers. (see Blomberg and Cronhjort, 1993). The system (10) as a set of ordinary differential equations provide oscillating cycles when the number of components, N is five or larger. With fewer than four components, the system goes to a steady state.

This has been discussed in terms of cellular automata (CA). Boerlijst and Hogeweg (1991) have studied a (random) CA model which is based upon the properties of a hypercyclic system with five components. Their model uses a two-dimensional lattice where each lattice point can have a state corresponding to one of the components of a hypercycle system. These components are reproduced by rules corresponding to the kinetics of (10), and also vanish corresponding to the decay term. There is also a possibility to move as a diffusing system. Rules for a cellular automaton based upon hypercycle kinetics are given in the appendix. With five or more components which provide oscillations in the differential equations (10), this leads to a structure with rotating spirals.

To understand properly the effect on parasites, it is appropriate to go in some detail about the development of the spirals. In these lattice systems, there are regions which are almost entirely filled with one component. A component can grow only if there is a catalyst in the neighbourhood. Thus, a component grows essentially along a boundary to the region of the catalyst that supports this reproduction. The component moves really towards that region. Its catalyst in its turn grows at another boundary towards its proper catalyst. This yields the cyclic structure and sustains the rotation. The components grow outwards, but this also means that they reach regions without any catalytic support and where there is no further growth. A parasite can be introduced in this model so that it competes with one of the components by having the same catalysing support, but it does not provide any catalytic support itself. If it grows faster or decays slower than its competing component, it may destroy the system. If it is introduced in the center of such a spiral structure, it will destroy that part. On the other hand, if it occurs outside a spiral, it cannot go towards the outgoing movement of the spiral but will be trapped at the non-growing regions and there vanish.

We have pursued this kind of considerations. Among other things, a clear model dependence is found. We have considered a reaction-diffusion system based on the reaction differential equations (10) but complemented with diffusion. In this, we also include the monomers to restrict growth (Cronhjort and Blomberg, 1994). In two dimensions, similar to the CA model, spiral patterns can occur. An example of such spiral is shown in figure 1 (for more details, see below). In this case, the parasites can enter the spirals and go towards the spiral centre. Normally they destroy the hypercycle. (Some strange patterns were found if there were significant differences between the diffusion constant of the parasites and the ordinary components in which cases there can be a coexistence).

In the calculations, the partial differential equation is treated as a discrete type of equations with the space (in the main case, a plane) divided into cells. For the figure, the space was divided in 20×20 squares, and each square is represented by its dominating component. The main difference to the CA is that the components are represented by continuous concentration variables, and that all components are present in all cells by their concentrations (although some may have very low concentrations). In the CA, there can only be one component in a cell. In each cell, there are oscillations as in the homogeneous

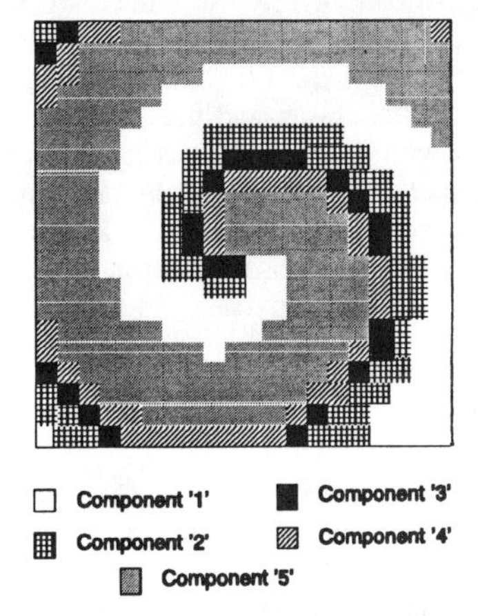

Figure 1.
A spiral resulting from the reaction-diffusion equation as described in the text.

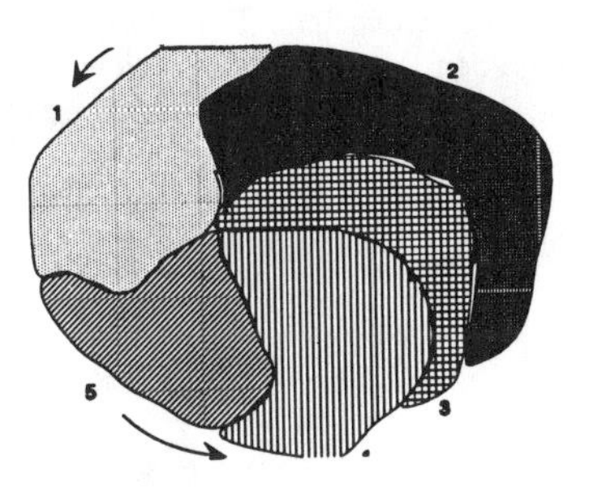

Figure 2.
A schematic picture of a rotating spiral type of cluster for the hypercycle cellular automaton with five components, see text.

situation, and these oscillations propagate as a kind of wave. In contrast to the CA, there is no true movement of regions. This, we believe makes the difference to the CA-model. As there is no outwards movement in the continuous model, a parasite can move into a spiral structure, reach the center and destroy it.

We have also considered three-dimensional patterns with the partial differential model. Normal structures in that case are composed by spiral rings, where the spirals rotate around an axis that may be formed as a circle. The spiral rings are not stable but their radii will decrease and eventually collapse. Spatial structures of this type do not seem to be stable in three dimensions. There is, however, a possibility to get stable straight cylindrical structures by proper boundary conditions. The result is similar to what is found in other reaction diffusion models, for instance of Belusov-Zhabotinsky type (Keener and Tyson, 1992).

There are further variations in this type of models. We have considered a (random) CA with what is called asynchronous updating, which is at each time step, one point is chosen at random and it is decided whether a component at that site will decay or lead to the production of a new component at a neighbouring site. Some details about the rules are presented in the appendix. The Boerlijst-Hogeweg model is based upon synchronous updating which means that the particles of the lattice are gone through in a systematic manner, and all states are updated simultaneously after this is done. Further, their model does not contain any explicit growth limitation. In a CA, there is always a growth limitation provided by the availability of empty space. We usually consider an explicit growth limitation, either provided by the introduction of monomers, which are consumed at each reproduction step and which then are restored towards an equilibrium value, or by letting the growth probability decrease with total concentration.

Under certain circumstances, we get clusters there are kept together by a kind of 'cage effect', essentially because components are reproduced close to existing ones and remain close to the place where they are created. With the hypercycle model, we get rotating clusters which can be in the form of spirals or components ordered in sectors which rotate as a wheel pattern. The density of the clusters is quite high, independent of the kinetic parameters. The latter influence the size of the cluster and then the overall, average density. A schematic view of such a cluster with five components is shown in figure 2. There are well-defined domains in the cluster which contain essentially one single component. The boundaries between clusters are, however, not smooth and this is also so for the cluster boundary (they are rather of fractal form). There are also clusters in the form of double spirals. Parasites will not penetrate into any of these clusters.

9. CLUSTERS IN CELLULAR AUTOMATA

The clusters provide some important aspects for these problems. They occur frequently in cellular automata with asynchronous updating and with a growth limitation that restricts the total concentration. If there is a tendency for a high concentration, clusters will occur. An interesting effect for the evolution problem is that these may compete as entities and form a kind of group selection. There may be slightly different kinds of replicating systems developed in different clusters. If they interact, they may interact by the full efficiency of the cooperativity, not merely by the growth features of the individual information carriers. This makes, in our opinions the clusters very interesting objects for the stability and development of early, primitive cooperative, replicating systems.

Clusters occur also in a simpler system than the many-component hypercyclic system. In particular, we have studied a system with only one component, which catalyses the replication of itself. With a general growth limitation and asynchronous updating, we get homogeneous clusters of this component. The growth limitation shall not be 'too local' which means that the growth is limited by a condition on the average density in a relatively large region around the growth point, not the local density of the immediate neighbourhood. In the latter case, we get continuous patterns, which may have a low density.

We may then introduce a parasite (see the appendix). We choose parameters so that the parasite should be fatal in a homogeneous system. Still, in the lattice clusters, the parasite is found to coexist with the host medium. The reason for this is interesting and worth to be discussed. A new component can be produced at an empty site if there is another component in its neighbourhood which acts as a catalyst. The main production occurs at the surface of the clusters, where empty sites are available. A parasite needs a main component as a catalyst to be reproduced and its growth therefore occurs close to the boundary between a region that is dominated by the main component and one dominated by parasites. A schematic view of such a main cluster and a surrounding parasite region is shown in figure 3. In the actual clusters, the boundaries are less smooth. *In that interface region, there are fewer empty places available than at the surface where the main component is reproduced.* In our model, the probability for a new parasite is twice that of a catalyst *per empty neighbouring site*. Parasites grow rather rapidly and form a layer at the surface of a cluster. (see the figure). If they encircle the entire cluster, they will then grow inwards and destroy the cluster. However, even if the parasite grows faster *per empty available site*, the main component will grow faster at the cluster surface than the parasite in the interface region. This appears to lead to an equilibrium situation with the parasites appearing at some parts of a cluster but not capable to encircle it. In our simulations, there will usually be only one surviving cluster. Other clusters have been encircled and killed by the parasites.

One important lesson from these models is that there are possibilities to manage the threat of the parasites. Further, clusters are entities that can show interesting evolution properties. However, we also see a clear model dependence and because of that, no general conclusions can be made. The models themselves are aimed to be illustrative and not really realistic. However, clusters may appear in many ways. We have not in our models made use of any tendency for polymers to attach to each other, a tendency that very often is seen in real systems. Thus, the clusters can certainly appear in realistic systems and the properties we speak about here could be relevant for the evolution of cooperativity.

Figure 3.
A main cluster with an autocatalytic component and the surrounding region of the parasite.

APPENDIX: THE HYPERCYCLE CA MODEL WITH ASYNCHRONOUS UPDATING.

We consider a hypercycle system as a random cellular automaton with asynchronous updating on a square lattice. Each point is either empty, in state '0', or filled and attains any of n states *'1'...'n'* which correspond to the n components of the hypercycle. At each time step, one point is selected at random. If that point has state 0, nothing happens and a new state is selected. If the point is in a state *'i'*, then

a) by a certain probability, the state changes to '0' (the molecule decays).

b) if it does not decay, then if there is a catalytic support for state 'i', that is a state 'i-1' at any of the eight neighbouring sites, an empty state among the eight neighbouring sites can be filled by a certain probability (reproduction). This probability is proportional to the number of empty sites among the neighbouring ones and it decreases with increasing overall concentration of filled places. In the figure, the chosen central state is '3', and it can be reproduced as there is a catalytic state 2 (3-1) in its neighbourhood. Other neighbouring filled states do not have any other relevance than diminishing the number of available empty sites where a new polymer can appear.

The dependence of the concentration can be introduced in several ways. There can be an explicit concentration dependence. It can also be treated by the introduction of monomers. For these, there is a separatc certain kinetic relation: at each time an empty site becomes filled, the number of monomers is decreased by a certain number. At each time step, the monomer pool is restored towards an equilibrium value attained in the absence of growth.

At certain intervals, points are selected and if in a filled state (not '0'), this state can be moved to an neighbouring, empty point with a certain probability (diffusion). The rate of diffusion can be varied by the probability and also by the interval at which the states are selected. Usually, diffusion selection is made after each step of the previous type.

A parasite can be introduced as a new state 'p'. It does support reproduction, but can be reproduced by one of the hypercycle components. If it is chosen at a time step, it can decay as the other states or be reproduced if there is a catalytic state at any of the neighbouring sites and one or more empty sites. The growth probability per empty place is higher for the parasite than for the ordinary state '1' (or the decay probability may be lower). The parasite in the figure can not help any growth but may be reproduced, for instance if it gets support from species 3.

In the one-component model mentioned in the text, there is only one component which provides catalytic support for itself. The model works exactly as described with one state '1' and the catalytic requirement is that there is a state '1' in the neighbourhood. The parasite can be reproduced if there a state '1' in the neighbourhood of a state 'p'.

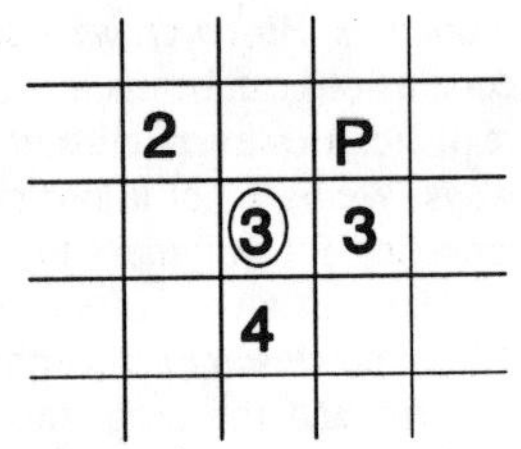

Figure 4.

A cellular automaton configuration around a central, chosen site with component '3', see text.

REFERENCES

Blomberg, C. and Cronhjort, M.: 1993, In *Cooperation and Conflict in Evolutionary Processes*, ed. J. Casti and A. Karlqvist, New York: J. Wiley & Sons.
Boerlijst, M.C. and Hogeweg, P.: 1991, *Physica D*, **48**, 17.
Bresch, C. , Niesert, U. and Harnasch, D.: 1980, *J. Theor. Biol.*, **85**, 399.
Cairns-Smith, A.G.: 1982, *Genetic takeover and the mineral origin of life*, Cambridge University press.
Crick, F.H.C.: 1968, *J. Mol. Biol.*, **38**, 367-379.
Cronhjort, M. and Blomberg, C.: 1994, *J. Theor. Biol.*, in press.
Dose, K. and Fox, S.: 1977, *Evolution and the origin of life.* Marcell Dekker.
Dyson, F: 1985, *Origins of Life*, Cambridge University Press.
Eigen. M.: 1971, *Naturwissenschaften*, **58**, 465-523.
Eigen, M. and Schuster, P.: 1979, *The hypercycle: A principle of natural self-organisation*, Springer, Berlin.
Joyce, G. F.: 1989, *Nature*, **338**, 217.
Kauffman, S.: 1986, *J. Theor. Biol.*, **119**, 1-24.
Keener, J.P. and Tyson, J.J.: 1992, *SIAM Review*, **34**, 1-39.
Kirkwood, T.B. and Holliday, R.: 1975, *J. Mol. Biol.*, **97**, 257.
Kontepudi., D.K. and Nelson, G.W.: 1983, *Phys. Rev. Lett*, **50**, 1023.
Maynard-Smith, J.: 1979, *Nature*, **280**, 445.
Oparin, A.I.: 1957, *The origin of life on earth.* 3rd edition, English translation by A. Synge. Oliver and Boyd.
Orgel, L. E.: 1970, *Proc. Nat. Acad. Sci. USA*, **67**, 1476.
Salam, A.: 1991. *J. Mol. Evol.*, **33**, 105.
Swetina, J. and Schuster, P.: 1982, *Biophys. Chem.*, **16**, 329.

ON ATTEMPTS TO CREATE LIFE-MIMICKING CELLS

F.R. Eirich
Polytechnic University
Brooklyn, New York 11201

ABSTRACT

Studies which try to show that the Origin of Life on this planet is explicable in physico-chemical terms, and may eventually be traceable as a form of Chemical Evolution, suffer from principal difficulties. The perhaps most formidable is that we know but one form of Living Matter, ours, that the transition form Non-living matter on this planet happened too long ago (ca. 3.8 x 10^9 years) to allow us to discern any clues, and that as the result of our bias, we might not even comprehend the steps of the transition as they occurred. It is, therefore, arguable that based on our present state of knowledge of the ability of lifeless matter to organize itself, and on the discoveries of Molecular Biology, experiments should be undertaken that aim at prompting the assembly of matter to form steady state units of measurable lifetime. In view of our lack of understanding of "what Life is", in general, we may use the non-living building blocks and the assembly modes of "our" Life, to obtain units with attributes of the living matter of which we are a part. A possible route to this end is imagined and presented.

1. INTRODUCTION

Our searches, so far, for reproducing credible scenarios for the Origin of Life on planet Earth suffer from such fundamental difficulties that any understanding of how inanimate matter transformed itself to become "alive", lies still in the indefinite future. One might thus think of approaches other than trying to duplicate the Genesis of "our" Life, and attempt to study the possibility of creating other, simpler, artificial forms of Life, in order to develop an understanding of the transition process.

2. PROPOSITION

Attempts to create life-mimicking cells are immediately confronted by the difficulties that we have no inkling of possible forms of Life other than our own, which means also that we do not know of any building blocks other than those produced by our extant biological apparatuses (contemporary living entities). Theoretical biologists, computer fans and engineers try different routes, in particular those of creating robots, cellular units, or systems with artificial intelligence that will perform in a manner analogous to the characteristics of our Life. I propose to look for ways in which we might observe simpler transitions from non-living to an "equivalent" of living matter. I emphasize chemistry and mimicking our biochemistry, because chemical units are more readily self-starting than mechanical or electrical ones, and because we may then derive information on a "non-living" to "living" transformation which can be understood in terms of the atomic (molecular) structure of matter and be akin to what happened on our young planet.

A rational approach along synthetic, constructionist lines requires a general definition of Life. Unfortunately, one can not get away from circular reasoning, since we do not know other life forms for a generalization. However, for the course advocated, it is sufficient to formulate a definition that can be derived from life as we observe it. Learning from other attempts to define Life (de Duve, 1991), the following can be postulated for our purposes: -"Living matter (on planet Earth) occurs in the form of identifiable cellular units based on CHNOPS chemistry, organized into a metabolism which maintains its molecular entities in a continuous, or cyclical fashion. These units regenerate their constituents with the help of catalytically acting multimers, and replicate themselves according to inherited blueprints (information), using material and extracting energy from their environment in the process, and responding (adapting) to changes in their surroundings in ways that help the units to survive with their identities intact."-.

Since we do not insist on creating artificial life by routes that may have been followed 4 billion years ago, we are not restricted by uncertainties as to the nature of our Earth's atmosphere at the time of Life's appearance, or whether Life arose in hot or cool, acid or neutral environments, and that only once, or any number of times. Implicit in this approach is, on the other hand, that the inanimate-animate transition and the character of Life on Earth were pretty well determinate, while the route which the transition and further evolution took was a matter of chance.

In any case, for rational modeling, one has to make an assumption about whether the essential chemistry evolved before or after cell formation, and how leaky and chemically active the enclosing membranes were. The choice proposed is that a leaky form of encapsulation occurred at an early state of metabolic evolution, allowing an early development of individuality. Under the circumstances considered, the membranes may form by accidental co-micellization of amphiphiles and multimers and their co-solutes, with subsequent liposome creation. While leaky, such membranes should retain small catalytically active polymers (the "multimers"). We propose to formulate the chemistry accordingly, and borrow further from reasonable assumptions about early life, that the enclosing membranes were at first chemically inert, but soon imbibed elements (rudimentary pigments) that responded to radiant energy (Morowitz et al., 1988). This constitutes a compromise between current views. Further advantages of utilizing our best knowledge of biochemistry and plausible thoughts on the evolution of membrane structure are, that the inevitable osmotic pressure to which our model cells will be exposed will be low, and that diffusive exchanges of small molecules through the membrane will be easy and permit a further evolution of the intracellular chemistry that leads to an equivalent of a primitive metabolism, see below, Fig. 2.

Combining the most plausible features of our present knowledge, this model addresses the central problem of the "Non-living to Living Transition" in the following way:-- Just as no amount of external operation will cause the parts of a clockwork, thrown haphazardly into a box, to assemble themselves accidentally to a working clock (requiring a watchmaker to assemble the parts according to a blueprint), or intact constituents identical with those of a living cell today, randomly thrown together, will never come to life, for units of chemical or mechanical nature to be collectively different and become more than the sum of their parts, the latter have to be narrowly confined while chemically reacting and organizing themselves within a dissipative flux of energy passing through cellular units.

Singular appearances of new properties and function occurred spontaneously before, during the evolution of our Universe. Instances are: the organization of photons into H, and of the latter into He and the heavier atoms in Suns and Novas, the combination of atoms into molecules (i.e., "materials"; it is not possible to derive, e.g., the properties of water from those of hydrogen and oxygen), the qualitative changes during polymerization and those due to polymer folding (the 10^6 -fold increase of their enzymatic-catalytic functions), the clustering of supra-macromolecules to become reaction centers or cellular scaffoldings*, the organization of multi-cellular creatures and, eventually, the

*Organs like tubules and ribosomes are likely to be later symbiotic additions.

evolution of the brain that allowed intelligence to appear in multi-organ animals. The appearance of Life may then be seen as another instance of a jump in properties and functions as the result of material organization in the course of irreversible processes (see Fig. 1). Property alterations due to phase changes do not belong into this class, since they occur during equilibrium processes at zero change in Free Energy. Self-assembly processes like those of the fragments of sonicated viruses are also different: they follow instructions already inherent in the parts (morphogenesis), instructions which millions of years of evolution, i.e., of synthesis and selection happened to build into long chain molecules, their supra-molecular structures, and eventually into a specific repository of information, i.e., DNA.

Thus, when as chemists we try to fashion life-mimicking units, we can not help but model them after the only life that we know. It seems then reasonable to provide, in solution, the reagents for spontaneous organic-chemical reactions known from biochemistry for their synthetic and energy-delivering value, a solution which contains also the ingredients for the synthesis of amphiphilic molecules (e.g., apolar peptides, fatty acids, glycerol and their phosphates) that allow an early encapsulation of a population of spontaneously formed cells exposed, e.g., to ultraviolet light, heat, or acidity, encapsulated reactive molecules will be subjected to a selection pressures: cells with a higher metabolic rate of regenerating damage-resistant multimeric molecules will survive better.

It lies in the nature of this model that it allows, in principle, for the participation of a myriads of compounds such as must have arisen on primitive Earth under geologically conceivable conditions, a number which though, under constant attack of destructive forces, must have become reduced to a, still very large, but finite number of compounds in dynamic equilibrium of synthesis and decay, as a function of the conditions that prevailed over the eons. In given locales, such as hot vents, acid pools, freezing or dehydrating exposures (including the effects of condensing agents), on surfaces or within interlayers, further selections to smaller numbers of compounds that crowded-out others must have taken place. It remains then for the personal experience and preference of todays' experimenter, which small numbers of self-synthesizing and themselves mutually catalytically influencing and self-organizing molecules he chooses for this analogous simulation of (terrestrial) chemical evolution.

3. SUGGESTED CHEMISTRY

Any combination of starting compounds that leads to a progressive

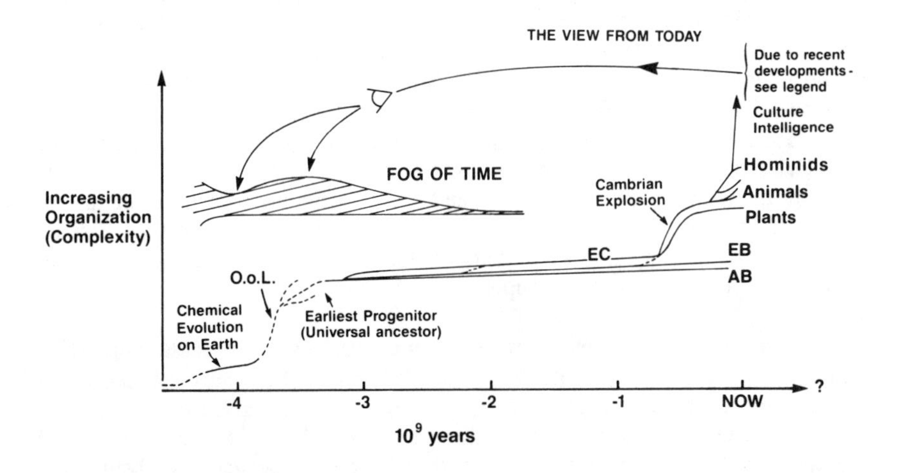

Figure 1. "The View from Today", in the Light of new Fossil Findings (Micropaleontology), of the results of our Planetary Missions (Planetary Geology), of work in Chemical Evolution, and of our progresses in Molecular Biology.- Ordinate: arbitrary units; Increasing Organization--Complexity. Abscissa: Time, in Billions (-10^9) of Years.
O.o.L. ...Origin of Life. EC...Eucaryotes. EB...Eubacteria AB...Archebacteria.
COMMENT: As our Universe follows its course of Dissipation and Aggregation, local fluxes of Energy gave rise to Dynamic Structures. In our Solar System, a Chemical Evolution led to the appearence of Life on Planet Earth. Life's three Kingdoms, their rising Organization shown schematically by three lines marked EC, EB, and AB, created progressively our Biosphere. The increase with Time in the complexity of Dynamic Structures including the Evolution of Life on Earth, proceeded steadily, though punctuated by periods of more rapid rises in Organization and consequent Quality of Function. The Development of the Hominid Brain in particular caused a practically vertical (on the Cosmic Scale) "Take-Off" in Complexity as a result of the creation of Language and Cultures.

synthesis of multimeric peptides, and of a parallel synthesis and consumption of keto- and thio-acids, of acyl-or carbamyl-phosphates, of PPi and/or ATP, may be tried. According to Dyson (Dyson, 1985), e.g., about 1000 copies per liposome of (activated) multimeric homo-or copolymers of the 8 or 9 original amino acids may be found suitable. The addition of primitive synthetic reducing enzymes like synthetic ferredoxins, plus ferrous salts as electron donors (de Duve, 1991; Waechterhäuser, 1991), should provide a suitable starting point for chemical reactions in which the multi-or-polymeric compounds undergo some kind of evolution observable by changes in multimeric composition and/or molecular weight. This amounts to, as far as the presumed history of Life on Earth is concerned, picking up at some midpoint of the continuous sequence of chemical evolution prior to the appearance of RNA.

The oligo-or-multimeric peptides proposed above execute ubiquitous functions in life as we know it. They are likely to be chemically suitable to act catalytically as "hardware", i.e. to have the capability to produce software and more hardware for our models; they are also likely to have appeared prebiotically. In molecular machines the distinction between hardware and software becomes blurred, and nucleotides have indeed been found to exhibit catalytic activity (Cech, 1986). However, while amino acids are easily synthesized prebiotically and have been found in meteorites (Kutter, 1987), the spontaneous synthesis of nucleotides presents up-to-now such problems (Shapiro, 1986) that their appearance before amino acids and peptides in the course of chemical evolution on Earth must be deemed unlikely (Dillon, 1986). The information contained in the molecular architecture of polypeptides was (contrary to the "Central Dogma") probably enough for the crude early forms of life, and should be enough for a fair reproduction of multimers of the chemical automatons of this study.

Historically, the most likely nature of the energy flux would have been caloric or radiant. One is free, though, due to the nature of this approach, to introduce "energy-rich" (-storing) molecules as energy sources, such as thioesters, or phospho- or -adenyl acylates, or one might choose phosphoryl sugars as source of energy from glycolytic decompositions, to allow desired anabolic and self-organizing reactions to start spontaneously. What is proposed here is really something like Spiegelman's (Spiegelman, 1967), Eigen's (Eigen, 1971), or Joyce's (Joyce, 1989) artificial evolution reactions with several important differences: (1) instead of biological enzymes, one employs in situ formed multimeric peptides, or introduces peptides that were surface-catalytically pre-synthesized (with their spontaneous indigenous sequences); (2) the synthesis aims at evolving higher polypeptides and their complexes and co-factors, and only later at nucleotides, eventually aiming at a peptide--nucleotide co-evolution, and (3) the reactions proceed within liposomes!

Utilizing Deamer's (Deamer et al., 1982) and Oro's (Deamer et al., 1980), method of including polymeric reagents into liposomes, and following this by dialysis and the regulated addition of "nutrients", it should be possible (e.g., by checking the consumption of the "nutrients", and/or the resulting wastes) to establish courses or cycles of reactions and self-organization within the fluxes of chemical reactions through liposomes. An important goal would be to establish the occurrence of replication and of its error rate.

Concerning the spontaneous formation of "leaky" membranes which allow the passage of select molecules, we observed distinct breaks in solution properties such as electrical conductivity or osmotic pressure (a measure of activity) during the progressive co-micellizations of amphiphiles, polymers and co-solutes. In the system studied by us in detail, dodecylsodiumsulfate, poly (vinylpyrrolidone) and, e.g., benzenesulfonic acid as "guest" molecule, the amphiphile becomes adsorbed in stages on the polymer until the latter gets to be embedded in an enlarged amphiphile micelle. The enclosing of the sulfonic acid could be followed by the changes in pH. We interpreted our observations (Fishman et al., 1975; Eirich, 1990), as shown on Fig. 2., as the result of the stepwise formation of mixed micelles.

Concerning the conversion of mixed micelles into liposomes, one can expect polymer and guest molecules to remain part of the membrane and to facilitate passive trans-diffusion in the course of partitioning of the guest molecules. Fig. 2, shows also the imagined enclosure in the liposome of a multimer and its presumed looping, for incipient catalytic action, around smaller co-solute molecules inside the liposomes. In such manner, one might expect inward diffusing ATP, imidazoles, sugars, etc., to be available inside of the liposomes for reactions towards an evolution of polypeptides and polynucleotides, and the latter to become instruments of enhanced information.

4. PROJECTIONS

Such developments are at this time potential, but not implausible, expectations inviting experimentation. They include anabolic reactions at first with the catalytic help of multimeric complementarity (de Duve's nomenclature; I dubbed it: "molecular messages", i.e., information by contiguity, (Eirich, 1989), which leads to a crude form of replication and to the formation of molecular complexes which then support selfsustaining reactions (fed by inward diffusing nutrients) or to their cycles (e.g., primitive carboxylic acid cycles). Unfortunately, we do not know any relevant cyclical reaction in non-biological organic chemistry, and in inorganic chemistry only the group

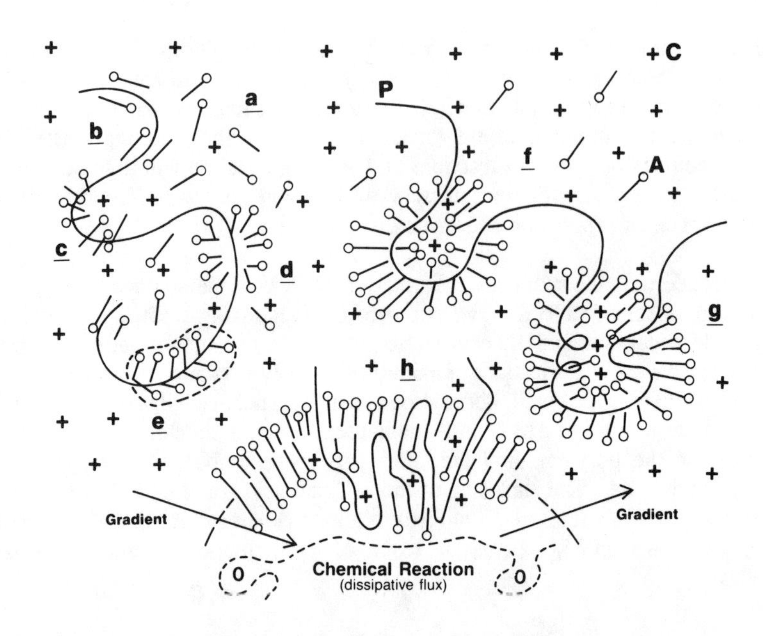

Figure 2. The Spontaneous Increase in Structural Entities, with increasing concentration, in combined aqueous solutions of Polymers(P), Amphiphile(A), and other Co-solutes(C). -a-, at lowest concentration, polymers, amphiphiles and co-solutes are free in solution; -b-, with rising concentration of one or both, A begins to become adsorbed on P; -c,d,e-, further rises in concentration lead to linear, spherical, and cylindrical micellization of A on P; -f,g-, further micellization leads to imbibing (micellar solubilization, mixed micellization) of C and P. -h-, on shaking, or after drying and rewetting, the mixed micelles may extrovert into Liposomes which hold the parent solution, including adventitious polymers of incipient catalytic capability. Convoluted sections of P, and some other Co-solutes, may become also part of the Liposome Membrane and then constitute locales for exchanges with the exterior solution. Chemical reactions within the Liposomic Cell, possibly catalytically enhanced*, create a Dissipative Flux of reagents and products into and out of the Cell.

*Observed for cells with proteinaceous walls by Fox (1977)

of Zhabotinski-Belousov (Winfrey, 1984) reactions which, though much quoted, at best point into the right direction. Knowing the rudimentary structural self-organization through templates,[15] and the formation of colloidal structures like micelles or liposomes, the most important quest must now be for self-organizing chemical reactions, following the mingling of peptide multimers and nutrients. It may well be that such reactions develop only in the confined space of small cells, originally under the influence of the inner surfaces of liposomic membranes. I refer to the possibility that membranes, the least specific of all cell constituents, self-assembling through the processes of micellization, exert an organizing influence on adjacent reactions.

CONCLUDING REMARKS

The upshot and conclusion of the above is the proposition that one should experiment with life-related organic reactions within liposomes of life-related composition. If at first heterotrophic, but eventually autotrophic in the wake of light absorbing pigments enclosed in the membranes, self-organizing reactions can be established, one would have in the realm of "our" chemistry synthesized self-maintaining cells and obtained pointers for the generalization of the transition from transitory to steady-state reactions in microscopic units, plus the concomitant possibility of an evolution towards life- mimicking cellular units.

The proposition above should be seen as principally different from other attempts to get closer to transgressing the inanimate-animate gap. Neuroscience, for all its recent progress is, by its methods, analytical and thus unlikely to produce a global understanding of (our) intelligence. The complexity of brain structure revealed is such that an attempt to produce intelligent machines based on analogues of brain tissue is too daunting to behold. The construction of thinking machines on the basis of computer technology, on the other hand, may ultimately lead to a brave new evolution of devices on a non-biological basis.

Compared to the above attempts, the proposition to prepare life-mimicking chemical automatons based on our knowledge of prebiotic chemical evolution, and of colloid and bio-chemistry, while still formidable, should lie more within our reach. If achieved, the lessons learned would yield valuable pointers for a more general understanding of how matter becomes "alive". The scientific enterprise of today can not provide answers to questions as "What is Life?". Science is not equipped to tell "what is", only "what works". More intrinsic answers will have to wait until we develop methods to look at Life from the outside instead, as now, from the inside of living beings of our kind.

In the forgoing, Chirality was left out of the scope of discussion. According to Ponnamperuma and al. (1994), prebiotic systems were able to choose from molecules of interstellar origin, where L-forms were somewhat more abundant. As an alternative, it may be assumed that prebiotic evolving systems became increasingly uni-chiral because this improved the course of life-friendly reactions and L-forms were more frequently available, or because L-forms were started with by chance and then selected for because of the advantages bestowed by uni-chirality. The reactions proposed above should thus be tried, from hindsight, with L-amino acids and D-sugars.

REFERENCES

Cech, T.R., RNA as an Enzyme, Sci. Amer. 255, 64 (1986).

Deamer, D.H. and G.L. Barchfeld, J. Mol. Evol., 18, 203 (1982).

Deamer, D.H. and Oro, J., Biosystems 12, 167 (1980).

de Duve, Ch., "Blueprint of a Cell" (1991). Neil Patterson Publ., Burlington, N.C.

Dillon, L.S., "The Genetic Mechanism and the Origin of Life" (1978), Plenum Press, New York.

Dyson, F., "Origins of Life" (1985); Cambridge University Press.

Eigen, M., Self-organization of Matter and the Evolution of Biological Macromolecules, Naturwissenschaften, 58, 465 (1971).

Eirich, F.R., On a Co-evolution of Polypeptides and Nucleic Acids, and Liposome Assembly, in: "Prebiological Self-organization of Matter", C. Ponnamperuma and F. Eirich, Eds., (1990); A. Deepak Publ., Hampton, Virginia.

Eirich, F.R. "Molecular Messages", Lecture, Gordon Conf. on Polymers, N.H., 1989.

Fishman, M., and F.R. Eirich, J. Phys. Chem. 79, 2740 (1975).

Fox, S.W., and K. Dose (1977), Molecular Evolution and the Origin of Life, Marcel Dekker, New York.

Joyce, G.F., RNA Evolution and the Origins of Life, Nature, 338, 217, (1989).

Kutter, G.S., "The Universe and Life" (1987); Jones and Bartlet Publ. Boston.

Morowitz, H., Heinz, B., and Deamer, D.H., The Chemical Logic of a Minimum Protocell, Origins of Life and Evolution of the Biosphere, 18, 281 (1988).

Ponnamperuma, C., and A.J. MacDermott, Chemistry in Britain, June, 1994.

Shapiro, R., "Origins" (1986); Summit Books, New York.

Spiegelman, S., An in-vitro Analysis of a Replicating Molecule; Amer. Scientist, 55, 221 (1967).

Waechterhäuser, G., "Order Out of Order"; in "Frontiers of Life", I. and K. Tran Thanh Van, I.C. Monolou, J. Schneider, C. McKay, Eds. (1991); Edition Frontiers, Gif-sur-Yvette, France.

Winfrey, A.T., Rotating Chemical Reactions, Scient. Amer. 233, 82 (1984).

COMPUTATIONAL SUPPORT FOR ORIGINS OF LIFE RESEARCH: A PERSONAL VIEW

Mitchell K. Hobish
University of Maryland
College Park, MD 20742 USA

1. INTRODUCTION

The language of computer science has been pervaded by terms derived from biology: such terms as *worm, virus, genetic algorithm,* and others are now common. There is a natural affinity between these seemingly disparate disciplines, such that many developments in modern molecular biology and biochemistry would not have been possible without computational support.

Computers can be of similar significant aid in investigations of the origins of life, with particular emphasis on the topic of self-organization. Both analytic and synthetic investigations have already felt the benefits of computational tools. Whether one moves from the present backwards in time to examine the evolutionary pathways for development of organisms from common ancestors, or forward in time from the first moments of the creation of the universe to that same point, computers will continue to allow the human mind to explore realms outside the boundaries of everyday experience, and extend human senses into realms heretofore inaccessible with standard laboratory techniques. Ultimately, it should be possible to combine the best features of computational and "bench" chemistry to understand the processes whereby abiogenic molecules combine to produce structures that, by still as-yet-to-be-defined criteria, would be classified as living systems.

In this paper, there is no pretension of covering all the literature in origins of life research that has computer-based relevance, but rather the author will endeavor to present to the reader a survey of the ways in which computers aid the general scientific process, and some examples of the ways in which computers support origins of life research particularly. In an effort to place this work in context, the author will explore the philosophical underpinnings of computer-supported origins of life research, discuss the state-of-the-art in such work, and project a bit into the future to see where this research could go, aided, to a very large extent, by the computer. Examples of existing research supported by computational techniques will be provided, and the rationale for incorporating advanced techniques into research on self-organizing systems will be discussed.

2. PHILOSOPHICAL UNDERPINNINGS

The question of "What is life?" has been asked by humans throughout recorded history. Some have said it's like former justice of the Supreme Court's Potter Stewart's definition of pornography, i.e., "I know it when I see it." We all have some inherent understanding of what it means to be alive, but can we formalize this "hunch," quantify it to enable us to apply a completely objective set of criteria to a new system to ascertain its claim to "aliveness?"

There is a tendency to see "life" as a physical *thing*, that is, we look at a physical entity and, based on several criteria, we say, "This thing is alive." However, we may do well to consider an alternate approach. To a first approximation, life may be thought of in terms of the flow of energy and information through a system. This flow may, for ease of interpretation, be considered as a series of interrelated but discrete *processes*. Given the ease with which life is, ends if one or more of these processes is disrupted, we must be careful to remember that this segmentation into discrete processes is merely a tool to allow us to consider the whole.

Having said this, we may consider "life" to be an embodiment of several processes, which, taken together, provide living systems with those observable criteria we use when we look at a system and bestow upon it the designation "living" or "not living." What are these processes? Can we find general principles embodied therein? When and how did these processes arise on, e.g., the planet Earth? Once we have defined these processes, can we find conditions that would, in a general sense, always give rise to such phenomena?

2.1 ECHOES OF THE PAST

Since we have not (yet) mastered movement through time, we must look to the echoes of past events, as encapsulated in the geophysical record (i.e., fossils, ancient gas bubbles, other planets in our Solar System) and in the existing result of those origins, i.e., today's living systems. For example, the energy budget of living systems is most often thought of in terms of the medium of energy storage and expenditure, the generation of chemical energy in the form of adenosine triphosphate (ATP) for later use in release of energy for metabolic processes. This may be misleading, however. ATP may be thought of as the embodiment of a concept, i.e., production and storage of activity for later use by living organisms. Indeed, even the terms with which we discuss the ATP-generating capabilities of living systems is filled with terms that deal with concepts and processes, i.e., production, storage, breakdown, etc. Similarly, we may look at such life criteria as irritability and reproductive capability as processes, i.e., transfer of information about the environment through an internal network to effectors, and the replication of genetic information into units that will be transmitted to progeny, rather than the existence of a nervous system, or the partitioning of nucleic acid copies into daughter cells. To understand exactly what life is and how it arose, we would do well to begin by looking for and at general principles, i.e., how is energy gathered, stored, and utilized, not, what proteins are used to generate and use ATP, and, what is it about the carbon-based, nucleoprotein architecture we find in terrestrial living systems that makes it suitable to support these systems?

2.2 INFORMATION TRANSFER

Life entails information transfer from one generation to another, i.e., genetics. Seen in its simplest form, genetics embodies several concepts common in the world of computers. It requires the long-term storage of information, and the expression of that information in a usable form. Generally, the long-term storage role is subsumed by deoxyribonucleic acid (DNA), and the expression role by ribonucleic acid (RNA). Given the inexorable flow of entropy into and through systems, we must also posit a mechanism for keeping the information intact, such that transfer of information between generations or its use within a given generation (in the individual) does not result in deleterious results. This is an energy-requiring activity, and so we must add another component to the list of attributes for life, some means of extracting energy from the environment and converting it to useful work. We may call this component metabolism. Other criteria have been suggested such as irritability (or, the ability to react to environmental stimuli), but we can subsume these under the metabolic and information-rectifying activities. Implicit in this discussion is the apparent need to keep the information-bearing entities and their protective mechanisms separate from the rest of the environment. Some segregation of matter is necessary. Given the apparent complexities of all these activities, we may be justified in calling this process self-organization, and include it in our list of life's defining criteria.

Computers may provide the first real chance humans have to understand processes that may have given rise to life as we know it. Their ability to amass, store, access, and assess data, to model environments of several types, make our approach to the origin of life question amenable to mathematical and computational experiments which later could be applied to at-the-bench, wet-chemistry experimentation, i.e., construction of a "living system."

3. GENERAL SUPPORT ROLES

Computational support for research activities has so pervaded the scientific realm that it is almost unthinkable to perform research without them. Whether for something as relatively simple as writing manuscripts, as specific as control of instrumentation, or as complex as global climate modeling, computers are now a basic utility in the researchers' toolbox.

Origins of life research covers a vast panoply of scientific and technical disciplines, from micropaleontology to cosmochemistry, and subsumes the development of techniques and instrumentation necessary to carry out this research. All of the disciplines under this broad umbrella can (and do!) make use of computers in a variety of ways. However, before we continue to address the specifics of computational support of origins of life research, several ways in which computers have begun to pervade the research realm generally will be addressed.

3.1 LARGE-SCALE INVESTIGATIONS

For several years, we have been in a period in which computers have further enhanced our abilities to perform research-related tasks. With the advent of personal computers whose capabilities now meet or exceed those which were hitherto available only with dedicated workstations, minicomputers, or even low-end mainframe computers, whole avenues of research have opened up. The ability to model complex environments and present them as "virtual reality" is but one such avenue which will likely bear great fruit in other areas such as telescience, i.e., the use of robots in exploration of the ocean depths on our own planet or the surfaces and atmospheres of other planets. The development of complex tools to program computers has similarly enhanced our research armamentarium. Indeed, in a wonderful example of recursion, we now even have computer-aided software engineering (CASE) tools to help us write increasingly complex programs to be applied to increasingly complex tasks such as the development of global circulation models, derivation of cosmological theories, and the design of spacecraft and instrumentation to enable such investigations.

3.2 SMALL-SCALE INVESTIGATIONS

At the other end of the spectrum, computers are increasingly being used to examine atomic and molecular-level activities. Pharmaceutical companies have developed protocols to facilitate the cost-effective design of new drugs, employing qualitative structure-activity relationships, taking advantage of tremendous amounts of data pertaining to molecular characteristics such as charge density, van der Waal's potentials, binding energies, pharmacological effects, and more. Efforts have been underway since at least 1982 to combine elements of virtual reality with force feedback manipulators to enable researchers to reach down into the molecular realm and actually manipulate the molecules of interest, and "feel" the effects of bringing two molecules into proximity.

In sum, then, the impact of computers is being felt throughout the vast panoply of research activities underway. Given the familiarity with which researchers view these tools, it may now be the time to examine new ways in which they may be utilized.

3.3 COMPUTER SUPPORT OF RESEARCH

Computers have become so much a part of scientific and technical work lately that it almost seems superfluous to describe some of the uses to which they are put in those environments. However, there are some sociological aspects of their use which in some way(s) mimic the kind of information transfer that must have been present in the early stages of life's origins, and so license to discuss the scientific process and the computer will be taken, and indulgence of the reader is requested during this little parenthetical journey.

The very nature of many of the experiments we can now undertake in the laboratory has changed due to the presence of dedicated and general computing power. Not only may we simulate experimental protocols in an effort to optimize acquisition of data (i.e., should more data points be taken in this region or in that region?) but computers are useful as control points in the use of increasingly complex instrumentation, and as measurement tools to take data, as well.

3.3.1 Experimental Design

Computers may be used in the very design of experiments. For example, through the development of mathematical models, many of the processes that exist in living systems may be investigated to so fine a level of detail that we may ascertain the minuscule interaction energies between hemoglobin and the protons, oxygen, carbonate, and small organic phosphate molecules that ultimately may govern such interactions. By using simulated data in conjunction with sophisticated nonlinear curve fitting (practicable only because of the availability of inexpensive computing resources), it is possible to design binding experiments that are exquisitely sensitive in those regions of the binding isotherms necessary for supportable characterization of the binding phenomena themselves. In the case of the binding of oxygen and organic phosphates to hemoglobin, for example, such an approach indicated the need for more data at extremes of the pH curve, i.e., below pH 6.5 and above pH 8.5, outside the physiological pH range.

3.3.2 Control of Instrumentation

For sophisticated structure determination of, e.g., proteins, high-field nuclear magnetic resonance (NMR) research would be impossible were it not for the use of embedded, dedicated microprocessors in the instrumentation itself to control parameters such as pulse width, frequency, nuclear Overhauser effects, etc. Even such relatively simple instrumentation control activities as liquid scintillation counting and high performance liquid chromatography or gas chromatography/mass spectroscopy benefit from computer control to maintain temperature, flow rates, and experimental conditions in general at levels that are impractical with mere human intervention. The result is, of course, more data, and more reliable data. This has the interesting secondary benefit of freeing up the researcher's time for activities that are as yet not completely supportable by computers, i.e., thinking, evaluating, and planning. However, these activities, too, have benefits derived from increasingly complex computational tools as neural networks, fuzzy logic, and visualization tools to enable the researcher to see their data from many different perspectives. As we fine-tune our efforts to understand phenomena at lower and lower signal-to-noise ratios, we are turning increasingly to the use of sophisticated statistical measures to bring our signals out of the noise floor. While statistics themselves should never be used as a criteria for drawing conclusions, they are useful when employed in conjunction with other criteria.

3.3.3 Facilitating Communications

Given the complexity of many experiments, it is natural that investigators look to each other to pool expertise in an effort to optimize the return on investment of their increasingly limited research resources. This may be something as formalized as a well-defined collaboration, or merely(!) the request that someone at a far remove from the principal investigator's laboratory examine some data, using their unique skills, and perhaps offer some insight as to the phenomena under discussion, and suggestions for future work. The use of computers and computer networks is almost an absolute requirement for such activity, as standard media, such as floppy disks, tapes and network file transfers may be used to transfer data between researchers' computing facilities. Indeed, the wide use of networks is having other effects, as well, as a new, exciting (or puzzling) result is often discussed in this virtual lecture hall long before a formal publication ever makes it to a journal editor's desk. In this way, significant synergy results.

Another aspect of network utilization is the access researchers can get to supercomputing facilities, thereby enabling them to perform calculations often unattainable in their local computing environment. Such calculations may be used to support visualization, i.e., the presentation of data in a graphical format that enables the research to see interrelationships in the data that may not have been apparent by examination of the raw data themselves.

4. SPECIFIC SUPPORT ROLES

If we are to understand the origins of life, we might do well to consider how existing organisms have descended from their ancestors. Whatever else may be said about processes common to living systems, certainly the role of *information transfer* by genetic mechanisms is key. In present-day terrestrial biosystems, this process is embodied in the nucleic acids and the proteins used to catalyze the transfer of the genetic information stored therein.

4.1 PATTERN RECOGNITION

Computers have virtually revolutionized the way we perform nucleic acid-related research. The very nature of nucleic acids—long-chain linear polymers made up of a limited number of subunits—allows the examination of the storage and manipulation of sequences via machines a suitable activity. Whole computer-based industries have grown up around the determination of nucleic acid sequences, their storage, analysis, and comparison for several purposes. New roles for nucleic acids in catalytic processes have been elucidated with the support of computation tools, which allow alignment of sequences for examination of both phylogenetic and catalytic processes.

Very often computational tools that deal with pattern recognition are employed in such pursuits. In a fascinating expansion of scale, the greatest advances in pattern recognition may come from the application of computers to origin of life investigations on a truly cosmic scale in the search for extraterrestrial intelligence, or SETI. While the United States Congress has recently voted to suspend funding for work in this area, the advances already made in the design of search strategies, and the design and construction of complex hardware to support advanced digital signal processing and pattern recognition cannot be made to go away. The all-sky Microwave Survey hardware and software, developed under grants and contracts from the U.S. National Aeronautics and Space Administration (NASA), will certainly find their way into other fields of endeavor in an excellent example of technology transfer.

4.2 PRIMITIVE PLANETS

Other cosmic-scale investigations that are pertinent to origins of life research include the search for planetary systems orbiting stars other than our Sun. Early work in this field was based on mathematical modeling, using at first simple Newtonian mechanics and later, utilizing increased computational power, embodying parameters that have more complex terms. Such work laid the foundation for understanding observational data of candidate systems, i.e., infra-red data of dust disks in the Vela system, and others. Indeed, even the observational data would not be so persuasive were it not for the utilization of computation tools to acquire and reduce data from orbiting spacecraft, including digital signal processing and image enhancement. To date, however, we still have no definitive statements as to the presence of specific planetary bodies around other stars, but the data are certainly indicative.

Once having found other planetary systems, we may begin to speculate on the nature of the planets themselves. Even within the confines of our own Solar System, examination of planets other then the Earth has begun to give us great insight into the processes that led to the formation of a planet that can support life. Observational data obtained by robotic explorers (themselves controlled largely by computers) have been used to support complex mathematical models that allow researchers to define, initiate, and follow the evolution of planetary atmospheres through aeons. These models can help origin of life researchers by providing boundary conditions for the circumstances that may have given rise to terrestrial biosystems. Data concerning gas evolution from planetary accretion and condensation, mixing ratios, gas escape, and more have already helped us understand what conditions were like on the primitive, abiotic Earth.

4.3 MATHEMATICAL MODELS

Mathematical models have found utility in many fields of endeavor, ranging from economics through other social and sociological activities and on to engineering, for example, to help design faster and more efficient cars, airplanes, and space transportation systems. The use of networked supercomputers now allows the speed and volume of computations necessary to calculate (and visualize) the complex hydrodynamic flows around rockets traveling through the atmosphere at the speeds necessary to achieve orbit or escape velocity.

As with physical models (i.e., the construction of clay simulacra in the design of automobiles) a mathematical model is an attempt to simulate structure, function, or phenomena. Of particular interest in these days of new-found environmental awareness, global circulation models (GCMs) are being used by Earth system scientists to help us understand our Earth and its processes. These complex models take into account such variables as sea- and land-surface temperature, concentration and distribution of gases in our atmosphere, the flow of energy through the Earth system, the distribution of water, and the effects of anthropogenic activities on so-called "natural" phenomena. Not only will such models give us a better understanding of the way the Earth system functions and the effects of humans on those processes, but their predictive ability will enable use to determine, e.g., when an El Nino event will occur, with its drastic effects on global weather, or when severe storms will form. This information may then be used to either adapt to or mitigate the effects of these phenomena, much, it is hoped, to the betterment of all. Indeed, we may now begin to further understand the interactions between the nascent Earth and the origins of life.

4.3.1 Cellular Automata

A particularly fascinating use of computers in modeling several aspects of "living" systems comes from their application to cellular automata. These mathematical constructs display several features ascribable to living systems, i.e., they grow, mutate, live, and die. In their simplest incarnation, the evolution of cellular automata may be calculated according to defined rules using paper and pencil; there is nothing inherent to cellular automata that makes their study accessible only via mechanical computational techniques. However, over the years since they were first envisioned, the field study has become increasingly complex, such that the more well-defined and refined rules and complex behaviors now used in their study are significantly facilitated by computers. By employing computers to investigate the responses of cellular automata to different initial conditions and different rules, we may be able to glean significant insight into processes that are embodied in living systems.

4.4 MOLECULAR MODELING

We are living and working in a time when we have demonstrated abilities to manipulate matter at many levels, down to the level of individual atoms. Given the extent to which humans find it an absolute requirement to manipulate their environment, it should come as no surprise that there have been numerous attempts to translate the biochemical phenomena and mathematical findings described above into concrete terms via molecular modeling. Indeed, this offspring of the marriage between biology, biochemistry, mathematics, theory, and computers may provide one of the happiest outcomes for the student of origins of life, in that it may prove entirely possible to first model and then subsequently "build" a self-replicating molecule, an auto-replicant. We will examine some of the tenets and operational practices of molecular modeling in an effort to determine if such an exciting event could indeed come to pass.

4.4.1 Catalytic Molecules

Recent advances in understanding the catalytic capabilities of molecules that heretofore were thought to have no such roles provides an opportunity to apply computational approaches to the basic questions surrounding the origins of life, be such origins on the Earth or elsewhere in the cosmos. With the maturity of the state-of-the-art in molecular design, information theory, category theory, and analytic and synthetic biochemistry, we should now be able to begin to use computers to help us more fully understand the processes and structures that give rise to living systems—indeed, to define just what is a living system—and to apply such knowledge to the design and, ultimately, the construction of a synthetic living system in the laboratory. Such a project may be decomposed into a number of phases, starting with an assessment of the roles that computers can play in this quest, continuing through a refinement phase, wherein would be assessed the kinds of computational tools that require development and the nature of the computing platforms, architectures, and software upon which the research should best be conducted, development of discrete "black boxes", each representing a component of the overall process, including energy-transducing ("metabolic") and information-transfer processes, definition of chemical structures which would embody these processes, and design of an auto-replicant, which would take raw materials from its environment to produce others of its kind.

4.4.2 Molecular Structure and Design

One area that will need addressing is the folding of biological macromolecules into functional conformations that will support self-replication. The details of protein folding have long been investigated with an eye toward understanding the forces and processes involved in turning a long-chain polymer into a precisely folded tertiary structure. There are many forces involved in such folding, and computers may be used to examine theoretical and practical considerations of interactions between amino acid side chains, hydrogen bonds, disulfide bonds, van der Waal's interactions, electrostatic interactions, and the interactions between proteins and prosthetic groups, such as metals and porphyrins. Indeed, several general phenomena must be dealt with, including protein "engineering" and dynamics. In this latter case, some attention has been paid to the overhead needed for molecular dynamical calculations, with the introduction of a new computer architecture to optimize such calculations. One approach is based on parallel processing using microcomputer architectures, linking parallel ports in Intel 80x8y microprocessor platforms for interprocessor communication. For a specific algorithm and application, this configuration works approximately 10 times faster than a VAX-11/780. Nucleic acid data, as well, must be made amenable to molecular modeling, with particular importance being given to the binding of various DNA-binding proteins and other nucleic acids involved in DNA replication and transcription.

4.5 VISUALIZATION

A key component to facilitating an investigator's interaction with increasingly complex data is visualization, discussed briefly above. This catch-all term subsumes the use of computers to transform numerical data into more-easily accessible graphical images, thereby to allow the investigator to use his/her own brain and visual acuity(?) to see relationships and structures that would take a digital computer (even with sophisticated programming) significant time to elucidate. It is when visualization tools are used in concert with molecular and mathematical modeling tools and databases of interactions energies and other thermodynamic parameters that the role of the computer in origins of life research–particularly design of self-replicating molecules–most attractive. As with most such endeavors, significant thought and planning must be applied "up-front," to create databases that are easily accessible, where data formats are standardized, and which therefore may be used by a broad range of applications programs, be they commercial or customized. Herein lies potential for problems, in that there is an increasingly large range of computing platforms and architectures that could be used for such investigations. Providing software and data in formats that could be used by all of them may be an intractable task. However, recent efforts by such industry giants as Apple and IBM to define, design, and build central processing unit (CPU) chips such as the PowerPC, and provide operating system software that will run applications across platforms will go far in alleviating this difficulty.

4.5.1 Data Volume and Accessibility

Another area of potential difficulty (but certainly not insurmountable) is the sheer volume of data required to fully describe the biological macromolecules of interest, whether extant or of as-yet-to-be-determined form and function. Researchers will require access to these large supporting databases, as it is unlikely that all such data will be available at the researcher's workstation (e.g., via compact disk/read-only memory (CD-ROM). High-speed computing networks will be required, with easy access to host machines that house the data. If, as seems plausible, given the stunning increases in the availability of on-line data, a researcher has to access several machines to access needed data, it may prove necessary to construct "intelligent" front-ends to facilitate the hunt for data across hosts and platforms. The parallels with Earth system science data access, as embodied in requirements and early plans for NASA's Earth Observing System Data and Information System are manifold, and researchers into biological, biochemical, and genomic phenomena should be made aware as early as possible of the tremendous amounts of work needed to bring such "one-stop shopping" into their own realms.

The tools to utilize the data, once acquired, are becoming increasingly available, for use on main-frames, workstations, and personal computers. An interesting twist is the availability of rigorous professional tools for use in this latter architecture and environment. For example, the makers of a well-known computer-aided design package have recently begun distribution of a molecular modeling package, HyperChem, by AutoDesk, that provides 2- and 3-D modeling, provides rendering and display routines for several forms, including sticks, disks, filled spheres, dot surfaces, stick and dot surfaces, and stereo images. The package also allows simulations based on several parameters, including energy, heat of formation, ionization potential, electron affinity, dipole moment, and UV-, visible-, and infrared spectra absorptions. For those of us who have watched the personal computer industry grow, the availability of a software package that integrates molecular mechanics, semi-empirical quantum mechanics, and molecular dynamics simulations into one package for a desk-top computer is nothing short of astonishing.

Visualization is a tremendously computation-intensive activity, so the more powerful the CPU and architecture in general, the better for the researcher. Visualization is more than "just" drawing on-screen the results of a calculation, especially when it comes to complex structures. Hidden-line removal, perspective, shading, rotation, etc., are all basic pieces of the puzzle. Of further complexity is to have the programmer provide tools for detailed manipulation of images, including the kinds of tools already familiar to users of word processing applications. Search tools, the ability to describe—and subsequently find—fragments of existing molecules that may have utility in modeling a self-replicant, will become increasingly important as origins of life researchers attempt to project back to early proto-organisms from existing molecules.

5. POTENTIAL SUPPORT ROLES

Given our new-found understanding of molecular motion and dynamics, we should be able to apply algorithms developed for macro-scale objects. Recent work in the modeling of multi-body systems (MBSs) may be applicable to mathematical modeling of molecular interactions. MBS models, which represent components as rigid or flexible bodies with inertia, and springs, dampers, and servomotors without inertia, interconnected by rigid bearings or supports, are of particular interest in molecular modeling because of their ability to include friction and contact forces. Strength considerations may be accounted for by equations of reaction. By modeling molecular candidates for autoreplicants in this manner, we may achieve a first approximation to molecular design which would then be modifiable using techniques designed for molecular modeling *per se.*

5.1 FRACTALS

The study of non-integer dimensional objects, or fractals, is likely to have applicability in mathematical and molecular modeling. First described by Benoit Mandelbrot, fractals have had enormous impact on the way many disciplines approach their subjects, and in the study of complex chemical system, as well. While fractals will likely have their greatest use in characterization of chemical systems larger than the size of small molecules, the utility of this approach may be applied to systems as a whole, and may provide insight that would not heretofore have been attainable. Of particular interest is the use of fractals to describe aggregation of colloids, heterogenic catalytic processes, the kinetics of heterogeneous reactions, and, given the role that electron distribution plays in many existing metabolic and energy-transducing phenomena, electron localization in crystals. Computers allow not only the calculation of fractal dimensionality, but support the visualization of the many structures and phenomena that may be characterized by fractals.

5.2 ELECTRON DISTRIBUTION

The role of electron distribution and other such phenomena should be addressed in studies of the origins of life. New understanding of molecular interactions brings quantum mechanics into to origin of life investigators' tool kit. Computer modeling of these phenomena and subsequent application of new findings to the mathematical and molecular modeling of primitive molecular structures will likely provide significant insight into the energy transduction processes necessary to the development and survival of self-replicating structures. In addition, polymers and biomolecules may have utility as computer components, i.e., storage and switching, and may presage give us the ability to see if existing biomolecules can (and do) support this kind of activity.

6. DESIGNING AND BUILDING AN AUTOREPLICANT

Given the broad outline above, it should come as no surprise that advocacy of the use of a combination of computational and chemical approaches to design (from first principles) and ultimately to construct a molecular assembly that displays several of the requirements of living systems. Specifically, the tools may well be at-hand to use computers to delineate and define those processes inherent in living systems, to transforms these inchoate processes into chemical and molecular terms, and design synthetic pathways which would be used by bio- and organic chemists to synthesize such molecular assemblies *in vitro*. Computer-based mathematical and molecular modeling tools, coupled with access to large amounts of structural and functional data stored in databases accessible across networks such as the Internet and the upcoming U.S. National Research and Educational Network and the National Information Infrastructure, and utilization of increasingly complex and comprehensive visualization tools will support such an endeavor. Whether such a molecular assembly has direct relevance to the specific assembly (or assemblies) that arose on the Earth is a moot point. That is, what we design in the laboratory may not (and probably would not) bear much resemblance to the first "natural" autoreplicants. But as embodiments of general principles, such a design and synthesis would tell us a great deal about the processes embodied in living systems in general, and, as such, may give us pointers to more fully understand the nature of the first bionts.

7. CONCLUSION

This paper has examined several aspects of the role(s) for computers in research in general, and in origins of life research specifically. It should be clear that there is a natural affinity for use of computational tools and techniques in existing origins of life research, and for extending investigations into realms hitherto inaccessible to the human researcher. Such extension could include, most notably, the design and synthesis of a self-replicating molecule, embodying principles and processes that are common to living systems.

The tools are at hand; we have but to commit ourselves to using them, or developing tools where needed, and to apply ourselves to the task.

ACKNOWLEDGMENTS

The author would like to thank Professor James Reggia of the Department of Computer Science, University of Maryland, for his support during the generation of this paper, and Professor Cyril Ponnamperuma, Laboratory of Chemical Evolution and the Department of Chemistry and Biochemistry, University of Maryland, College Park, for the invitation to address the Conference. Happy Birthday, Cyril! Janice Hobish provided critical commentary and moral support.

This work was supported, in part, by NASA Award NAGW-2805 to Professor James Reggia.

SUGGESTIONS FOR FURTHER READING

Borman, S., 1991: Fractals Offer Mathematical Tool for Study of Complex Chemical Systems, *Chemical and Engineering News*, April 22, 28-35.

Colwell, R.R., Ed., 1989: *Biomolecular Data: A Resource in Transition*, Oxford University Press, Oxford, UK, 367 pp.

Forrest, S., 1993: Genetic Algorithms: Principles of Natural Selection Applied to Computation, *Science*, **261**, 872-878.

Freedman, D., 1993: AI Helps Researchers Find Meaning in Molecules, *Science*, **261**, 844-845.

Hopfield, J.J., J.N. Onuchic, D.N. Beratan, 1988: A Molecular Shift Register Based on Electron Transfer, *Science*, **241**, 817-819.

Kasparek, S.V., 1990: *Computer Graphics and Chemical Structures: Database Management Systems*, John Wiley & Sons, New York.

Lesk, A.M., Ed., 1988: *Computational Molecular Biology*, Oxford University Press, Oxford, UK, 254 pp.

Marinucci, M.R., F. Giorgi, 1992: Regional Climate Modeling. *Proceedings of the International School of Physics "Enrico Fermi: Course CXV, "The Use of EOS for Studies of Atmospheric Physics*, Gille, J.C., and G. Visconti, Eds., Italian Physical Society, North-Holland, Amsterdam, 231-251.

Ratzlaff, K.L., 1989: *Introduction to Computer-Assisted Experimentation*, John Wiley & Sons, New York, 438 pp.

Reggia, J.A., S.L. Armentrout, H.-H. Chou, and Y. Peng, 1993: Simple Systems that Exhibit Self-Directed Replication, *Science*, **259**, 1282-1287.

Scheraga, H.A., 1989: "Some Computational Problems in the Conformational Analysis of Polypeptides and Proteins." In *Computer-Assisted Modeling of Receptor-Ligand Interactions*, R. Rein and A. Golombek, Eds., *Progress in Clinical and Biological Research*, **289**, Alan R. Liss, Inc., New York, 3-18.

Schiehlen, W., Ed., 1990: *Multibody Systems Handbook*, Springer-Verlag, Heidelberg.

Solomon, E.I., and M.D. Lowery, 1993: Electronic Structure Contributions to Function in Bioinorganic Chemistry, *Science*, **259**(5101), 1575

Venkatarahgavan, G., and R.J. Feldmann, Eds., 1985: Macromolecular Structure and Specificity: Computer-assisted Modeling and Applications, *Annals of the New York Academy of Sciences*, **439**.

MEMBRANE PHASE SEPARATIONS, ASYMMETRY AND IMPLICATIONS IN THE ORIGIN OF LIFE

Michael O. Eze
International Centre for Theoretical Physics,
Trieste 34100, Italy
and
University of Nigeria,
Nsukka, Nigeria

ABSTRACT

Membrane lipids, by affecting membrane physical state, influence solute transport as in *Escherichia coli*, and play a prominent role in homeoviscous (homeophasic) adaptation whereby cells adapt to varying temperatures. Thus, on prebiotic earth, lipid–doped precells were possibly stabilized. Studies to investigate this hypothesis are advocated.

1. INTRODUCTION AND BIOMEMBRANE STRUCTURE

What lessons can be learned from the *Escherichia coli* membrane to help in the design of experiments probing the interface between the living and the non–living? It is to address part of this question that this essay is all about.

A biological membrane surrounds every prokaryotic and eukaryotic cell (cell, or plasma membrane), as well as every organelle (organellar, or intracellular membrane) in a eukaryotic cell. It has been argued that the first living cell could not have evolved in the absence of this functionally vital membrane (Oro and Lazcano, 1990; De Duve, 1991).

The "Fluid Mosaic" model (Singer and Nicolson, 1972) describes the biomembrane (Fig.1) as a bimolecular leaflet (bilayer) of membrane lipids in which float the membrane integral (intrinsic) proteins, and other molecules like sterols. These components are amphiphilic (amphipathic), i.e., at least one part of the molecule is hydrophobic, and the other hydrophilic. The hydrophobic parts of the integral proteins (and the other molecules) are inside the hydrophobic core of the bilayer. The peripherial (extrinsic) proteins are hydrophilic, and remain in the inside and outside aqueous phases, interacting with the polar parts of the membrane lipids and integral proteins. The carbohydrates (receptors and antigenic determinants) extrude to the outside of the cell, from the polar parts of the lipids and integral proteins. The disposition of these components thus imposes the inherent asymmetry on the membrane architecture, vital for all the membrane functions (Singer and Nicolson, 1972).

2. ARTIFICAL, AND NATURAL LIPID BILAYERS, AND THE THERMOTROPIC GEL–TO–LIQUID CRYSTALLINE PHASE TRANSITION

Each typical membrane lipid, e.g., the phospholipid, dipalmitoylphosphatidylcholine (di16:0PC), contains two acyl chains on one polar head. In water (at the right conditions) the molecules spontaneously self–assemble into bilayer vesicles. In the bilayer, the acyl chains sequester into the hydrophobic core away from water, while the polar heads are in contact with water on the inner and the outer surfaces (Singer and Nicolson, 1972; McElhaney, 1976; Eze, 1991). Prebiotic amphiphiles might have formed bilayers in the primeval broth (Oro and Lazcano, 1990; De Duve, 1991).

The bilayer undergoes a thermotropic, cooperative, and reversible gel–to–liquid crystalline phase transition, in response to temperature (Fig.2). For a pure lipid, containing only one lipid head type (e.g., di16:0PC), the transition is sharp and occurs at the transition temperature, T_c. During the transition, there is long range order, i.e., bilayer configuration is preserved. Above T_c, the bilayer is in the liquid crystalline state in which there is short range disorder, with acyl **chains** melted and mobile. Below T_c, is the gel state with rigid and immobile acyl chains.

A heterogeneous mixture of phospholipid heads and acyl chains (as is the biomembrane) undergoes a broad transition having an onset T_S, an end T_L, and a midpoint T_M. At T_M, 50% gel and 50% liquid crystalline lipids coexist (McElhaney, 1976; Eze, 1990, 1991). Biomembranes (which also contain proteins, sterols, etc.) do exhibit lateral phase separations, manifested as overlap of two or more broad transitions (Linden *et al.*, 1973; McElhaney, 1976; Eze and McElhaney, 1981; De Rosa *et al.*, 1986; Eze, 1990, 1991). Phase transitions may be measured with physical techniques (Linden *et al.*, 1973; Sinensky, 1974; Eze and McElhaney, 1981; McElhaney, 1976, 1984; De Rosa *et al.*, 1986; Dey *et al.*, 1993).

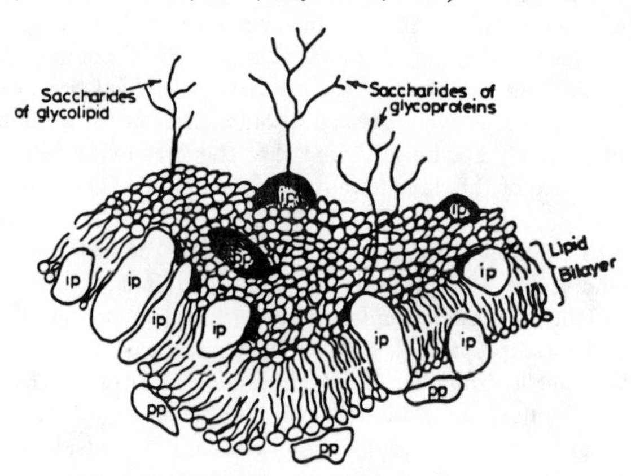

Figure 1. The Fluid Mosaic Model of Biomembrane Structure.
(∝≈) = membrane lipid molecule;
(ip) = integral (intrinsic) protein;
(pp) = peripheral (extrinsic) protein.

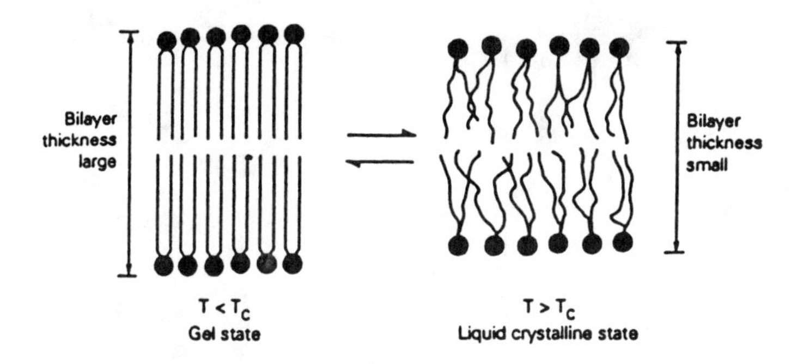

Figure 2. Gel–to–liquid crystalline phase transition. T = temperature; T_c = transition temperature. (After Eze, M.O., 1991). Reprinted from Biochemical Education with kind permission from Pergamon Press, Oxford, UK

3. CELLS WITH DIFFERING MEMBRANE FATTY ACIDS AND FLUIDITY AND PHYSICAL STATE

Escherichia coli K1060 is an unsaturated fatty acid (UFA) auxotroph incapable of synthesizing or catabolizing UFAs, and requires UFA to grow. *Escherichia coli* K1060 whose membranes have differing acyl chains, and thus varying fluidity and physical state were prepared by growing cells in UFAs, namely, linoleate (18:2c,c), palmitoleate (16:1c), oleate (18:1c), palmitelaidate (16:1_t), and elaidate (18:1_t). The membrane fluidizing potency decreased thus: $18 : 2c, c > 16 : 1c > 18 : 1c > 16 : 1_t > 18 : 1_t$. The physical state was determined with Differential Thermal Analysis (DTA), and as shown in Fig.3, they exhibited lateral phase separations (Eze and McElhaney, 1981).

4. SOLUTE TRANSPORT STUDIES AND INFORMATION THEREFROM

Active transport of ^{14}C – L–proline and ^{14}C – L–glutamine across these membranes clearly revealed a dependence on the fluidity and physical state of the membranes, seen as breaks in Arrhenius plot of temperature–dependence data for each UFA enrichment. As revealed in Fig.4, in 18:1_t cells, L–glutamine transport maximum velocity ($V_{\max}$) responded predictably at lower temperatures. However, starting from high temperatures, at the point (37.5°C), i.e., close to T_L of 40°C where gel phase lipid begins to form in coexistence with liquid crystalline lipid, there is a discontinuity in the Arrhenius plot for this transport system in favour of increased $V_{\max}$.

A related phenomenon was earlier observed (Linden *et al.*, 1973) for active transport of sugars in 18:1_t–grown *Escherichia coli* strain $30E\beta ox^-$, another

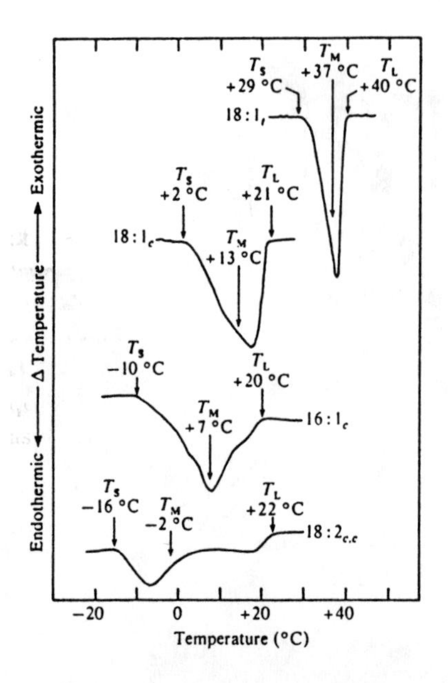

Figure 3. Differential Thermal Analytical) (DTA) thermograms of lipids from membranes of *Escherichia coli* K1060 enriched in various unsaturated fatty acids. T_S = onset of transition; T_M = temperature at which 50% gel and 50% liquid crystalline lipid coexist; T_L = end of transition (After Eze and McElhaney, 1981).

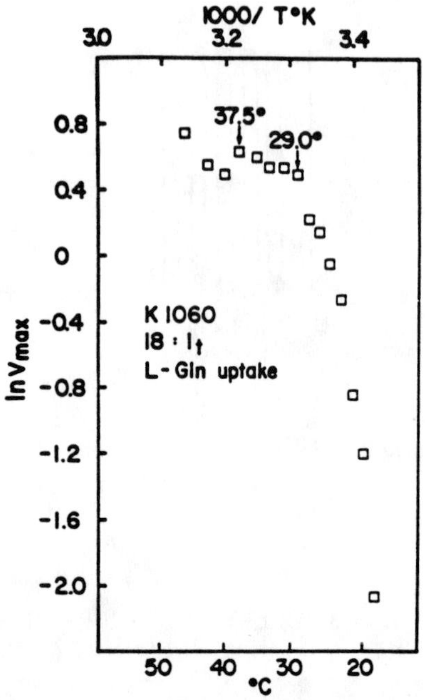

Figure 4. Arrhenius plot of V_{max} of *L*–glutamine (Gln) transport in *Escherichiacoli* K1060 enriched in elaidate (18:1$_t$). (Adapted from Eze and McElhaney, 1987).

UFA auxotroph. The authors explained that such enhancement of active transport occurred because of enhanced isothermal lateral compressibility in the membrane milieu in which gel and liquid crystalline lipids coexist. This signalled the importance of lateral phase separations in membrane lipids during cellular function, and directly applies also to our observations.

5. HOMEOVISCOUS (OR HOMEOPHASIC) ADAPTATION

To acclimatize to varying temperatures, micro–organisms resort to hemeoviscous (Sinensky, 1974), or homeophasic (McElhaney, 1984) adaptation. This involves the incorporation into the membrane of low–melting fatty acids at low temperatures, and high–melting ones at high temperatures. This has been observed also in some eukaryotes (Thompson, Jr., 1989), some organs of hibernating mammals (Aloia and Raison, 1989), and in fish liver (Dey *et al.*, 1993). Thus each organism ensures that its growth temperature lies within a borad thermotropic phase transition of its membrane lipids, displaying lateral phase separations (Eze, 1991). The implicit asymmetry in the molecular architecture of membrane components optimizes these effects.

6. IMPLICATIONS TO CHEMICAL EVOLUTION AND THE ORIGIN OF LIFE

In the investigations on chemical evolution and the origin of life, the role of the lipid membrane has always been implicated (Oro and Lazcano, 1990; De Duve, 1991; Luisi, 1993). Also Academician A.I. Oparin had suggested that phase separations (in the general sense, I suppose) would play a crucial role (Fox *et al.*, 1974). Membrane lipids, a special class of amphiphilic lipids, can self–aggregate by a thermodynamically driven process to form the bilayer (liposomes) as the configuration of least free energy (Bangham *et al.*, 1974; Oro and Lazcano, 1990: De Duve, 1991; Luisi, 1993). Thus, the message for bilayer formation is inherent in the lipid molecule itself. Additionally, the functional attributes of many membrane proteins and enzymes are intricately linked to their presence in, and interactions with the membrane hydrophobic milieu (Eze, 1990).

It is hereby hypothesized that membrane–type lipid molecules played a vital role in the natural selection of those precells (or protocells) which survived the harsh conditions (temperature, etc.) that existed on early earth. The type and amount of such lipids associated with any pre–, or protocellular structure (polynucleotides–plus–polypeptides–plus–lipids) would determine within which temperature ranges there could be a broad, reversible phase transition accompanied by lateral phase separations. Structures allowing for wide temperature range, and marked phase separation, would be favourably selected for, and should be able to survive wide fluctuations in temperature, in the primordial ocean surface, in the hydrothermal vent, or elsewhere. Such pre–, or protocells would therefore have the chance to evolve into higher forms.

7. POSSIBLE INSIGHTS FROM THE ARCHAEBACTERIA

Archaebacteria constitute a primitive group of organisms. Some of their characteristics betray them as prokaryotes whereas some others make them resemble eukaryotes (De Rosa *et al.*, 1986; De Duve, 1991; Runnegar, 1992) (see also Prof. Oshima's chapter in this book). They have been referred to by De Rosa *et al.* (1986) as " ... a distinct primary kingdom, ... contributing to a better understanding of the universal ancestor". Derivatives of archaebacterial membrane lipids have been found in sedimentary rocks as old as Precambrian Era (De Rosa *et al.*, 1986). These primitive organisms still inhabit rather harsh ecological niches, akin to conditions that could have existed on prebiotic earth, including saturated brine (extreme halophiles), hydrothermal vents (methanogens), hot springs (extreme thermophiles) (De Rosa *et al.*, 1986). Their lipids (with ether, rather than ester bonds to glycerol) suit their environments. They are mostly bipolar with a continuous hydrocarbon (isoprenoid) chain between the two polar ends, rather like two regular membrane lipids joined covalently at their methyl ends. Thus, the membrane is a monolayer of these bipolar isoprenoid lipids (De Rosa *et al.*, 1986; De Duve, 1991). In the thermophiles, e.g., *Sulfolobus sulfataricus*, the hydrophobic chains contain cyclopentane rings which increase in number as temperature increases. Physical studies of the membrane reveal possibilities of lateral phase separation (De Rosa *et al.*, 1986), suggesting the importance of hoeoviscous (homeophasic) adaptation in cells on early earth.

8. PROSPECTS FOR EXPERIMENTATION

Experiments specifically designed to probe the possible role of membrane-type lipids in the stabilization of precellular molecular assemblies (microsystems) during the process of chemical evolution of the first cell, should involve the doping of the microsystems with different types and levels of lipid. The stabilities of these assemblies over time, and under various conditions would suggest their survival potential. The stability should be correlated with lipid content, and the ability to exhibit lateral phase separations during response to incubation temperature. The relevant physical parameters, e.g., phase separations, in the microsystems should be obtained by suitable physical techniques (*vide supra*).

ACKNOWLEDGMENTS

The author would like to thank Professor Abdus Salam, the International Atomic Energy Agency and UNESCO for hospitality at the International Centre for Theoretical Physics, Trieste, where this work was prepared. He would also like to thank the Swedish Agency for Research Cooperation with Developing Countries, SAREC, for financial support during his visit at ICTP under the Associateship scheme.

REFERENCES

Aloia, R.C., and J.K. Raison, 1989: Membrane Function in Mammalian Hibernation, *Biochim. Biophys. Acta*, **988**, 123–146.

Bangham, A.D., M.W. Hill, and N.G.A. Miller, 1974: Preparation and Use of Liposomes as Models of Biological Membranes. Chapter 1, in *Methods in Membrane Biology*, Vol.I, Plenum Press, New York, 1–68.

De Duve, C., 1991: *Blueprint for a Cell: The Nature and Origin of Life*, Neil Patterson Publishers, Burlington, North Carolina.

De Rosa, M., A. Gambacorta, and A. Gliozzi, 1986: Structure, Biosynthesis and Physicochemical Properties of Archaebacterial Lipids, *Microbiol. Rev.*, **50**, 70–80.

Dey, I., C. Buda, T. Wiik, J.E. Halver, and T. Farkas, 1993: Molecular and Structural Composition of Phospholipid Membranes in Livers of Marine and freshwater Fish in Relation to Temperature, *Proc. Natl. Acad. Sci.* (USA), **90**, 7498–7502.

Eze, M.O., 1990: Consequences of the Lipid Bilayer to Membrane-Associated Reactions, *J. Chem. Education* (Amer. Chem. Soc.), **67**, 17–20.

Eze, M.O., 1991: Phase transitions in Phospholipid Bilayers: Lateral Phase Separations Play Vital Roles in Biomembranes, *Biochemical Education*, **19**, 204–208.

Eze, M.O., and R.N. McElhaney, 1981: The Effect of Alterations in the Fluidity and Phase State of the Membrane Lipids on the Passive Permeation and Facilitated Diffusion of Glycerol in *Escherichia coli*, *J. Gen. Microbiol.*, **124**, 299–307.

Eze, M.O., and R.N. McElhaney, 1987: Lipid and Temperature Dependence of the Kinetic and Thermodynamic Parameters for Active Amino Acid Transport in *Escherichia coli* K1060, *Biochim. Biophys. Acta*, **897**, 159–168.

Fox, S.W., G.A. Deborin, K. Dose, and T.E. Pavlovskaya, 1974: Historical Introduction: A.I. Oparin and the Origin of Life. In *The Origin of Life and Evolutionary Biochemistry*, K. Dose, S.W. Fox, G.A. Deborin and T.E. Pavlovskaya (Eds.), Plenum press, New York, 3–6.

Linden, C.D., K.L. Wright, H.M. McConnell, and C.F. Fox, 1973: Lateral Phase Separations in Membrane Lipids and the Mechanism of Sugar Transport in *Escherichia coli*, *Proc. Natl. Acad. Sci.* (USA), **70**, 2271–2275.

Luisi, P.L., 1993: Defining the Transition to Life: Self–Replicating Bounded Structures and Chemical Autopoiesis. In *Thinking About Biology*, W. Stein and F.J. Varela (Eds.), SFI Studies in the Sciences of Complexity, Lecture Note Vol.III, Addison–Wesley Publishers, 3–24.

McElhaney, R.N., 1976: The Biological Significance of Alterations in the Fatty Acid Composition of Microbial Membrane Lipids in Response to Changes in Environmental Temperature. In *Extreme Environments: Mechanisms of Microbial Adaptation*, M.R. Heinrich (Ed.), Academic Press, New York, 255–281.

McElhaney, R.N., 1984: The Relationship Between Membrane Lipid Fluidity and Phase State and the Ability of Bacteria and Mycoplasmas to Grow and Survive at Various Temperatues. In *Membrane Fluidity*, M. Kates and L.A. Manson (Eds.), Plenum Press, New York, 249–278.

Oro, J., and A. Lazcano, 1990: A Holistic Precellular Organization Model. In *Prebiological Self Organization of Matter*, C. Ponnamperuma and F.R. Eirich (Eds.), A. Deepak Publishers, Hampton, Virgina, 11–34.

Runnegar, B.N., 1992: The Tree of Life. Section 9.3 in *The Proterozoic Biosphere: A Multidisciplinary Study*, J.W. Schopf and C. Klein (Eds.), Cambridge University Press, UK, 471–475.

Sinensky, M., 1974: Homeoviscous Adaptation – A Homeostatic Process that Regulates the Viscosity of Membrane Lipids in *Escherichia coli*, *Proc. Natl. Acad. Sci.* (USA), **71**, 522–525.

Singer, S.J., and G.L. Nicolson, 1972: The Fluid Mosaic Model of the Structure of Cell Membranes, *Science*, **175**, 720–731.

Thompson (Jr.), G.A., 1989: Lipid Molecular Species Retailoring and Membrane Fluidity, *Biochem. Soc. Trans.*, **17**, 286–289.

GENERAL CRYSTAL IN PREBIOTIC CONTEXT[1]

I. Simon
Institute of Enzymology, BRC,
Hungarian Academy of Sciences
Budapest P.O.Box 7, H-1518, Hungary
and
International Centre for Theoretical Physics,
Trieste, Strada Costera 11, I-34100, Italy

ABSTRACT

General crystal is an extension of the crystal concept to any form of matter which exhibits neighbour structure determination. This extension makes many results of solid state physics applicable to heterogeneous matter. Among others it includes the description of phase transition from random to unique structure. The advantage of the general crystal approach is demonstrated on globular protein, one of the most important macromolecules of life, which is capable to adopt unique 3D structure spontaneously, regardless of the heterogeneous character of its chemical structure and conformation. It is suggested that the use of general crystal concept may help to find candidates among heterogeneous matters capable of spontaneous self-organization in the same way as crystallisation results in unique structure of homogeneous matter, and to apply some of the results of solid state physics to describe the phase transition and other behaviour of this matter.

1. INTRODUCTION

Folding of proteins and nucleic acids into their unique, biologically active forms is one of the most important unsolved problems of today's biochemistry. It is probable that the folding problem will not be solved until we understand what makes a certain sequence of amino acids capable of folding to a unique structure, since polypeptides with random sequences do not adopt unique conformations. Self-organisation of protein, RNA and DNA is crucial to their biological functions. It is obvious that self-organisation played even more important roles in the early days of evolution, when no splicing enzymes nor molecular chaperones were present.

It is evident that under certain conditions, every homogeneous matter undergoes self-organisation into a unique structure. The procedure is called crystallisation, and solid state physics established several principles and laws about this phase transition as well as properties of the matter in the solid phase. Unfortunately, biochemistry teaches us that only heterogeneous matter can play active biological

[1] This work was performed under the Associate Membership of the author, who also acknowledge additional financial support from OTKA 1361.

roles, while the homogeneous ones, like homopolymers play only passive supporting or storage roles.

To find criteria that make heterogeneous matter capable of adopting unique structures spontaneously and to learn about laws governing the behaviour of this matter may be important from the viewpoint of the origin of life for the reason we discuss below.

2. CONCEPT OF GENERAL CRYSTAL

Every crystal has a common feature, called Neighbour Structure Determination (or NSD) since in a crystal the position of every atom or ion is determined by the structure of any neighbouring unit-cell by translation and other symmetry. For homogeneous matter, symmetry and NSD are equivalent. On one hand, symmetry is a perfect form of NSD. On the other hand, in homogeneous matter, consisting of identical elements, the way of NSD is identical for all elements and this identical determination results in structures with translation (and other) symmetries. However, symmetry is not the only form of NSD. It was suggested that it is worth extending the crystal concept in such a way that any form of matter which shows NSD should be recognised as a general crystal. The advantage of this extension is that all results of solid state physics which are based on the NSD and not on the symmetry of the crystals can be applied in a much larger set of matter than the homogeneous ones (Simon, 1986).

For example, let us consider a sequence of 100 letters, in which every overlapping 5 letter segment is an English word. To create even a much shorter such sequence is so difficult that it is evident that these sequences, if they exist at all, are extremely rare, forming a relatively small subset of the 26^{100} possible sequence of letters. Since there are only about 10^5 five letter words which make sense in English out of the about 10 million possible combinations of the letters, it is very hard to change any letter, outside the few terminal ones, in such a way that all the five words containing the letter in question, change into another meaningful word. Therefore here every letter is determined by its neighbours. Consequently, this sequence can be created from any of its parts into a unique one by successive fitting of all letters to the growing sequence. It was shown by a simple probability calculation, that even if in some point more than one letter would fit to the preceding four letters, in general, the propagation cannot be continued in any wrong directions further than a few steps (Simon, 1981). Note, that here NSD is implemented by probability, therefore it is not as perfect, and rigid, as in a real crystal.

3. PROTEIN AS GENERAL CRYSTAL

A protein is a more realistic example. Due to the way of the *de novo* folding (Simon and Asbóth, 1980; Simon 1979) and the natural selection for extended lifetime (Simon, 1981), all native protein structures have a common feature, namely, they are the only conformation of the polypeptide chain in which all of the overlapping short segments appear in one of the low energy conformations of the respective oligopeptides (Simon, 1985; Simon et al., 1991). Since an average pentapeptide has about ten million conformations which correspond to local minima in the conformational energy surface; 99% of these correspond to high energy local minima and only about 10^5 have really low energy structure. Therefore the case is very similar to that discussed for the overlapping English words. Thus the conformation of an amino acid residue is more or less determined by the conformation of the adjacent one, to the same extent whether the residue in question is part of a symmetrical structure, like alpha helix or beta strand, or belongs to non-symmetrical structures like coil or turn. One should note however, that the restriction, that all overlapping short segments are in low energy conformation is so strong, that only a relatively small subset of the possible amino acid sequences can adopt such a structure. (A simplified calculation suggests that from the possible $20^{100}=10^{130}$ different 100 residue long polypeptide 10^{45} belong to this subset (Simon, 1985)). Thus polypeptides in general do not fall into the category of a general crystal, only a relatively small (but in absolute value, extremely large) subset which, among others, contains all naturally edited, or successfully designed protein sequences.

Since proteins are general crystals there has to be an analogy between behaviour of solid states governed by NSD, and that of globular proteins.

Crystallisation from supercooled liquid or from supersaturated solution is based exclusively on NSD and it has nothing to do with the symmetry. The procedure starts with random nucleation and subsequent fast growth by adjusting new elements to the already existing part of the crystal. Protein refolding, self-organisation of the polypeptide chain, takes place according to the same pattern. Adding even a very small piece of crystal nucleus or a surface which catalyses the formation of the crystal nucleus to the supercooled liquid or to the supersaturated solution can speed up the crystallisation. Likewise, any trace of native structure due to incomplete unfolding, the surface of molecular chaperones do the same kinetic effect on folding. In general, the same thermodynamics and kinetics which describe crystallisation can be applied to protein folding as well.

A phase transition between states of matter is a similar case. The well-defined structure of solid state and the random mass distribution in a gas phase correspond to the unique structure of the folded protein and the unfolded one respectively. In a certain interval of temperature and pressure there is a third phase called liquid. This may correspond to the molten globule state of protein. The phase diagrams of the three phases, including the critical point where the three phases can coexist generally used in solid state physics can be applied to describe the folded - unfolded equilibrium of proteins too.

The last example is about influence of foreign atoms on the properties of the solid state. Their influence on the transport properties is mainly based on their influence on the symmetry feature of the matter. However, their influence on the overall stability (melting temperature) and the local stability (movements of vacancies and dislocations under external pressure) are based on NSD. It is known that while in low concentration foreign matter always reduce the melting temperature, they can influence the local stability in both directions, independently from the global stability (Kittel, 1976). Likewise, it was shown that ligand binding or amino acid replacement influence the overall stability (measured by thermal unfolding) and the local stability (measured by H-D exchange) are independent (Privalov, 1989; Woodward et al., 1982; Simon et al., 1984).

These examples on the general crystal character of proteins also show that NSD can take place in a probability level. This is a great advantage of general crystals over the real ones from the viewpoint of evolution. As we know a protein can undergo certain limited structural changes without loosing its overall conformation. This happens when protein binds any other molecule, including the substrate which transferred into product by a protein as an enzyme. Therefore a general crystal adopts its structure according to the environment, and is capable of active interaction with the environment. Another important feature is the tolerance of foreign matter. Crystallisation is often used for purification, because a growing crystal hardly tolerates foreign matter. This is due to the deterministic (not probabilistic) way of NSD. When NSD takes place by probability, the matter can be more tolerant. The evolution of protein sequences results in many slightly different chemical structures in various species, which adopt the same 3D structure, but they have different stability and other biochemical properties. So a general crystal can evolve.

4. CONCLUSION

Turning back to the question of prebiotic evolution, the general crystal approach may help to overcome difficulties emerging from the fact that while crystallisation is a perfect way of self-organisation and it works properly on simple inorganic matter, it works only on homogeneous ones which are not of interest from the viewpoint of evolution. Therefore, when we are looking for heterogeneous matter capable of adopting unique structures spontaneously, it is worth checking the NSD ability of the candidates.

ACKNOWLEDGEMENTS

The author would like to thank Professor Julian Chela-Flores for his valuable comment on this work. The author also would like to thank Professor Abdus Salam and the ICTP for hospitality.

REFERENCES

Kittel, C. (1976)
Introduction to Solid State Physics, 6th ed.
John Wiley and Sons Inc. New York

Privalov, P.L. (1989)
Thermodynamic Problems of Proteins Structure
Ann. Rev. Biophys. Biophys. Chem. 18, 47-69

Simon, I. (1979)
Investigation of Protein Folding: Uneven Distribution of Point Mutations along Polypeptide Chains
J. Theor. Biol. 81, 247-258

Simon, I. (1981)
Possible Mechanism for the Dynamic Stability of Proteins Structure
J. Theor. Biol. 90, 487-493

Simon, I. (1985)
Investigation of Protein Refolding: a Special Feature of Native Structure Responsible for Refolding Ability
J. Theor. Biol. 113, 703-710

Simon, I. (1986)
Proteins as General Crystals
J. Theor. Biol. 123, 121-124

Simon, I. and Asbóth, B. (1980)
An Additional Argument in Favour of Continous Folding during Biosynthesis of Proteins
J. Theor. Biol. 82, 685-688

Simon, I., Tüchsen, E. and Woodward, C. (1984)
Effect of Trypsin Binding on the Hydrogen Exchange Kinetics of Bovine Pancreatic Trypsin Inhibitor B-Sheet NH's
Biochemistry 23, 2064-2068

Simon, I., Glasser, L. and Scheraga, H.A. (1991)
Calculation of Protein Conformation as an Assembly of Stable Overlapping Segments: Application to Bovine Pancreatic Trypsin Inhibitor
Proc. Natl. Acad. Sci. USA 88, 3661-3665

Woodward, C., Simon, I. and Tüchsen, E. (1982)
Hydrogen Exchange and the Dynamic Structures of Proteins
Mol. Cell. Biochem. 48, 135-160

SELF-ORDERING AND POLYMERIZATION IN BIOLOGICAL MACROMOLECULES AND SYMMETRY BREAKING

Giuseppe Vitiello *

Dipartimento di Fisica, Università di Salerno, 84100 Salerno, Italy

In the study of the origin of life puzzling questions concern the mechanisms producing not only the self-ordering of living matter, but also the high efficiency of its chemical activity and the high stability of its functional properties. Such questions are relevant to understand the origin of life and at the same time to understand how life *still now* is sustained and *survives*; in some sense, as Cyril Ponnamperuma observed in his lecture, one should ask not only "how *did* life start", but also "how *does* life start".

Living matter presents self-ordering in space *and* in time, namely it presents pathways of biochemical reactions sequentially interlocked. One crucial problem of chemistry and of molecular biology is that such pathways cannot be expected to occur in a random chemical environment and even "the embedding of the simplest biochemical pathway in a random chemical environment will cause the pathway to collapse" (Rasmussen et al., 1992). Chemical efficiency and functional stability at the degree of the one observed in living matter seem to be out of reach of any probabilistic approach based on microscopic *random kinematics*: self-ordering efficiency and stability thus must be properties emerging from a microscopic *dynamics* underlying the biochemical activity. Statistical kinematical arguments, although necessary ingredients, are not sufficient tools to fully understand living systems. The great body of knowledge produced by molecular biology appears as the phenomenological manifestation of basic dynamical laws. In other words, the dynamics, and not only the kinematics, must be investigated in the study of living matter. The situation is similar to the one for non-living (inert) matter: chemistry is the phenomenological level emerging from the quantum mechanical laws ruling the interactions among electrons, atoms and molecules. Before the discovery of Quantum Mechanics chemistry was a rich collection of observations, however unable to explain periodicities, affinities among elements, the strength of the molecular bonds, etc.. For example, chemistry was able to produce only a *catalog* of different ordering patterns in crystals, or, at most, it could *watch* and *register* the sequence of steps in the growth of the crystal, and it has been even possible to assemble it since

* Electronic mail address: vitiello@sa.infn.it

the molecular *engineering* has been enough developed. Nevertheless, one had to wait the developments of quantum theories to understand the basic physical laws prescribing the atoms to sit in their positions in the crystal ordered pattern(Anderson,1984).

Since many years we are therefore pursuing the formulation of a microscopic dynamical scheme for living matter, which could account for those questions which require physical, more than chemical, tools (Del Giudice et al., 1982-1988; Vitiello, 1992,1993). Of course, there is a large territory, say an *interface*, where the contact between the physics of living matter and molecular biology is of crucial importance. In the same way there are concepts and tools of quantum physics which do not belong to biochemistry and vice versa; but this is typical of any interdisciplinary effort. In the following I will shortly summarize the main qualitative aspects of our approach. The mathematical apparatus and more detailed discussions may be found in the above quoted references.

We study some aspects of living matter in the framework of Quantum Field Theory (QFT) since it is the available theory successfully describing elementary constituents as electrons, atoms, molecules and their interactions. We consider in fact the living matter as just one of the possible *phases* or *states* in which matter appears and therefore the dynamical laws for its elementary constituents must be the same ones as in other (inert) states.

The first problem to which we address ourselves is the one of the self-ordering of living matter. QFT is tailored for systems with an infinite number of degrees of freedom and a preeminent role is played by symmetry properties of the dynamical equations for the fields. The equations (and the observables quantities) are symmetric under some group of transformations, e.g. translations, rotations, etc., if they remain invariant when such transformations are performed on the fields. Thus, in some sense, one cannot distinguish among dynamical configurations (states) of the fields each other related by those transformations, which therefore are called symmetry transformations. An ordered state of the system is, on the contrary, characterized by the possibility of distinguishing it from other states with different (or none) ordering. Order is therefore lack of symmetry; to generate from a set of symmetric states an ordered state one has to single out one of them and then to *break* its symmetry. To break the symmetry means that the elementary constituents must be organized according to a certain configuration, e.g., of their spatial positions, like in the crystal lattice, or of their time evolution, like in the laser which is generated by *in phase*, or *coherent*, emission of photons by the laser source. The probability of generating such a *coherent* or ordered behaviour of the full set of elementary constituents out of a random kinematics is of course practically zero and also believing in miraculous fluctuations one may obtain correlations at most in very small domains, never ranging over the whole system volume: this is the reason why light is not emitted in laser state. Coherence requires that the ordering *information* (the characterizing arrangement or configuration) be carried over the whole system volume so to be shared by the full set of elementary constituents; in QFT such an ordering information is carried by a wave mode generated *by the dynamics*, as shown by a general theorem, and ranging over the whole system. The quantum associated to such a wave mode is therefore a *collective* mode, also called the Goldstone particle, and is actually observed in crystals, where is called phonon, in ferromagnets, where is called magnon, etc.. QFT is therefore able to describe the ordering of the system as induced by *dynamically* generated wave modes extending over the whole system.

In the case of living matter the elementary constituents are molecules with electrical dipole moment, namely water molecules and biomolecules of various species, but always carrying an electrical dipole. Laboratory observations by biologists have shown that water surrounding proteins appears in organized domains with non-zero density of electrical polarization (see e.g. Celaschi and Mascarenhas, 1977). The dipolar property of water is generally recognized to be a relevant parameter affecting the biochemical activity. Fröhlich suggested that electrical polarization could be taken as a parameter characterizing the living matter state(Fröhlich, 1968-1980). We have then assumed in our QFT approach that the relevant symmetry to be broken is the rotational electrical dipole symmetry of the field equations for the water molecule excitations(Del Giudice et al.,1983-1986). By general theorems of QFT and by using also a specific model we have shown that, in the presence of some dipolar *impurity*, as e.g. a protein chain, water may indeed present domains with coherent electrical dipole vibrations. In our scheme therefore water is considered as the *matrix* for living matter where a first level of *dynamical* ordering is achieved: in the presence of some kind of molecules, or macromolecules, carrying electrical dipole, the water gets organized in coherent domains with non-zero polarization density, as indeed is observed for water surrounding biomolecules. It is interesting to observe, in connection with the problem of the origin of life, that such a first level of organization might well be realized in ponds of quiet water, or in other situations where temperature was not so high to interfere with and destroy the dipole ordering, and on the other hand not so low to freeze the freedom of water molecules dipoles to rotate or to vibrate so to form a coherent domain. Similarly, strong thermal fluctuations could have a negative effect in the coherent domain formation; however, they could also have a positive effect since the dipole-dipole electrostatic bounds of neighbour water molecules must be broken in order for the dipoles to enter in the coherent vibrational mode. We in fact stress that the mechanism of spontaneous breakdown of symmetry sets in *long range* correlations (forces) extending over domains large with respect to the elementary components scale (over the system volume in the absence of surface effects); therefore local, short range forces among the constituents do not enter in the ordering mechanism; at most they set in after order is achieved; more often they act as background *noise* producing local distortions, deformations and defects in the ordered pattern. This is a crucial point where the dynamical origin of the ordering is clearly seen; it is also the crucial point which shows that ordering obtained by assembling *molecule by molecule* with short range forces (*hooks*), as hydrogen bonds in water, has negligible probability to occur, even more negligible if the *one by one* molecule assembling should relay on random collisions: ordering is long range correlation of dynamical origin. In the case of water the long range electrical dipole-dipole interaction is of radiative nature and for this reason we talk also of superradiance(Del Giudice et al.,1988b).

We observe that in biological systems ordering is not imported from some external source, but is indeed of dynamical origin as described in the above QFT scheme; in this sense we speak of self-ordering. An important feature is the one related with the nonlinear dynamics of wave propagation on biochains, as pointed out by Davydov (Davydov, 1979-1982). The external energy input (not carrying information), induces ATP reactions at one of the end-points of the biochains. A soliton wave is then generated and it propagates without dissipation on the biochain; the resulting electrical dipole excitation may thus trigger the polarization of the surrounding water molecules, i.e. the spontaneous breakdown of dipole rotational symmetry: the incoherent informationless energy input is thus used by the system dynamics to generate ordering. The nonlinear character of the system-environment interaction is rich of consequences and a discussion on weak feedback and environment effects may be found in Del Giudice et al.,1988c; and Celeghini et al.,1990.

Let me now briefly discuss the propagation of electromagnetic field in the ordered water domains, which leads to a further level of ordering and is strongly related with functional properties of living matter.

There is a competition between electromagnetic (em) field and long range correlations in ordered media so that either the em field is prevented from penetrating the ordered medium, or it breaks the ordering correlations, provided its strength is above certain threshold; in such a case the em field remains confined in the uncorrelated regions thus propagating inside tubes or filaments, whose diameter is controlled by a factor depending on the inverse of strength of the medium correlation. Outside the filament the em field is zero; inside the filament, where the em field is confined, the correlation is destroyed. This mechanism is known as the Meissner effect and controls the magnetic field propagation in superconductors (Parks, 1969). The associated propagating regime of the em field is also called self-focusing propagation: the em field is allowed to propagate in the correlated medium *only within a network of filaments* (Del Giudice et al., 1986).

Self-focusing is a known phenomenon in nonlinear optics(Askar'yan, 1974) where it may happen provided some dynamical conditions are met, which do not include the coherence of the medium. In living matter the ordering of the medium is a pre-requisite for the self-focusing to occur. In nonlinear optics the field strength must be strong enough to supply energy to create correlations. In the case of living matter the field strength may be as weak as the one of the medium correlations. The diameter of the filament in a completely aqueous medium with maximum of polarization is computed to be of the order of 15 nanometers (which is a figure very near to the inner diameter of microtubules(Clegg, 1981, 1988)).

The self-focusing propagation of em field may have relevant effects on polymerization mechanisms. The gradient forces which are present on the side boundary of the filament may be shown to selectively act on the surrounding molecules attracting or repelling them according to a definite resonant frequency pattern (Del Giudice et al, 1986).

Molecules attracted on the filament may contribute to a change of the field frequency in the filament thus modifying the resonant pattern. Consequently, attraction or repulsion of molecules of different oscillatory frequency will occur. In this way the filament gets coated by a molecular pattern which may stabilize into a polymeric structure when the coating molecules may form stable chemical bonds. In such a case, even if the filament field disappears the polymeric structure will survive. Otherwise, if the chemical bonds are not stable, the polymeric structure will disassemble as soon as the filament field will vanish. We thus see how the em filamentary network may originate a polymeric net structure. The cytoskeleton structure in living systems may be described by such a network with its continuous creation or depletion of branches and with its movements and intricate geometry.

It is also interesting to observe that the medium coherence also plays the role of *protecting* the system from unwanted (*noisy*) em weak disturbances, the strength of the correlations acting as a threshold. Chemical activity involving em molecular interactions weaker than the coherence strength are indeed shielded from unwanted em perturbations of very low strength with respect to the ordering correlations which therefore are not allowed to propagate. The stability of delicate biochemical activity (which would collapse in a random chemical environment) is thus insured by the coherence of the medium.

The self-focusing propagation of the em field with the consequent polymerization process may have been the basic mechanism generating a number of interesting polymerization phenomena in primordial environments presenting ordered domains. Once the formation of specific molecular sequences was triggered by self-focusing em propagation, further mechanisms of modular duplication and chemical evolution may have taken place.

REFERENCES

Anderson,P.W. 1984: *Basic Notions of Condensed Matter Physics*, Benjamin, Menlo Park, California.

Askar'yan, G.A., 1974: The self-focusing effect, *Sov. Phys. Usp.* **16**, 680.

Celaschi, S., and S.Mascarenhas, 1977: Thermal stimulated pressure and current studies of bound water in lysozime, *Biophys. J.* **20**, 273.

Celeghini, E., E. Graziano and G. Vitiello, 1990: Classical limit and spontaneous breakdown of symmetry as an environment effect in quantum field theory, *Phys. Lett.* **145A**, 1.

Clegg, J.S., 1981: Intracellular water metabolism and cell architecture, part 1, *Collect. Phenomena* **3**,389.

Clegg, J.S., 1988:Intracellular water metabolism and cell architecture, part 2. In *Coherent Excitations in Biological Systems*, H. Fröhlich and F. Kemmer (Eds.), Springer, Berlin, p.189.

Davydov, A.S., 1979: Solitons in molecular systems, *Physica Scripta* **20**, 387.

Davydov, A.S.,1982: *Biology and quantum mechanics*,Pergamon, Oxford.

Del Giudice, E., S. Doglia, and M. Milani, 1982: A coherent dynamics in metabolic active cells, *Physica Scripta* **26**, 232.

Del Giudice, E., S. Doglia, M. Milani and G. Vitiello, 1983: Spontaneous symmetry breaking and boson condensation in biology, *Phys. Lett.* **95A**, 508.

Del Giudice, E., S. Doglia, M. Milani and G. Vitiello, 1985: A quantum field theoretical approach to the collective behaviour of biological systems, *Nucl. Phys.* **B251** [**FS 13**], 375.

Del Giudice, E., S. Doglia, M. Milani and G. Vitiello, 1986: Electromagnetic field and spontaneous symmetry breaking in biological matter, *Nucl. Phys.* **B275** [**FS 17**], 185.

Del Giudice, E., S. Doglia, M. Milani and G. Vitiello, 1988a: Structures, correlations and electromagnetic interactions in living matter: Theory and applications. In *Biological coherence and response to external stimuli*, H. Fröhlich (Ed.), Springer-Verlag, Berlin, p.49.

Del Giudice, E., G. Preparata and G. Vitiello,1988b: Water as free electric dipole laser, *Phys. Rev. Lett.* **61**, 1085.

Del Giudice, E., R. Mańka, M. Milani and G. Vitiello,1988c: Non-constant order parameter and vacuum evolution, *Phys. Lett.* **206B**, 661.

Fröhlich, H., 1968: Long range coherence and energy storage in biological systems, *J.Quantum Chemistry* **2**, 641.

Fröhlich, H., 1977: Long range coherence in biological systems, *Riv. Nuovo Cimento* **7**, 399.

Fröhlich, H., 1980: The biological effects of microwaves and related questions, *Adv. Electron. Phys.* **53**, 85.

Parks, R.D. (Ed.),1969: *Superconductivity*, Marcel Dekker, Inc. N.Y..

Rasmussen, S., R. Feldberg and C.Knudsen, 1992: Self-programming of matter and the evolution of proto-biological organizations, LA-UR 92-1722, Los Alamos.

Vitiello, G., 1992: Coherence and electromagnetic field in living matter, *Nanobiology* **1**, 221.

Vitiello, G., 1993: Living matter Physics. In *Coherent and emergent phenomena in biomolecular systems*, P.A. Hanson, S. Hameroff, S. Rasmussen and J. Tuszynski (Eds.), MIT-Press, in print

PART IV
BIOPHYSICAL ASPECTS OF SELF-ORGANIZATION

RETURN TO DICHOTOMY: *BACTERIA* AND *ARCHAEA*

A. Yamagishi, and T. Oshima
Tokyo Institute of Technology
Nagatsuta, Yokohama 227, Japan

ABSTRACT

We briefly reviewed the argument about the phylogenetic tree of life on the earth, and the validity of the group archaebacteria or *Archaea*. We explained our proposal about dichotomic division of the general phylogenetic tree of life. We also proposed the way of handling eukaryotes in the prokaryotic phylogenetic tree.

1. PHYLOGENETIC TREES OF LIFE

In 1977 Woese and Fox (1977) have reported the cataloguing analysis of small subunit ribosomal RNA of methanogenic bacteria. They found that methanogenic bacteria belong to an unique group different from eukaryotes and common bacteria (eubacteria). They proposed a group archaebacteria to represent the third group. Thermophilic sulfur-dependent bacteria, and extremely halophilic bacteria were also included in archaebacteria (Fox et al., 1980). They suggested that the life on the earth should be divided into three not two nor five (Woese and Fox, 1977).

Until 1989, there was no way to determine the root of the general phylogenetic tree. In 1989 two groups have reported the way to determine the root of the global phylogenetic tree (Gogarten et al., 1989; Iwabe et al., 1989). They used duplicated genes to determine the root in the tree. Several genes have been used to determine the root. It was concluded that the root is between eubacteria and archaebacteria with eukaryotes on the archaebacterial branch (Iwabe et al., 1989; Miyata et al., 1991). The root was adapted to the phylogenetic tree derived from rRNA sequences (Woese et al., 1990, Fig. 1). Woese et al. proposed the names of these three groups: *Archaea, Bacteria* and *Eucarya* for archaebacteria, eubacteria and eukaryotes, respectively (Woese et al., 1990).

2. RELATION BETWEEN EUKARYOTES AND ARCHAEBACTERIA

On the other hand, the relation between eukaryotes and archaebacterial groups is an unsettled question. Sequence analysis of rRNAs suggests that archaebacteria are monophyletic (Gouy and Li, 1989) and the eukaryotes have separated before the division of archaebacterial groups. Protein sequences did not give conclusive results. Recently Lake analyzed the elongation factor sequences (Rivera and Lake, 1992; Lake, 1994). They suggested that the eukaryotes are more closely related to eocytes (thermophilic archaebacteria) than halophiles.

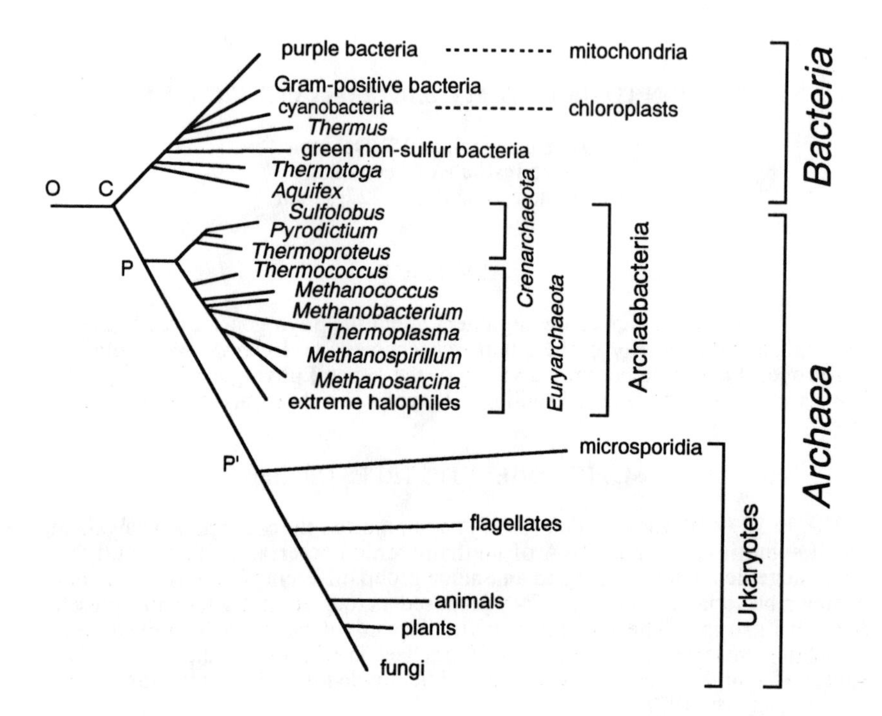

Figure 1. Phylogentic tree and the dichotomic division of life.

The results suggest that the nuclear genes of eukaryotic cell like rRNA are closely related to archaebacteria and may be included in archaebacteria. It is rather natural to include eukaryotes in archaebacteria.

3. DICHOTOMIC DIVISION OF LIFE ON THE EARTH

We pointed out that the most primitive division of the life is the separation between eubacteria and archaebacteria at the position indicated by the letter c in Fig.1 (Yamagishi and Oshima, 1993). We proposed the dichotomic division of life: Bacteria (eubacteria) and Archaea. The dichotomic division is shown in Fig.1. We proposed to include the nuclear genomes or genes of eukaryotes in Archaea (Yamagishi and Oshima, 1993). Because the separation of archaebacterial group and urkaryotes is so small, these two groups should not be taken as the most primitive division. Thus life on the earth should be divided into *Archaea* and *Bacteria*

4. POSITION OF EUKARYOTES IN THE PROKARYOTIC TREE

The next problem is how to handle eukaryotes. Phylogenetic analysis of rRNA and protein sequence of chloroplasts and mitochondria supported the symbiotic theory of these organelles in eukaryotic cells. Chloroplasts of green plants are closely related to and included in cyanobacteria (Giovannoni et al., 1988; Turner et al., 1989). Mitochondria are included in alpha subdivision of proteobacteria (Purple bacteria, Yang et al., 1985; Olsen et al., 1994). Each component of the eukaryotic cells is assigned to the prokaryotic phylogenetic tree.

We also propose that the eukaryotes should not be positioned in the phylogenetic trees as a whole. Instead, each component of the eukaryotic cells, or each gene of the eukaryotic cells should be placed at each suitable position in the prokaryotic phylogenetic tree (Yamagishi and Oshima, 1993). We also think it appropriate to use the word urkaryotes proposed by Woese et al. (Woese and Fox, 1977) to represent nuclear genomes or genes of eukaryotes.

What about eukaryotes as a whole, then? We think it appropriate to treat eukaryotes like molecules, which are the counterparts of atoms (Yamagishi and Oshima, 1993). Each prokaryotic species is accepted as an atom or element. Thus the eukaryotic cells can be expressed as the combination of each component of prokaryotes. For example, green plants contain nucleus, mitochondria and chloroplasts. The nucleus probably consists of cells or Archaea whose precise positions in the phylogenetic tree are not known. Though, there are still many possibilities including that the nucleus is made of several prokaryotic components from *Archaea* and *Bacteria*. Then the nucleus itself must be expressed as the combination of prokaryotic components. Accordingly, the universal phylogenetic tree contains only prokarytoes and prokaryotic component of eukaryotes. Eukaryotes should not appear in the phylogenetic tree as a whole.

Accepting these ideas on eukaryotes, life on earth should be divided into two groups, *Archaea* and *Bacteria*. In Fig.1, we used the word archaebacteria to represent the groups of *Crenarchaeota* and *Euryarchaeota* (Woese et al., 1990). The phylogenetic nomenclature, *Archaea* should be used to mean that the group consists of three subgroups: *Euryarchaeota*, *Crenarchaeota* and Urkaryotes. It is also convenient to use the common name archaebacteria, though it is not a taxon name, for the group contains *Crenarchaeota* and *Euryarchaeota*.

REFERENCES

Fox, G.E., E. Stackebrandt, R.B. Hespell, J. Gibson, J. Maniloff, T.A. Dyeer, R.S. Wolfe, W.E. Balch, R.S. Tanner, L.J. Magrum, L.B. Zablen, R. Blakemore, R. Gupta, L. Bonen, B.J. Lewis, D.A. Stahl, K.R. Luehrsen, K.N. Chen, and C.R. Woese 1980: The phylogeny of prokaryotes, *Science* **209**, 457-463.

Giovannoni, S.J., S. Turner, G.J. Olsen, S. Barns, D.J. Lane, and N.R. Pace 1988: Evolutionary relationships among cyanobacteria and green chloroplasts, *J, Bacteriol,* **170**, 3584-3592.

Gogarten, J.P., H. Kibak, P. Dittrich, L. Taiz, E.J. Bowman, B.J. Bowman, M.F. Manolson, R.J. Poole, T. Date, T. Oshima, J. Konishi, K. Denda, and M. Yoshida 1989: Evolution of the vacuolar H^+-ATPase: Implication for the origin of eukaryotes, *Proc. Natl. Acad. Sci., USA,* **86**, 6661-6665.

Gouy, M. and W.-H. Li 1989: Phylogenetic analysis based on rRNA sequences supports the archaebacterial rather than the eocyte tree, *Nature* **339**, 145-147.

Iwabe, N., K. Kuma, M. Hasegawa, S. Osawa, and T. Miyata 1989: Evolutionary relationship of archaebacteria, eubacteria, and eukaryotes inferred from phylogenetic trees of duplicated genes, *Proc, Natl, Acad, Scie, USA,* **86**, 9355-9359.

Lake, J.A. 1994: Reconstructing evolutionary trees from DNA and protein sequences: Paralinear distances, *Proc. Natl. Acad. Sci., USA,* **91**, 1455-1459.

Miyata, T., N. Iwabe, K. Kuma, Y. Kawanishi, M. Hasegawa, H. Kishino, Y. Mukohata, K. Ihara, and S. Osawa 1991. Evolution of archaebacteria: Phylogenetic relationships among archaebacteria, eubacteria, eukaryotes. In *Evolution of life: Fossils, Molecules, and Culture*, S. Osawa and T. Honjo (Eds.), Springer-Verlag, Tokyo, 337-351.

Olsen, G.J., C.R. Woese, and R. OverBeek 1994: The winds of (Evolutionary) change: Breathing new life into microbiology, *J. Bacteriol,* **176**, 1-6.

Rivera, M.C. and J.A. Lake 1992: Evidence that eukaryotes and eocyte prokaryotes are immediate relatives, *Science* **257**, 74-76.

Turner, S., T. Burger-Wiersma, S.J. Giovannoni, L.R. Mur, and N.R. Pace 1989: The relationship of a prochlorophyte *Peochlorothrix hollandica* to green chloroplasts, *Nature* **337**, 380-382.

Woese, C.R., O. Kandler, and M.L. Wheelis 1990: Towards a natural system of organisms: Proposal for the domains Archaea, Bacteria, and Eucarya, *Proc. Natl. Acad. Sci., USA,* **87**, 4576-4579.

Woese, C.R. and G.E. Fox 1977: Phylogenetic structure of the prokaryotic domain: The primary kingdoms, *Proc. Natl. Acad. Sci., USA,* **74**, 5088-5090.

Yamagishi, A. and T. Oshima 1993. Proposals on the group of archaebacteria and naming of the last common ancestor. In *The Abstracts of the International Workshop on Molecular Biology and Biotechnology of Extremophiles and Archaebacteria* 40-41.

Yang, D., Y. Oyaizu, H. Oyaizu, G.J. Olsen, and C.R. Woese 1985: Mitochondrial origins, *Proc. Natl. Acad. Sci., USA,* **82**, 4443-4447.

RNA: Genotype and Phenotype

Peter F. Stadler
Institut für Theoretische Chemie, Universität Wien
Währingerstraße 17, A-1090 Wien, Austria
Santa Fe Institute
1660 Old Pecos Trail, Santa Fe, NM 87501, U.S.A.

ABSTRACT

RNA folding is a combinatory map assigning (secondary) structures to sequences. The frequency distribution of structures is highly non-uniform: there are few abundant and many rare structures. The distances between sequences folding into the same structure are distributed randomly, while extensive neutral networks of nearest neighbors with common structures percolate through the entire sequence space, and hence only a small number of mutations is necessary in order to reach a desired structure from an arbitrary initial sequence. These features make RNA an optimally evolvable polymer.

1. INTRODUCTION

The dichotomy of genotypes and phenotypes is the basis of biological evolution. The information for the unfolding of an organism, i.e., a phenotype, is contained in (DNA) genes and all inheritable variations are, as far as we know today, the effects of changes in the nucleotide sequence of the genome. Selection, on the other hand, acts by enhancing the relative frequency of "fitter" phenotypes, causing the eventual extinction of less fit ones. This is known as the Darwinian principle of evolution. Fitness, i.e., the rate of reproduction (and the probability of survival of the off-spring until the age of its own reproduction), is a property of the phenotype (and the environment, of course). A crucial step in the understanding of evolutionary dynamics on molecular level is therefore the development of a theory for the mapping from genotype to phenotype and the subsequent evaluation of the phenotype by the environment.

2. WHY RNA?

RNA molecules provide a unique model for the interplay of genotype, phenotype, and fitness. They are replicated template directed in a suitable environment, and replication errors occur at the level of the nucleotide sequence (genotype). Spiegelman (1971) pointed out that such an RNA replication system shows all ingredients of the Darwinian scenario: The replication rate depends on the three-dimensional structure (phenotype) of the molecule, which must be recognized by the replicase, and which must be melted during the replication process. The underlying mechanism and its reaction kinetics are known in detail (Biebricher and Eigen 1988).

The major problem with RNA in chemical evolution is to be seen in its highly elaborate molecular structure: for RNA template induced reactions molecules of high stereochemical purity are required, and chiral compounds such as ribose have to be present as pure single antipods (Racemic mixtures of D- and L-ribose incorporated into polynucleotides are prohibitive for template induced replication). It was suggested therefore that less complicated molecules which could nevertheless act as templates preceded an RNA world (Orgel 1986; Joyce 1989) Recent progress in template-chemistry (Orgel 1992) makes this suggestion more and more plausible.

Replication errors, *mutations*, lead to new molecular species whose replication efficiency is evaluated by the selection mechanism. The higher the error rate the more mutants appear in the population. The stationary mutant distribution is characterized as *quasispecies* (Eigen 1971), an extension of the notion of a species in biology. A broader spectrum of mutants makes evolutionary optimization faster and more efficient in the sense that populations are less likely caught in local fitness optima. There is, however, a critical *error threshold* (Eigen 1971; Eigen and Schuster 1979; Swetina and Schuster 1982): if the error rate exceeds the critical limit heredity breaks down, populations are drifting in the sense that new RNA sequences are formed steadily, old ones disappear, and no evolutionary optimization according to Darwin's principle is possible. In the non-catalyzed case the error-rate per nucleotide, p, is very large, on the order of 10^{-2} to 10^{-1}. The maximal chain length (Eigen 1971; Eigen and Schuster 1979)

$$n_{max} = \frac{\ln \sigma}{p} \qquad (1)$$

that a master sequence can adopt and still allow for stable replication over many generations is thus very unlikely to exceed tRNA-size. The superiority parameter σ expresses differential fitness between the master sequence and the average of the remaining population; in the limit $\sigma \rightarrow 1$ we are dealing with *neutral evolution* (Kimura 1983).

The discoveries of Thomas Cech and Sidney Altman (Cech 1986; Guerrier-Takada *et al.* 1983; Guerrier-Takada and Altman 1984; Symons 1992) have shown that, in contrast to previous belief, RNA molecules, "ribozymes", can catalyze several classes of reactions with similar high efficiency and specificity as protein enzymes. Thus RNA molecules are unique in chemistry and biology: they can not only act as templates for their reproduction but are also catalysts for several reactions involving other RNA molecules or oligopeptides. The catalytic activities of RNA molecules are all related to cleavage or formation of nucleotide or peptide bonds. The capability of RNA to catalyze peptide bond cleavage and formation (Noller 1991; Noller *et al.* 1992; Piccirilli *et al.* 1992, Bartel and Szostak 1993) is of particular interest since it can be interpreted as a hint that primordial ribosomes might have worked without proteins. Although the catalytic repertoire of ribozymes is rather poor compared to that of protein enzymes, it comprises the key reactions that are necessary to produce polynucleotides from oligomers and monomers. It is possible therefore that there was a period in the origin of life on Earth when RNA, or a chemically related precursor, served as both genetic material and specific catalyst (Gilbert 1986; Joyce 1991). In other words, RNA once was both genotype and phenotype.

Catalytic networks of interacting RNAs catalyzing each others replication have been proposed by Eigen and Schuster (1979, long before the catalytic properties of RNA had been discovered) as a means to overcome the limitations on information content imposed by the error threshold. In the purest form of this model, the *hypercycle*, one RNA species catalyzes the replication of its successor in a cyclic arrangement. The growth rates of a species depends now on the presence and concentration of other members of the cycle. Frequency dependent replication in the absence of mutations has been studied extensively in the past (For a comprehensive survey see Hofbauer and Sigmund (1988). All kinds of complex dynamical behavior including chaotic attractors (Schnabl *et al.* 1991) were observed. Introduction of mutation into these dynamical systems provides substantial difficulties for qualitative analysis and very few investigations with explicit and exact consideration of mutations were performed so far (For a general perturbation approach to the problem see Stadler and Schuster (1992). Very recent results (Stadler and Nuño 1993) indicate that an error-threshold exists for catalytic networks as well. Catalysed replication, however, is very likely to be much more accurate than uncatalysed replication.

Parasites that use a network's catalytical activity for their own reproduction but do not improve the viability of any other member in the network present a serious threat to such an organization, since they may overgrow and eventually kill the entire network (Eigen and Schuster 1979, Stadler and Happel 1993). It is very reasonable to assume that such parasites will arise as mutants of members of the network. This problem can be resolved by assuming that the reactions take place in surface layers with the transport of matter being controlled by diffusion (Boerlijst and Hogeweg 1992), or by assuming that the replication RNAs are included in lipid vesicles or the like.

It is worth noting in this context that RNA plays an active role in the translation of genetic information in all present-day organisms: RNA is the major catalytic component of the ribosomes and maybe even the catalyst for protein synthesis (Noller 1991). The recognition of the correct amino acid is not direct: The anticodon of a specific tRNA carrying the proper amino acid is used for recognition at the ribosome. Maturation of the tRNAs involves RNaseP, a riboprotein in which again RNA is the catalytically active component (Kole *et al.* 1980, Pace and Smith 1990). Furthermore, DNA is primed by RNA in the initial stage of DNA replication. These facts are consistent with the view that (i) translation was an invention made in a world of self-replicating catalytic RNAs, and that (ii) DNA was a later addition, introduced as a permanent storage device for the increasing amount of genetic information needed for an increasingly complex protein machinery. It seems that despite the fact that proteins allow for much more efficient and specific catalysis life never found a way to completely replace the ancient ribozymes at the very interface between nucleic acids and poly-peptides. Information transfer from DNA to RNA and vice versa (transcription and reverse transcription) is routine even in present day biochemistry. We can speculate therefore that a phylogenetic extrapolation of present day genetic information may go back as far as to the first efficiently replicating species of a primordial RNA world.

It is a lucky coincidence that among all biopolymers RNA is the easiest case for a computational approach. The 3D structure is dominated by its secondary structure, the Watson-Crick and **GU** base pairs. The free energy of folding can be explained almost completely by the energy of secondary formation (see, e.g., Freier *et al.* 1986). Usually, triple helical regions and so-called pseudo knots are, per definition, considered part of the tertiary interactions folding the two-dimensional (planar) secondary structure into the full 3D shape of the molecule. Secondary structure is well conserved in evolution (Sankoff *et al.* 1978, Cech 1988, Waterman 1989, Le and Zuker 1990, Gutell 1992). Consequently, the secondary structure is a reasonable first approximation to the true shape (phenotype) of an RNA molecule. RNA secondary structure can be efficiently predicted from the sequences by efficient computer programs (Zuker and Sankoff 1984, Martinez 1984, Zuker 1989, McCaskill 1990, Hofacker *et al.* 1993a).

3. LANDSCAPES AND COMBINATORY MAPS

The Hamming distance is defined as the number of positions in which two sequences differ. This metric arranges the sequences space, $\mathcal{C}$, i.e., the set of all sequences of length n, in a highly symmetric graph. A quite natural distance between secondary structures can be defined in a similar way. For the details of the definition of distances between secondary structures we refer to Hogeweg and Hesper (1984), Shapiro (1988), Shapiro and Zhang (1990), Fontana *et al.* (1993a). In the following we will denote the distance in sequence space between the configurations (sequences) x and y by $d(x,y)$. For the distance of two structures a and b in the shape space $\mathcal{S}$ we will write $D(a,b)$.

Both the sequence space $\mathcal{C}$ and the shape space $\mathcal{S}$ are discrete spaces of combinatorial complexity; mappings between such spaces have been termed *combinatory maps* (Fontana *et al.* 1993a). For the special case of objects being mapped to numerical properties the term (combinatory) *landscape* is widely used. The mathematical object "landscape" occurs in many different contexts. Classical examples are spin glass Hamiltonians and the cost functions of combinatorial optimization problems. For a recent overview see

Kauffman (1993) and Schuster and Stadler (1993). It is clear that the dynamics of evolution is closely linked to the structure of the fitness landscape (Eigen 1971, Eigen *et al.* 1989, Fontana *et al.* 1987, 1989, 1993a, Amitrano *et al.* 1989, Bonhoeffer and Stadler 1993, Huynen and Hogeweg 1994), and to the structure of the combinatory map f of phenotype formation in a more general context (Fontana *et al.* 1993b, Schuster 1993, Schuster *et al.* 1993).

In the RNA case we have landscapes of numerical properties as functions of secondary structures $p : \mathcal{S} \to \mathbb{R}$. The composition of the combinatory map $f : \mathcal{C} \to \mathcal{S}$ describing the folding process with these landscapes yields again a family of landscapes $p(f) : \mathcal{C} \to \mathbb{R}$ relating the sequences directly to the physical properties of the unfolded phenotype. In fact, the free energy ΔG of the folding process is obtained as a by-product of the calculation of the secondary structure.

RNA free energy landscapes, $\Delta G : \mathcal{C} \to \mathbb{R}$ have been studied in detail (Fontana *et al.* 1991, 1993ab, Bonhoeffer *et al.* 1993). Landscapes of kinetic properties as functions of the sequences are described in Tacker *et al.* (1993a). The mapping from genotypes, i.e., sequences, to phenotypes, i.e., secondary structure graphs, and further-on to properties of the phenotype defining its fitness are schematically depicted as follows

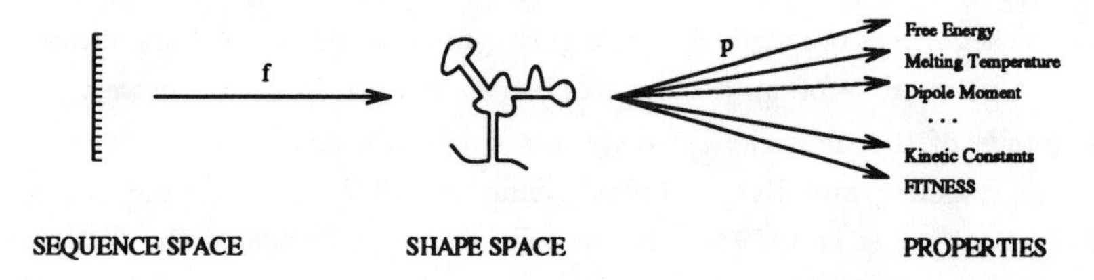

It has been shown recently (Tacker *et al.* 1993b) that the structure of f does not strongly depend on the algorithm or parameter set used to predict the secondary structures. The choice of the distance measure D in shape space has also only a minor influence on the results. We will focus on the data that have been obtained from a minimum free energy approach with a parameter set taken from Freier *et al.* (1986). All calculation reported here have been performed using the `Vienna RNA Package` (Hofacker *et al.* 1993a).

4. THE COMBINATORY MAP OF RNA FOLDING

4.1. THE DISTRIBUTION OF STRUCTURES

There are many more sequences than structures. Counting only planar secondary structures without isolated base pairs and with at least three unpaired bases in hairpin loops one finds that there are approximately

$$S_n \approx 1.4848 \times n^{-3/2} (1.8488)^n \tag{2}$$

secondary structures for sequences of length n for the 4^n sequences (Stein and Waterman 1978, Hofacker *et al.* 1993b). Hence the question arises how these relatively few structures are distributed over the sequence space $\mathcal{C}$.

We find that the frequency distribution of secondary structures obtained by folding a random sample of sequences follows roughly a generalized Zipf law (Mandelbrot 1982),

$$f(r) = a(b + r)^{-c}, \tag{3}$$

where r is the rank (by frequency) of the structure S and $f(r)$ is the fraction of occurrences of S in the sample (Schuster *et al.* 1993). The exponent c describes the distribution of rare sequences, the constant b is a rough measure for the number of frequent structures. Our data suggest that there are few frequent structures and many very rare ones. The few thousand most common structures cover already more than 90% of the sequence space for **AUGC** sequences with chain length $n = 30$ and 40.

For much longer sequences a direct computation of the frequency distribution becomes too costly. One can, however, investigate the distribution of coarse grained structure representations (see e.g. Schuster *et al.* 1993, Hofacker *et al.* 1993a). It turns out that a frequency distribution of the generalized Zipf type occurs at all resolutions of the secondary structure representation.

The parameters b and c depend strongly on the chain length. Not unexpectedly, the number of the most frequent structures, b, increases exponentially with chain length, see figure 1a. The parameter c describing the scaling of the power law tail of the distribution decreases with chain length, indicating that a larger fraction of sequences folds into rare structures for longer chains. The data in figure 1b are consistent with $c \to 1$ for $n \to \infty$.

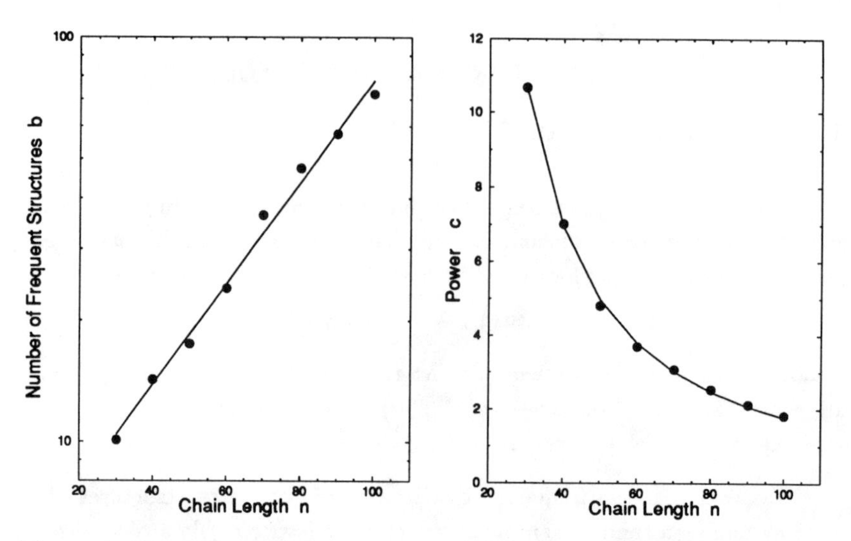

Figure 1. Dependence of the parameters of the generalized Zipf's law, equation (3), on the chain length. Data are for coarse grained structures are shown.

4.2. CORRELATION

A basic property of combinatory landscape is *ruggedness.* A landscape is rugged if it has lots of local optima, if adaptive (up-hill) walks are short, and if the correlation between nearest neighbors is small. Adaptation and optimization is harder on more rugged landscapes. While the notions of local optima and adaptive walks do not have counter-parts in general combinatory maps (their definition require the image set to be ordered), we can generalize the definition of pair-correlation to mapping from one metric space into another one (Fontana *et al.* 1993a):

$$\rho(d) = 1 - \frac{\langle D^2(f(x), f(y)) \rangle_{d(x,y)=d}}{\langle D^2(f(x), f(y)) \rangle_{\text{random}}} \tag{4}$$

The average in the enumerator runs over all pairs of configurations with fixed Hamming distance d while the average in the denominator runs over all pairs of configurations. An empirical correlation length ℓ is obtained from the definition $\rho(\ell) = 1/\mathrm{e}$, which can be used to compare landscapes (see, e.g., Fontana *et al.* 1993a and b).

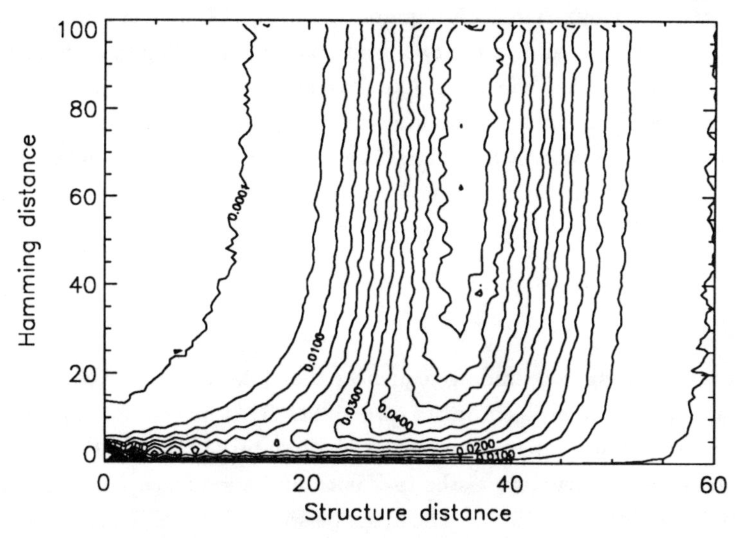

Figure 2. Density surface for chain length 100. The distance between structures is the tree edit distance (Fontana *et al.* 1993b).

The *density surface* $\wp(D|d)$ is the conditional probability that two secondary structures have distance D in shape space provided their underlying sequences have distance d in sequence space (Fontana *et al.* 1993b, Huynen *et al.* 1993). Density surfaces contain more information then the autocorrelation functions. For instance, the probability for finding a neutral neighbor, i.e., a sequence in Hamming distance 1 which folds into the same structure is $\wp(0,1)$. In fact, the autocorrelation function can be obtained from the density surface:

$$\rho(d) = 1 - \frac{\sum_D D^2 \wp(D|d)}{\sum_d \sum_D D^2 p(d) \wp(D|d)} \qquad (5)$$

where $p(d)$ is the probability for two randomly chosen sequences to have distance d.

The particular form of the density surfaces, figure 2, suggest that fairly small balls in sequence space already show the global characteristics of the folding map f — in other words (nearly) all structures can be found in a small neighborhood around *every* sequence. Not all sequences can fold into a particular secondary structure even if we consider all suboptimal structures as well. The set of sequences *compatible* with a secondary structure S can be characterized as follows: On each unpaired position of the structure we can have an arbitrary base, and on the two positions corresponding to a base pair we must have bases that can pair. Our "shape space covering" conjecture states then, that in a small neighborhood of any sequence compatible with a common structure we will find almost certainly a sequence actually folding into this structure. Small means here of the order of once or twice the correlation length ℓ.

4.3. INVERSE FOLDING

Inverse folding, i.e., finding sequences that fold into prescribed structures is an important problem in its own right. This inverse folding problem can be translated into an equivalent combinatorial optimization problem: Given a target structure t^* the task is to find a sequence x minimizing $D(f(x), t^*)$, where $D(\,.\,,\,.\,)$ is a metric in shape space. If there is a sequence x folding into t^*, i.e., $f(x) = t^*$, then the inverse folding problem has a solution for t^*. A heuristic algorithm for this optimization problem is included in the `Vienna RNA Package`; it is described in detail in Hofacker *et al.* (1993a).

Simulations with this algorithm strongly support the shape space covering conjecture (Schuster *et al.* 1993). Starting with a compatible sequence it is not far to the nearest sequence which folds into the desired secondary structure. Furthermore, large samples of sequences folding into the same secondary structure have been generated using the inverse folding algorithm. We find that these sequences cannot be distinguished from a random sample of compatible sequences. The distribution of sequences folding into a common secondary structure is therefore as random as can be, see figure 3.

4.4. NEUTRAL NETWORKS

The definition of a set $C(S)$ of sequences compatible with a secondary structure S suggest to use a modified definition of neighborhood on these sets. We will say that two sequences are neighbors in $C(S)$ if they differ either in a single unpaired base or in the type of a single base pair of S. This definition makes $C(S)$ again to a connected graph. A ***neutral network*** is a connected subgraph $\mathcal{N}$ of $C(S)$ such that all sequences in $\mathcal{N}$ fold into the same structure; if x and y are neighbors and fold into the same structure, $f(x) = f(y)$, then

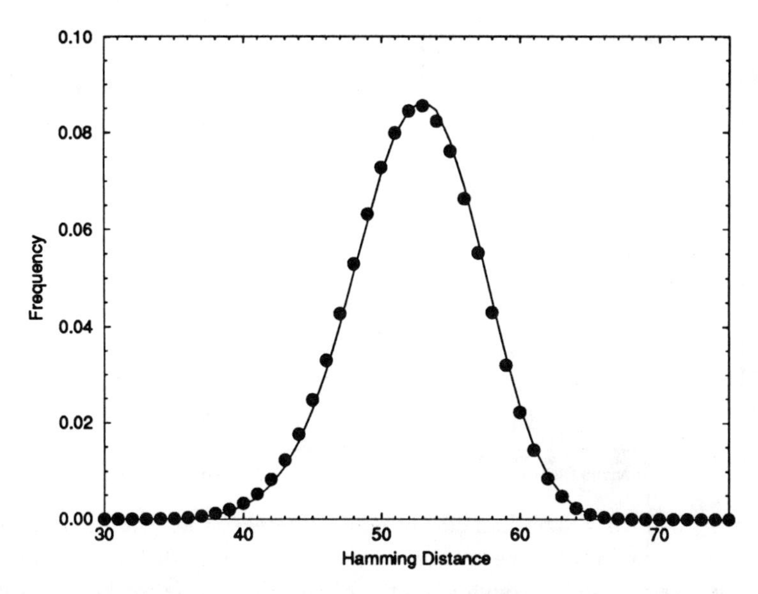

Figure 3. Distribution of pair distances of a more than 2000 sequences folding exactly into the $tRNA^{phe}$ clover-leaf structure, •. The full line is the distance distribution of random compatible sequences without **G-U** pairs which have been used as initial guesses of the inverse folding algorithm. The slight shift between the two curves is due to the incorporation of a few **G-U** pairs into the clover-leafs.

x and y belong to the same neutral network $\mathcal{N}$. In other words: A neutral net is a connected component of the set of all sequences folding into a given secondary structure.

Computer simulations show a surprising result: neutral networks are very large, in fact in general they are much too large to be listed exhaustively. We have performed the following computer experiment in order to determine how far neutral net reach out in sequence space. Starting with an arbitrary initial sequence we search for neutral neighbors in $C(S)$ subject to the constraint that the distance for the starting point increases after each step. Figure 4a shows the probability that such a neutral walk reaches a Hamming distance of at least d. Note that since only a single walk is performed in each network we calculate only a lower bound on the diameter of neutral networks. We find that most neutral networks reach through essentially the entire configuration space.

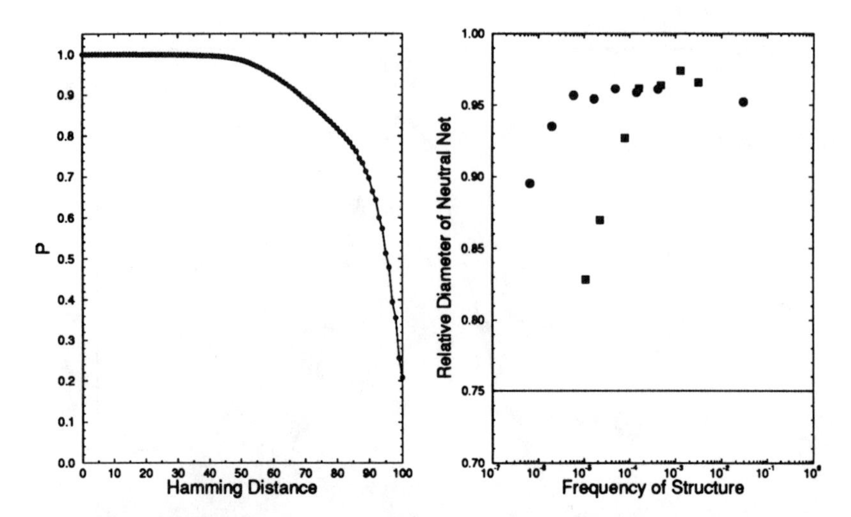

Figure 4. Left: Average diameter of neutral networks at chain length 100. More than 50% of the networks exceed the average distance of random sequences, and a roughly a fifth of the neutral networks percolates through the entire sequence space.
Right: The diameter of neutral networks (here scaled by the chain length) depends on the frequency of the underlying secondary structure. Squares refer to chain length 30, circles refer to chain length 40. The dotted line marks the average distance of random pairs of sequences.

Neutral networks of rare structures are, as expected, smaller than the neutral networks of the most frequent structures, see figure 4b. Even for the rarest structures that we have encountered in our computer experiments (frequencies near 10^{-6}) we observe an average diameter of the neutral networks, $\bar{d}$, well above the the average distance of two random structures, $d = 3n/4$. The average distance between two randomly chosen compatible sequences is even slightly smaller that $3n/4$.

5. DISCUSSION

Our data show that optimization of structures by evolutionary trial and error strategies is much simpler than often assumed. In fact, the combinatory map of RNA secondary structures is *optimally suited* for evolutionary adaptation. Exploration is easy because of huge neutral networks that are nearly neutral, and optimization is not a hard task since a molecule with desired secondary structure is just a few mutations away from almost anywhere in sequence space. Populations replicating with sufficiently high error rates will readily spread along these networks and can reach more distant regions in sequence space. A reduced mutation rate causes the population to concentrate in the most favorable part of the neutral network and to adapt locally in this region (Fontana *et al.* 1987, 1989).

Computer simulations allow us to vary parameters that are hardly accessible to an experimental approach. Systematic studies of the effects of base composition (alphabet) and the influence of variation of the thermodynamic parameters of the folding procedure have been performed (Bonhoeffer *et al.* 1993, Fontana *et al.* 1991, 1993a and b, Schuster 1993, Schuster and Stadler 1993, Tacker *et al.* 1993ab). The results are surprising. The natural **AUGC** alphabet exhibits the largest correlation length, the most extensive neutral networks and the smallest shape space covering radii of all investigated alternatives. Has the repertoire of nucleic acids been evolved for optimal evolvability on RNA landscapes?

The consequences of our results for natural and artificial selection are immediate. We predict that there is no need to systematically search huge portions of the sequence space, nor does one need specially designed initial conditions. These properties provide further support to the idea of widespread applicability of molecular evolution (Eigen 1971, Eigen and Schuster 1979, Eigen *et al.* 1989).

ACKNOWLEDGEMENTS

The results on RNA landscape are the outcome of joint work with Peter Schuster, Walter Fontana, and Ivo Hofacker. Stimulating discussions with M.A. Huynen are gratefully acknowledged. I thank the German *Max Planck Society* for supporting my stay at the Santa Fe Institute.

REFERENCES

Amitrano, C., L. Peliti, and M. Saber, 1989: Population Dynamics in Spin Glass Model of Chemical Evolution. *J. Mol. Evol.*, **29**, 513-525.

Bartel, D.P. and J.W. Szostak, 1993: Isolation of New Ribozymes from a Large Pool of Random Sequences, *Science*, **261**, 1411-1418.

Biebricher, C.K. and M. Eigen, 1988: Kinetics of RNA replication by Qb replicase. In *RNA genetics*, Vol.I, E. Domingo, J.J. Holland, and P. Ahlquist (Eds.), CRC Press, Boca Raton (Fl.), 211-245.

Boerlijst, M. and P. Hogeweg, 1992: Spiral Wave Structure in Pre-Biotic Evolution. Hypercycles Stable Against Parasites. *Physica D*, **48**, 17-28.

Bonhoeffer, S., J.S. McCaskill, P.F. Stadler, and P. Schuster, 1993: RNA Multi-Structure Landscapes. A Study Based on Temperature Dependent Partition Functions, *Eur. Biophys. J.*, **22**, 13-24.

Bonhoeffer, S. and P.F. Stadler, 1993: Errorthreshold on Complex Fitness Landscapes. *J. Theor. Biol.*, **164**, 359-372.

Cech, T.R., 1986: RNA as an enzyme, *Sci. Am.*, **255/5**, 76-84.

Cech, T.R., 1988: Conserved sequences and structures of group I introns: building an active site for RNA catalysis. *Gene*, **73**, 259-271.

Eigen, M., 1971: Selforganization of matter and the evolution of biological macromolecules, *Naturwissenschaften*, **58**, 465-523.

Eigen, M. and P. Schuster, 1979: *The hypercycle — a principle of natural self-organization*, Springer-Verlag, Berlin.
The book is combined reprint of three papers in *Naturwissenschaften* **64**, 541-565 (1977); **65**, 7-41 and 341-369 (1978).

Eigen, M., J.S. McCaskill, and P. Schuster, 1989: The molecular quasispecies. *Adv. Chem. Phys.*, **75**, 149-263.

Fontana, W., and P. Schuster, 1987: A computer model of evolutionary optimization, *Biophys. Chem.*, **26**, 123-147.

Fontana, W., W. Schnabl, and P. Schuster, 1989: Physical aspects of evolutionary optimization and adaptation. *Phys. Rev. A*, **40**, 3301-3321.

Fontana, W., T. Griesmacher, W. Schnabl, P.F. Stadler, and P. Schuster, 1991: Statistics of Landscapes Based on Free Energies, Replication and Degradation Rate Constants of RNA Secondary Structures. *Mh. Chem.*, **122**, 795-819.

Fontana, W., P.F. Stadler, E.G. Bornberg-Bauer, T. Griesmacher, I.L. Hofacker, P. Tarazona, E.D. Weinberger, and P. Schuster, 1993a: RNA Folding Landscapes and Combinatory Landscapes. *Phys. Rev. E*, **47**, 2083-2099.

Fontana, W., D.A.M. Konings, P.F. Stadler, and P. Schuster, 1993b: Statistics of RNA Secondary Structures. *Biopolymers*, **33**, 1389-1404.

Freier, S.M., R. Kierzek, J.A. Jaeger, N. Sugimoto, M.H. Caruthers, T. Neilson and D.H. Turner. (1986). Improved free-energy parameters for predictions of RNA duplex stability. *Biochemistry* **83**, 9373-9377.

Gilbert, W, 1986: The RNA World, *Nature*, **319**, p.618.

Guerrier-Takada, C., K. Gardiner, T. Marsh, N. Pace, and S. Altman, 1983: The RNA moiety of ribonuclease P is the catalytic subunit of the enzyme, *Cell*, **35**, 849-857.

Guerrier-Takada C. and A. Altman, 1984: Catalytic activity of an RNA molecule prepared by transcription *in vitro*, *Science*, **223**, 285-286.

Gutell, R.R., 1992: Evolutionary Characteristics of 16S and 23S rRNA Structures, In *The Origin and Evolution of the Cell*, H. Hartman and K. Matsuno (Eds.), World Scientific, Singapore, 243-309.

Hofacker, I.L., W. Fontana, P.F. Stadler, L.S. Bonhoeffer, M. Tacker, and P. Schuster, 1993a: Fast Folding and Comparison of RNA Secondary Structures (The Vienna RNA Package). *Mh. Chem.* (1993) in press. SFI-preprint 93-07-44, Santa Fe Institute, Santa Fe, New Mexico.

Hofacker, I.L., P. Schuster, and P.F. Stadler, 1993b: Combinatorics of Secondary Structures. Submitted to *SIAM J. Disc. Math.*

Hofbauer, J., and K. Sigmund, 1988: *The theory of evolution and dynamical systems*, Cambridge University Press, Cambridge, U.K.

Hogeweg, P. and B. Hesper, 1984: Energy directed RNA folding rule, *Nucl. Acid. Res.*, **12**, 67-74.

Huynen, M.A., D.A.M. Konings, and P. Hogeweg, 1993: Multiple coding and the evolutionary properties of RNA secondary structure, *J. Theor. Biol.* in press.

Huynen, M.A. and P. Hogeweg, 1994: Pattern Generation in Molecular Evolution; Exploitation of the variation in RNA landscapes. Submitted to *J. Mol. Evol.*

Joyce, G.F., 1989: RNA evolution and the origins of life, *Nature*, **338**, 217-224.

Joyce, G.F., 1991: The rise and fall of the RNA world, *The New Biologist*, **3**, 399-407.

Kauffman, S.A., 1993: *The Origins of Order*. Oxford University Press, Oxford UK, New York.

Kimura, M., 1983: *The neutral theory of molecular evolution*, Cambridge University Press, Cambridge (UK).

Kole, R., M.F. Baer, B.C. Stark, and S. Altman, 1980: *E. coli* RNAase P has A Required RNA Component, *Cell*, **19**, 881-887.

Le, S-Y. and M. Zuker, 1990: Common structures of the 5' non-coding RNA in enteroviruses and rhinoviruses: thermodynamical stability and statistical significance, *J. Mol. Biol.*, **216**, 729-741.

Mandelbrot, B.B., 1982: *The fractal geometry of nature*, Freeman, New York.

Martinez, H.M., 1984: An RNA folding rule. *Nucl. Acids Res.*, **12**, 323-334.

McCaskill, J.S., 1990: The equilibrium partition function and base pair binding probabilities for RNA secondary structure. *Biopolymers*, **29**, 1105-1119.

Noller, H.F, 1991: Ribosomal RNA and translation. *Ann. Rev. Biochem.*, **60**, 191-227.

Noller, H.F., V. Hoffarth, and L. Zimniak, 1992: Unusual resistance of peptidyl transferase to protein extraction procedures, *Science*, **256**, 1416-1419.

Orgel, L.E., 1986: RNA catalysis and the origins of life, *J. Theor. Biol.*, **123**, 127-149.

Orgel, L.E., 1992: Molecular replication. *Nature*, **358**, 203-209.

Pace, N.R. and D. Smith, 1990: Ribonuclease P: Function and Variation, *J. Biol. Chem.*, **265**, 3587-3590.

Piccirilli, J.A., T.S. McConnell, A.J. Zaug, H.F. Noller, and T.R. Cech, 1992: Aminoacyl esterase activity of the *tetrahymena* ribozyme, *Science*, **256**, 1420-1424.

Sankoff, D., A-M. Morin, and R.J. Cedergren, 1978: The evolution of 5S RNA secondary structures, *Can. J. Biochem.*, **56**, 440-443.

Schnabl, W., P.F. Stadler, Ch. Forst, and P. Schuster, 1991: Full Characterization of a Strange Attractor. Chaotic Dynamics in Low Dimensional Replicator Systems. *Physica*, **48D**, 65-90.

Schuster, P., 1993. RNA based evolutionary optimization. *Origins of Life* in press.

Schuster, P., Fontana W., Stadler P.F., and Hofacker I.L., 1993: From Sequences to Shapes and Back: A Case Study in RNA Secondary Structures, *SFI Preprint 93-07-045*. Submitted to *Proc. Roy. Soc. B* (1993).

Schuster, P. and P.F. Stadler, 1993: Landscapes: Complex Optimization Problems and Biopolymer Structures, *SFI Preprint 93-11-069*. Submitted to *Computers and Chemistry*.

Shapiro, B.A., 1988: An algorithm for comparing multiple RNA secondary structures. *CABIOS*, **4**, 381-393.

Shapiro B.A. and K. Zhang, 1990: Computing multiple RNA secondary structures using tree comparisons. *CABIOS*, **6**, 309-318.

Spiegelman, S., 1971: An approach to the experimental analysis of precellular evolution. *Quart. Rev. Biophys.*, **4**, 213-253.

Stadler, P.F. and P. Schuster, 1992: Mutation in Autocatalytic Networks — An Analysis Based on Perturbation Theory, *J. Math. Biol.*, **30**, 597-631.

Stadler, P.F. and R. Happel, 1993: The Probability for Permanence, *Math. Biosc.*, **113**, 25-50.

Stadler, P.F. and J.-C. Nuño, 1993: The Influence of Mutation on Autocatalytic Reaction Networks, *Math. Biosc.*, in press.

Stein, P.R., and M.S. Waterman, 1978: On some sequences generalizing the Catalan and Motzkin numbers, *J. Discr. Math.*, **26**, 261-272.

Swetina, J., and P. Schuster, 1982: Self-replication with errors - A model for polynucleotide replication, Biophys. Chem., **16**, 329-345.

Symons, R.H., 1992: Small catalytic RNAs, *Ann. Rev. Biochem.*, **61**, 641-671.

Tacker, M., W. Fontana, P.F. Stadler, and P. Schuster, 1993a: Statistics of RNA Melting Kinetics. SFI-preprint 93-06-043, Santa Fe Institute, Santa Fe, New Mexico. *Eur. Biophys. J.*, in press.

Tacker, M., P.F. Stadler, E.G. Bornberg-Bauer, I.L. Hofacker, and P. Schuster, 1993: Robust Properties of RNA Secondary Structure Folding Algorithms, SFI-preprint, Santa Fe Institute, Santa Fe, New Mexico.

Waterman, M.S., 1989: Consensus Methods for Folding Single-Stranded Nucleic Acids, In *Mathematical Methods for DNA Sequences*, M.S. Waterman (Ed.), CRC Press, Boca Raton (fl.), 185-224.

Zuker, M., 1989: On Finding All Suboptimal Foldings of an RNA Molecule. *Science*, **244**, 48-52.

Zuker M., and D. Sankoff, 1984: RNA secondary structures and their prediction. *Bull. Math. Biol.*, **46**, 591-621.

GRADUAL RISE OF CELLULAR TRANSLATION

M. Rizzotti
Dipartimento di Biologia, University of Padova, 35121 Padova, Italy

ABSTRACT

A hypothesis is presented for the gradual increase of the role of polynucleotides in the biosphere, from mononucleotide-like molecules driving the condensation of amino acids, to polyribonucleotides acting as gene-messengers and transferers. The molecules involved are supposed to have changed from short to long, but also from variable and relying on chemical affinity, to constant and stereospecific. Accordingly, the genetic code changed from approximate to accurate.

1. INTRODUCTION

"RNA worlds" still come up against unresolved difficulties when attempts are made to devise a realistic pathway for the synthesis of mono- and polyribonucleotides (de Duve, 1988). Simpler polynucleotide precursors (Joyce et al., 1987) meet almost the same difficulties. In this paper, a pathway in five stages is depicted which does not start from RNA (Gesteland and Atkins, 1993) but ends with it, attention being paid not to introduce any phenomenological discontinuity. This pathway relies on three explicit assumptions.

First, it is assumed that primordial macromolecules were largely random. This means that, although some regularities may have occurred (Fox, 1988), it was improbable that two equal individual macromolecules would have been found in the same protocell population. (In the following, the names of four kinds of macromolecules are italicized in order to stress the fact that their constituent micromolecules, and the bonds among them, were variable.)

Second, it is assumed that the reproductive property was present in the protocells before the occurrence of replicating polynucleotides (DNA or RNA) and proteins (Morowitz et al., 1991). In fact, reproduction does not imply the other modern properties of a cell: it essentially implies that the protocells produced by the division of a parent protocell had the capability of taking new material from the outside and organizing it in a way similar to that of the parent protocell. This second assumption appears now the strongest of the three, although it was taken for granted until the widespread conversion to the idea of a necessary link between reproduction and polynucleotide replication.

The third assumption, that RNA preceded DNA in the particulate organic world, is accepted by nearly everybody (Lazcano et al., 1988). No special assumption is necessary about temporarily related phenomena, such as single aspects of metabolism (e.g., autotrophy or heterotrophy) or nature of the defining structures (membranes).

2. EMERGENCE OF OLIGOPEPTIDES AND OLIGONUCLEOTIDES

In order to avoid too many conditional sentences, the five stages of this hypothetical pathway are told like sequential parts of a story, with some comments.

STAGE 1. Primordial protocells were composed of heteropolymers, formed in turn of amino acids and other organic micromolecules (*amino acid-containing heteropolymers*), together with many other macro- and micromolecules without any dimensional discontinuity. They reproduced slowly, because the catalytic properties of their constituents were very low. Their interior was heterogeneous, from the point of view of chemical polarity, because there were hydrophobic as well as hydrophilic microzones.

The *amino acid-containing heteropolymers* were partially recruited from the environment and partially condensed inside the protocells (Rizzotti, 1991). Condensation reactions produced water molecules, so they were favoured in the hydrophobic microzones, because these zones tended to expel the water. In addition, condensation reactions may have been performed with the participation of other molecules called condensing agents. Among the condensing agents were oligophosphates, the simplest of which was pyrophosphate. But the oligophosphates most likely to "patrol" the hydrophobic microzones were those linked to hydrophobic compounds such as adenine. This was because these hydrophobic compounds were anchoring the oligophosphates to the hydrophobic microzones of the protocells. The link between hydrophobic compound and oligophosphate was mediated by suitable molecules such as small carbohydrates. Although the condensing agents were something like ADP or ATP, they were not a uniform chemical species but a collection of similar molecules. Let us call them conventionally *organic oligophosphates* (see figure). Simultaneously, many other molecules were undergoing condensation reactions and many other condensing agents were also acting.

STAGE 2. The *organic oligophosphates* whose hydrophobic portion was aromatic had a better chance to become more complex. This is because aromatic compounds are flat, so they can approach each other for short periods thanks to weak stacking: this contact facilitated condensation reactions, thus leading to variable oligomers, more often with alternating purine and pyrimidine bases (Brack and Orgel, 1975).

Also these *oligonucleotides* were involved from time to time in amino acid condensation, and in this case a clearer specificity could emerge (Navarro Gonzales et al., in press), because the contacts between the two molecules (amino acid and *oligonucleotide)* were much more numerous than those between amino acid and *organic oligophosphate*. The first distinction was basically between the different polarity of hydrophilic and hydrophobic amino acids: some *oligonucleotides* linked more frequently to hydrophilic amino acids, while others linked more frequently to hydrophobic amino acids. The monomer of the *oligonucleotide* which linked directly to the amino acid almost always carryied an adenine (A) or a very similar molecule. It was followed by one or two or more monomers of the *oligonucleotide*, mainly carrying A or U (or similar molecules). The *oligonucleotides* with U as their second base (apart from the initial A) preferred the hydrophilic amino acids while the *oligonucleotides* with A as their second base (apart from the initial A) preferred hydrophobic ones. This corresponds to the present-day relationship between a few amino acids and their anticodons (Woese, 1967). The possible bases of the *oligonucleotides* following the first two or three were not important for the loose specificity, because they had less chance to come into contact with the side chain of the amino acid. However, a longer *oligonucleotide* could wrap all round the amino acid, thus protecting the bond between the acyl group of the amino acid and the acidic or alcoholic hydroxyl group of the initial "adenin-nucleotide" of the *oligonucleotide* from hydrolysis (Hopfield, 1978).

STAGE 3. The *oligonucleotides* did not show a preference only for amino acids; they also showed preferences among themselves. An *oligonucleotide* could embrace an amino acid, but could also twist round a complementary *oligonucleotide* in a short double helix. As the *oligonucleotide* did not embrace the amino acid too strongly, unlike many modern anticodonic oligonucleotides (Balasubramanian, 1982), it could easily dissociate from the amino acid and associate in a double helix with another *oligonucleotide*. In doing so, protection from deacylation disappeared, thus allowing quicker formation of a peptide bond (Hopfield, 1978). The double helix itself was short and hence weak, so it easily freed the *oligonucleotide* for the next cycle of condensation. The result of this process was an alternating oligopeptide: it carried hydrophilic residues alternating with hydrophobic ones (Brack and Orgel, 1975). Variability of the alternating oligopeptide was lower than that of the primitive *amino acid-containing heteropolymers*, but still higher than that of modern polypeptides; for this reason, let us write *oligopeptides*. Summarizing, alternating *oligonucleotides* led to alternating *oligopeptides*.

Alternating *oligonucleotides* have another interesting feature: they can act as templates for their own replication. Because of its efficiency, template-directed replication gradually replaced spontaneous oligomerization. As a consequence, a split began to take place between activation by means of an "adenin-nucleotide" and transfer by means of an *oligonucleotide*. The bonds are different, being anhydridic in the former and esteric in the latter: the constant feature is the link of the amino acid to an adenin-nucleotide.

The growing importance of double helices for the approach of the amino acids towards each other and replication of the *oligonucleotides* led to an enhancement of the properties that gave rise to more effective double helices. The complementary association favoured the most effective pyrimidine for the most abundant purine, thus favouring U (Ishigami and Nagano, 1975), and ribose emerged as the preferred carbohydrate that linked oligophosphates and aromatic bases together.

At the same time one of the two configurations of the carbohydrate prevailed, because of enantiomeric cross-inhibition in double helix formation (Joyce et al., 1984). Moreover, the interaction between amino acid and *oligonucleotide* was stereospecific enough to determine a reciprocal configurational constraint (Lacey et al., 1993). However the actual choice prevailed (Bonner, 1991), there had been no reason for any choice before.

STAGE 4. The alternating oligonucleotides (already more defined than the *oligonucleotides*) also condensed among themselves from time to time, leading to longer polymers: let us call them conventionally polynucleotides. Because of their origin, these polymers were mostly alternating too. They replicated themselves whenever internal conditions weakened the stability of the double helices, thus enabling their strands to undergo another replication cycle (Orgel, 1992). However, replication was carried out with low efficiency, so that many oligonucleotides were still produced in the place of full-length polynucleotides.

Polynucleotides also helped in synthesizing oligopeptides: this is because the oligonucleotides which carried amino acids began to establish short double helices with them, besides among themselves. With regard to the previous stage, this was an alternative path to an identical result. The polynucleotides were the precursors of genes and messengers at the same time, while the oligonucleotides were the precursors of tRNAs.

Of course, the resulting oligopeptides were still alternating, provided the double helix tracts involved an odd number of bases and there were no gaps. Moreover, two bases did not establish a complementary association strong enough to give the amino acid time to become condensed with the growing oligopeptide. At least, the frequency of condensation reactions with a two-base association was much lower. Four bases gave too strong an association, so the turnover was lower. Moreover, the distance between the new amino acid and the growing oligopeptide was excessive (with five bases, the new amino acid was almost carried to the opposite side of the double helix, as 11 bases make a complete turn in standard double-helix RNA). So the preferred length of the double-helix tracts turned out to be three bases. The emergence of the preferential size of the codon constituted the formal premise for the rise of a translation process, but it did not yet accomplish a true translation as long as only monotone alternating polymers were synthesized.

STAGE 5. Deriving from previous stage, the entrance of new bases and distinctions among preexisting ones became important. This fact, together with other aspects of the internal milieu, reduced the tendency for backfolds to form double-helix segments in the longest polynucleotides (Orgel, 1992). In parallel, the mutual specificity between amino acids and anticodons improved in the direction of steric constraints. This was achieved by means of a more stable anticodonic structure, namely by means of a rigid double-helical segment that stabilized the anticodon loop of the oligonucleotides (Orgel, 1968). Later, the single rigid segment split into two rigid segments of roughly the same length, the precursors of the amino acid and anticodon stems. For their stability a GC composition prevailed (Eigen and Winkler-Oswatitsch, 1981). These two segments were connected by a flexible corner, so that the amino acid and the anticodon could come into contact with each other. This was the basis for a complete translation machinery, although still primitive because of the low catalytic performance that implied noise (low specificity, low stability, and hence low accuracy) and slowness (low rate of polymerization).

Polymerization rate and specificity took advantage of simple proteins. When the first true aminoacyl-tRNA synthetases appeared, and the direct interaction between amino acid and anticodon was no longer necessary, the evolution of some anticodonic structures began to disengage from the related amino acid structures. Moreover, tRNAs no longer needed to undergo U bending (Shimizu, 1982; Hou, 1993); they retained their L shape, adding to the first leaf the other two of their modern cloverleaf structure (Kinjo et al., 1986).

The different model of a primitive tRNA proposed by Hopfield (1978) can be derived from an [anticodonic] oligonucleotide by the growing of a stemmed loop (corresponding to the T loop of the present-day tRNAs) between the 3' amino acid end and the anticodon. However, if the anticodon was responsible for a modest selectivity, as also Hopfield assumes, its selectivity remained too loose, as it had no definite tridimensional conformation. Moreover, the specificity of primitive aminoacyl-tRNA synthetases for primitive tRNAs could only be directed to the T loop, because the amino acid stem was precisely the part of the model which was unstable, as it was dissociated and reassociated; when it was reassociated, the amino acid was also associated with the anticodon, so that it was not detectable as such by the synthetase. At the moment, it seems that the T loop is much more neglected by synthetases than the anticodon or the amino acid stem, or even than all the other portions of the molecule (Normanly and Abelson, 1989). This is probably due to the fact that the T loop is relatively conservative as it should associate with a sequence of rRNA. In evolutionary terms, the question could be raised why, in Hopfield's opinion, the anticodon is expected to have remained conformationally indeterminate while a relatively secondary feature of tRNA, such as the T loop, acquired a definite and modern structural arrangement so early.

As for other relevant features of modern tRNAs, the flanking base at the 3' end of the anticodon is always a purine, resembling the A in the former *oligonucleotides*. So, that A has two records: the purine just before the anticodon and the A at the 3' end of tRNA. The invariant flanking U at the 5' end of the anticodon might well be, unlike what is expected in tight models (Balasubramanian, 1982), a record of the pairing with the A of the "aminoacyl adenilate" which helped in bringing the side chain of the amino acid in contact with the anticodon (Ralph, 1968).

3. CONCLUDING REMARKS

No special role is reserved here for rRNAs. As a matter of fact, they are not formally essential to the translation process and the same is true for the many proteins involved in the whole translation machinery. Even in their cooperation with modern mRNAs, the only formally essential role of ribosomes is the correct positioning of the first codon. Ribosomes may be conceived as complex heteromultimeric enzymes: as they deal with polynucleotides, their cofactors are largely polynucleotides as well. The same happens for other enzymes that deal with polynucleotides, such as the pre-tRNA processing enzymes (e.g., ribonuclease P) of prokaryotes or the pre-mRNA processing enzymes (snRNPs) and the telomerases of eukaryotes. All these RNAs may be regarded as remains of an RNA world, but also as efficient coenzymes which originated in particular lineages of the biosphere.

DNA itself is not formally essential. Probably, soon after the first cell, DNA collected all the genes into one (or a few) molecules, so that the distribution of its two single strands in the two daughter cells was easily controlled. Previously, the preservation of the numerous RNA gene-messengers was assured during cell fission only on a probabilistic basis, provided each gene-messenger was represented by many copies.

The author is grateful to P. Omodeo (University of Siena, Italy) for discussions, to J. M. Hennessy and P. Kenway for language revision and to C. Friso for computer drawing.

ESSENTIAL REFERENCES

Balasubramanian, R., and P. Seetharamulu, 1985: *J. theor. Biol.*, **113**, 15-28.

Bonner, W. A., 1991: *Origins Life,* **21**, 59-111.

Brack, A., and L. E. Orgel, 1975: *Nature,* **256**, 383-387.

de Duve, C., 1988: *Nature*, **336**, 209-210.

Eigen, M. and R. Winkler-Oswatitsch, 1981: *Naturwissenschaften,* **68**, 217-228.

Fox, S. W., 1988: *The emergence of life*, Basic Books, New York, 208 pp.

Gesteland, R. F., and J. F. Atkins (Eds.), 1993: *The RNA world*, Cold Spring Harbor Laboratory Press, Plainview, 630 pp.

Hopfield, J. J., 1978: *Proc. Natl Acad. Sci. USA,* **75**, 4334-4338.

Hou, Y.-M., 1993: *TIBS*, **18**, 362-364.

Ishigami, M., and K. Nagano, 1975: *Origins Life,* **6**, 551-560.

Joyce, G. F., Schwartz, A. W., Miller, S. L., and L. E. Orgel, 1987: *Proc. Natl Acad. Sci. USA,* **84**, 4398-4402.

Joyce, G. F., Visscher, G. M., van Boeckel, C. A. A., van Boom, J. H., Orgel, L. E., and J. van Westrenen, 1984: *Nature*, **310**, 602-604.

Kinjo, M., Hasegawa, T., Nagano, K., Ishikura, H., and M. Ishigami, 1986: *J. Mol. Evol.*, **23**, 320-327.

Lacey, J. C. Jr, Wickramasinghe, N. S. M. D., Cook, G. W., and G. Anderson, 1993: *J. Mol. Evol.*, **37**, 233-239.

Lazcano, A., Guerrero, R., Margulis, L. and Oro, J., 1988. J. Mol Evol., 27, 283-290.

Morowitz, H. S., Deamer, D. W., and T. Smith, 1991: *J. Mol. Evol.*, **33**, 207-208.

Navarro Gonzales, R., Khanna, R. K., and C. Ponnamperuma, in press: In *Chemical evolution: origin of life, C.* Ponnamperuma and J. Chela-Flores (Eds.), Deepak, Hampton, 135-155.

Normanly, J., and J. Abelson, 1989: *Ann. Rev. Biochem.*, **58**, 1029-1089.

Orgel, L. E., 1968: *J. Mol. Biol.*, **38**, 381-393.

Orgel, L. E., 1992: *Nature,* **358**, 203-209.

Ralph, R. K., 1968: *Biochim. Biophys. Res. Comm.*, **33**, 213-218.

Rizzotti, M., 1991: *Materia e vita* (Matter and life), UTET, Torino.

Schimizu, M., 1982: *J. Mol. Evol.*, **18**, 297-303.

Woese, C., 1967: *The genetic code*, Harper & Row, New York, 200 pp.

STAGE. 3. Alternating oligonucleotides can embrace (a) amino acids and (b) each other via complementary association, possibly leading to the synthesis of alternating oligopeptides (c).

STAGE 4. Alternating polynucleotides and the rise of translation by means of a three-letter code (left). A two-letter code (right) would lead to a monotone oligopeptide, instead of an alternating one.

STAGE 5. Left: beginning of the definite three-dimensional structure of tRNAs: the third base of the anticodon and the amino acid itself are not yet precisely defined. Right: hypothetical bending of primitive tRNAs that allowed contact (and reciprocal stereochemical recognition) between anticodon and amino acid to be linked to the 3' end.

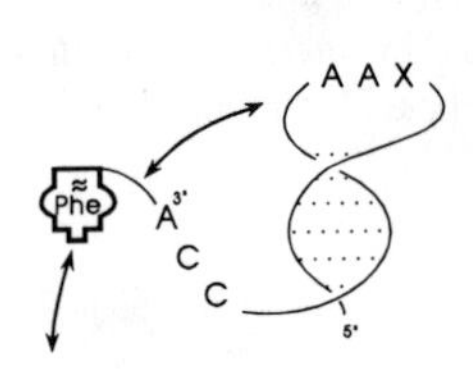

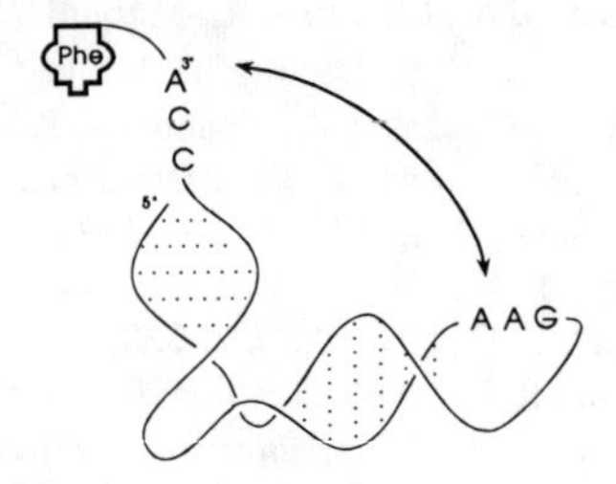

STAGE 1. One of many possible examples of an *organic oligophosphate* which may have participated in primordial condensation reactions. Every hydroxylic group of the carbohydrate could bind to an adenine or an oligophosphate, but there were many more possible variations: they could arise from involvement of other groups of the adenine in the formation of the *organic oligophosphate*, from other bases that assumed the same role, from the sharing of the base among two or more carbohydrate molecules, from length and branching of the oligophosphate, or reactions at the other hydroxylic groups of the carbohydrate, to list only a few.

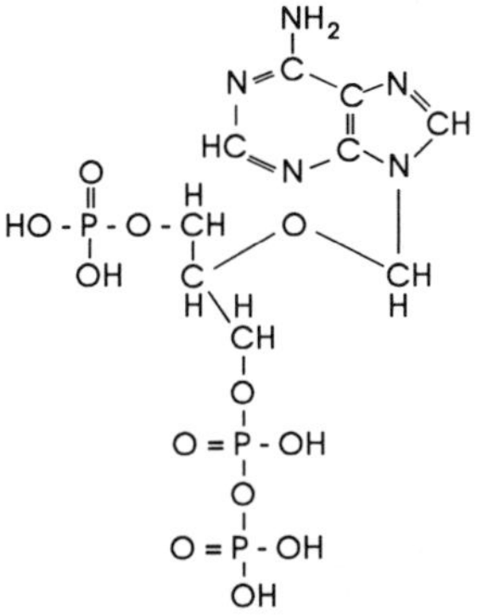

STAGE 2. Diagrams of the embracement of (left) a hydrophobic amino acid by a modern oligonucleotide with sequence AUA... and (right) a hydrophilic amino acid by a modern oligonucleotide with sequence AAU.... The two bases following the initial A establish loose contact with the amino acid residue, forming a sort of micelle and leaving outside, on the surface of the micelle, the more polar carbohydrate and phosphate moieties. A more schematic representation (as used for subsequent stages) is given under both diagrams.

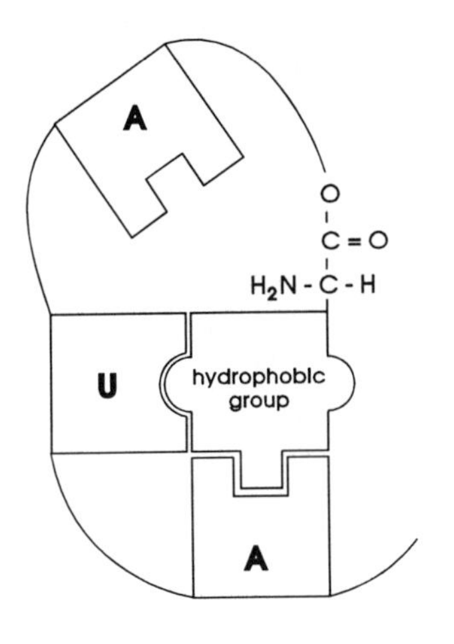

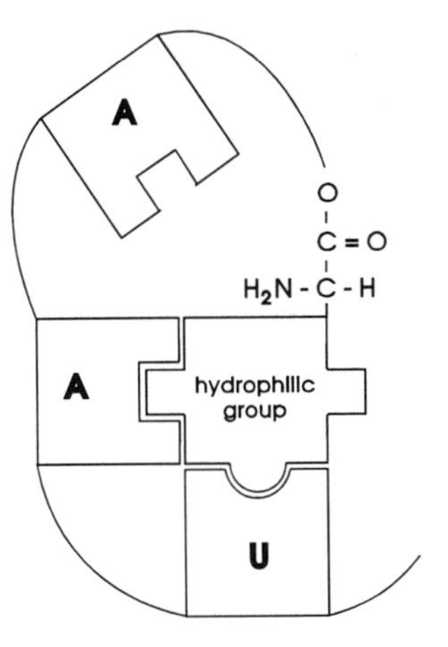

MOLECULAR RELICS FROM CHEMICAL EVOLUTION AND THE ORIGIN OF LIFE

Julian Chela-Flores
International Centre for Theoretical Physics,
Miramare P.O. Box 586; 34100 Trieste, Italy
and
Instituto Internacional de Estudios Avanzados,
Apartado 17606 Parque Central, Caracas 1015A, Venezuela

ABSTRACT

The main hypothesis proposed in this work intends to remove the difficulty that arises from the conjecture that the RNA world may have left molecular relics that may still be extant in the angiosperms.

We discuss whether it is possible to envisage a possible evolutionary pathway of the RNA replicators spanning the vast time span separating the first appearance of the angiosperms, late in the Mesozoic era (the Lower Cretaceous), from the most likely suberas in which the RNA world may have occurred, namely the Hadean/Early Archean.

In order to address this question we suggest that through *horizontal gene transfer,* as well as through a *series of symbioses* of the precursor cells of the land plants, the genes of the replicases (RNA-directed RNA polymerases) associated with putative DNA-independent RNA replicators may have been transferred vertically, eventually becoming specific to the angiosperms.

1. INTRODUCTION

It is widely believed that the chirality of amino acids and nucleic acids are relics of the earliest stages of chemical evolution (Chela-Flores, 1991; Salam, 1991; Mason, 1992). Once the macromolecules of life were formed, the evolution of the earliest life forms ('the early replicators') enhanced the importance of chirality. This led to the highly asymmetric environment of the macromolecules of the living cell - the hallmark of life itself. Chirality is not only a molecular relic of great interest in basic science, but currently it is being used in the design of synthetic drugs (Crossley, 1992).

We discuss, and evaluate, the general question of whether it is possible for further relics of the earliest stages of evolution to have survived till the present This question has been raised in the past by various authors. Some well known examples are the antiquity of introns, based on the finding that some group I introns are ancient (Saldanha *et al,* 1993) and, more recently, the question of molecular relics from the RNA world has been raised in the context of the origin of RNA plant pathogens, such as viroids (Diener, 1993).

Some of the above examples of the DNA-independent RNA replicators are largely specific to the angiosperms, which are recent additions (in geologic time) to the biota. We, therefore, address the paradox that arises from the following conjecture:

The RNA world may have left molecular relics that may still be extant, in spite of the vast time span separating the Mesozoic era from the most likely subera in which the RNA world may have occurred.

We recall that the first appearance of the angiosperms was in the Lower Cretaceous (cf., Sec. 3.1 below).

The main hypothesis proposed in this paper intends to remove the above paradox: we suggest that through horizontal gene transfer, as well as through a series of symbioses in the precursor cells of the land plants, the genes of the replicases associated with RNA plasmids and other putative DNA-independent RNA replicators may have been transferred vertically, eventually becoming specific to the angiosperms.

The series of symbioses postulated in this hypothesis implies that the gene for a putative RNA-directed RNA polymerase (RdRPase), the replicase responsible for the replication of the molecular relics, was transferred to a subsequent host cell during the early Archean, when a prokaryote (possibly a cyanobacterium) first became the host of primordial RNA replicators.

Such replicases have already been suggested to exist in some angiosperms (Fraenkel-Conrat, 1983; 1986) such as cabbage (*Brassica oleracea capitata*), cowpea (*Vigna sinensis*), cucumber (*Cucumis sativus*), and tobacco (*Nicotiana tabacum*).

2. ARE THERE MOLECULAR RELICS FROM THE RNA WORLD?

2.1 ASPECTS OF THE RNA WORLD.

We should look for further insights into the RNA world by considering the possible existence of molecular relics.The discussion of whether such relics are extant is the main topic of this work, which we now proceed to consider in some detail.

The antiquity of introns, or DNA sequences that are not represented in the mRNA, has been discussed in several works (Cavalier-Smith, 1985; Lambowitz, 1989). Their discovery led to the general question of whether introns are ancient, and whether they are related to viroids, putative relics of the RNA world, a topic that we discuss in the next subsection.

Diener has suggested that small circular pathogenic RNAs of plants may be relics of precellular evolution (Diener, 1989). Viroids have been well studied, particularly those with agricultural importance, since they may even affect, amongst others the Solanaceae family, notably *Solanum tuberosum* (the potato). Viroids have been shown to be specific to monocots (cf., Table 1).

TABLE 1: An example of a viroid specific to a monocot

ANGIOSPERM: (MONOCOT)	VIROID SPECIFIC TO THE GENUS	SUBCLASS/ ORDER/ FAMILY.
Coconut	Coconut cadang-cadang, CCCVd (Diener, 1991)	Arecidae/ Arecales/ Palmae

Viroids have also been shown to be specific to a wide variety of dicots (cf., Table 2).

TABLE 2: A list of some viroids specific to dicots

ANGIOSPERM (DICOT)	VIROID SPECIFIC TO THE GENUS (Diener, 1991)	SUBCLASS/ ORDER/FAMILY.
Avocado	Avocado sunblotch,ASVBd	Magnoliidae/ Laurales/ Lauraceae
Cucumber	Cucumber pale fruit, CPSVd	Dilleniidae/ Cucurbitales/ Cucurbitaceae
Apple	Apple scar skin, ASSVd	Rosidae/ Rosales/ Rosaceae
Citrus	*Citrus exocortis*, CEVd	Rosidae/ Rutales/ Rutaceae
Grapevine	Grapevine yellow speckle, GYSVd	Rosidae/ Rosales/ Rosaceae
Potato	Potato spindle tuber, PSTVd	Asteridae/ Scrophulariales/ Solanaceae
Tomato	Tomato apical stunt, TASVd	Asteridae/ Scrophulariales/ Solanaceae
Chrysanthemum	*Chrysanthemum* stunt, CSVd	Asteridae/ Asterales/ Compositae

The credits in favour of interpreting viroids as molecular fossils are abundant (Diener, 1993).

2.2 ARE RNA PLASMIDS MOLECULAR RELICS OF THE RNA WORLD?

In order to complete this line of discussion we should also evaluate whether other RNA molecules, which may replicate independent of its DNA, may be considered as molecular relics. The usual form of DNA used in bacterial transformation is plasmid DNA. Plasmids are small autonomously replicating pieces of DNA found in the cytoplasm; if, besides, the DNA is capable of inserting itself reversibly into the main chromosome, the plasmid is called an episome.

A line of male sterile maize (*Zea mays* L.) is called S type according to the fertility restoration pattern, characterized by the mediation of a single gene locus Rf_3 (Pring *et al.*, 1977), other types of male sterility require more than one gene to recover the normal phenotype.

It is for these reasons that the name "RNA plasmid" has been given to an episomal system consisting of two species of autonomously replicating RNAs that were reported to exist in the cytoplasm of mitochondria of healthy (i.e., virus free) maize cells (Finnegan and Brown, 1986): there are two species in this episomal system, the lenghts of which are 2850 bps and 900 bps, respectively. These RNA plasmids are synthesized in a DNA-independent way.

The origin and gene products of these plasmids remains unknown. However, there is a variety of cellular RNA-dependent RNA polymerases of no known function (Levings III and Brown, 1989). Double-stranded RNAs have been isolated from leaves of the variety of dicots, as shown in Table 3:

TABLE 3: Some examples of dicotyledonean double-stranded RNAs (replicases) of no specific function (Fraenkel-Conrat, 1983; 1986; Ikegami and Fraenkel-Conrat, 1979, 1978) that may have been observed in a broad sample of dicots

PLANT	BINOMIAL SYSTEMATICS	FAMILY/ ORDER	SUBCLASS
Cabagge	*Brassica oleracea*	Cruciferae/ Capparales	Dilleniidae
Cucumber	*Cucumis sativus*	Cucurbitaceae/ Cucurbitales	Dilleniidae
Cowpea	*Vigna sinensis*	Leguminosae/ Fabales	Rosidae
Tobacco	*Nicotiana tabacum*	Solanaceae/ Scrophulariales	Asteridae

2.3. POSSIBLE EVOLUTIONARY PATHWAY FOR RNA REPLICATORS TO BECOME SPECIFIC TO THE MONOCOT MITOCHONDRIAL CYTOPLASM.

During evolution the process of horizontal gene transfer (cf., Sec. 4.3 below for an introduction) may have played a role of continually restricting the habitat of the early RNA replicators:

First the replicators may have been imported into a cyanobacterium. Subsequently, during the Proterozoic, the RNA replicators may have been imported into an early purple bacterium by bacterial-bacterial conjugation. Mitochondria may have arisen by the invasion of aerobic or anaerobic photosynthetic bacteria into ancestral prokaryotic cells (Kuntzel and Kochel, 1981). Later, at an advanced stage of protistan diversification algae emerged close to the metazoa-metaphyte radiation (Perasso *et al.*, 1989).

2.4 ON THE DESCENT OF THE ANGIOSPERMS

In particular we shall discuss below (cf., Sec. 4.3 for references and further details) some of the spore-fossil evidence that land plants may have emerged in the Ordovician from charophycean algae.

It is generally agreed that the angiosperms descended from the progymnosperms, whose first appearance goes back to at least 370 million years before the present (Mybp). Finally, it is plausible that the monocot-dicot divergence may have occurred some 200 Mybp (Wolfe *et al.*, 1989).

In this manner, we introduce the following hypothesis:

By a combination of repeated symbiosis and horizontal gene transfer in the ancestral lineage of the angiosperms, the DNA-independent RNA replicators may have become specific to the dicots and monocots.

However, this point is discussed more fully for all putative RNA replicators in Sec. 4, below.

3. A PARADOX ARISES FROM THE HYPOTHESIS THAT MOLECULAR FOSSILS FROM THE RNA WORLD MAY BE REPRESENTED BY EXTANT RNA REPLICATING INDEPENDENTLY OF THE NUCLEAR DNA

3.1 DNA-INDEPENDENT RNA REPLICATORS ARE SPECIFIC TO ANGIOSPERMS, WHOSE FIRST APPEARANCE WAS LATE IN GEOLOGIC TIME.

A paradox arising from the assumed existence of molecular fossils may be put in a clearer form in the following terms already discussed briefly in preliminary form (Chela-Flores, 1994a, 1994b):

Where were the survivors of the Archean era - the DNA-independent RNA replicators - before they became specific to higher plant taxa in the comparatively recent Mesozoic era?

3.2 THE FIRST EXPERIMENTAL CHALLENGE: A SEARCH FOR RdRPases IN CYANOBACTERIA AND PURPULE BACTERIA.

The means of importing the RNA plasmid into the cytoplasm of a precursor of the mitochondria is not entirely clear, but a possible mechanism is discussed in Sec. 4.4.3 below. However, what has been shown is that viroids and RNA plasmids are specific to plant cells, in which their chloroplasts are organelles arising from cyanobacteria by symbiosis.

We recall that according to the serial endosymbiosis theory (Margulis, 1991, 1993), some chloroplasts originated through polyphyletic associations between various cyanobacteria and the precursors of eukaryotic cells.

3.3 ON A PUTATIVE CYANOBACTERIUM RNA REPLICASE

A property of viroids - passive replication due to a host enzyme - suggests that a cyanobacterium RNA replicase be searched for from this perspective (i.e., an RdRPase), which would be responsible for the replication of the DNA-independent RNA replicators.

Specificity should also be studied in the various classes of cyanobacteria: Coccogoneae and Hormogoneae.

These taxa should also be considered from this point of view, since there are many examples of laminated stromatolites that were constructed by filamentous organisms (cf., Sec. 5 below).

There are also examples of laminated fossil stromatolites where the constructing organisms apparently were coccoid rather than filamentous (Walter, 1983).

Finally, specificity of the RNA replicators should also be tested in members of Chloroxybacteria, since this class includes a putative cyanobacterium genus ancestral to chloroplasts *(Prochloron).*

3.4. SOME LESSONS MAY BE LEARNT FROM *Cyanophora paradoxa.*

As mentioned in Sec. 1.1 the existence of RdRPase has been suggested in cowpea, cucumber, and tobacco, but otherwise, we have little information on the transcription repertoire of a cyanobacterium.

Some insight, nevertheless, was gained some time ago with the discovery of the histone-like protein of size analogous to the *Escherichia coli* HU protein (Kornberg and Baker, 1992).

We may assume that holobionts (symbionts and hosts) are new targets of natural selection due to the interactions of the partners in the symbiosis. In this manner, symbiosis may be an important driving force in evolution.

Such interactions between host and symbiont may be, for instance, horizontal gene transfer (defined below in Sec. 4.3), or the adoption of whole symbionts as organelles. Some intracellular symbiotic bacteria have substantially reduced genomes.

For instance, *Cyanophora paradoxa* is a flagellated protist, a euglenoid harboring cyanelles, which are cyanobacterium-like symbionts lacking cell walls.

Cyanelles are functional chloroplasts (Margulis, 1993) and are known to have only 10% the DNA content of a nonsymbiotic cyanobacterium (Maynard-Smith, 1991). These endosymbiotic prokaryotes may approach the stage reached by chloroplasts, which retain their protein-synthesis apparatus, but many of whose proteins are coded for by nuclear genes.

One possibility that is worthy of attention is that, in our postulated RNA replicator-containing cyanobacterium, the replication of the relic molecule is due to the prokaryotic host's putative RdRPase: After symbiosis the symbiont may have a much smaller genome being constrained to rely on its host's genomic repertoire, as in the above well-studied case of the cyanelles.

4. ORGANELLES MAY LOSE THE ABILITY OF THEIR PROKARYOTIC PRECURSORS TO CODE FOR RdRPases

4.1 ON THE EVOLUTION OF ANGIOSPERMS.

What is already known about the integration of the cyanelle genomes with their hosts may be a model of what to expect in our present case of interest:

Cyanobacteria, being the putative hosts of DNA-independent RNA replicators, such as viroids, became chloroplasts of plant cells, the most likely precursor genus being *Prochloron.* In this case the cyanobacterium becomes a *de facto* organelle. In so doing, the chloroplast may lose the ability of its precursor (i.e., the cyanobacterium) to code for the enzyme RdRPase.

The proposed work remaining in plant biology is not an easy task for a number of reasons:

It is evident that bryophytes and tracheophytes arose from green algae, and that the particular ancestral group was similar to extant Charophyta (Lewis, 1991).

Some support for this view is that flavonoid compounds, such as anthocyanin, are probably present in all angiosperms, but are mostly absent from seedless vascular plants.

Anthocyanin is a particularly significant biochemical indicator of angiosperms. This is underlined by the observation that the expression of a monocot anthocyanin-specific transcriptional activator R in dicots increases their anthocyanin biosynthesis.

The monocot is maize and the dicots are members of the subclass Dilleniidae (*Arabidopsis*) and of the subclass Asteridae (*Nicotiana*) (Lloyd *et al.*, 1992).

In this context it is interesting to remark that some flavonoid compounds are present in charophycean green algae (Swain, 1991). This supports the hypothesis that this group of seaweeds may have been ancestral to land plants.

On the other hand, chloroplasts may have evolved from photosynthetic bacteria more than once. This is a consequence of considering the distribution and comparison of chlorophylls.

Grass green plastids (chlorophyll a and b) indicate that they were acquired from organisms of the *Prochloron* group of bacteria, whereas blue-green and red plastids probably derived from another class of cyanobacterium (Coccogoneae, coccoid cyanobacteria). This, however, does not alter our hypothesis.

4.2 A SECOND EXPERIMENTAL CHALLENGE: TESTING THE VERTICAL TRANSFER OF THE RdRPase GENE IN CHAROPHYCEAN GREEN ALGAE.

A possible rationalization of the pathway of RNA replicators from the RNA world to the angiosperms may be seen in the following terms:

RNA replicators may have been specific to coccoid cyanobacteria, then to *Prochloron*. According to the above argument, some members of this cyanobacterium genus later became organelles of higher eukaryotes, probably of charophycean green algae that led to the origin of land plants.

One consequence of these comments is that specificity of the RNA replicators should be studied in Coleochaete. This is the extant genus of charophycean green algae which, as already emphasized in Sec. 4.1, most closely resembles the now extinct ancestors of the land plants.

4.3. CAN WE RATIONALIZE THE SEQUENCE OF GENE TRANSFERS FROM THE EARLY REPLICATORS TO THE ANGIOSPERMS?

Some additional difficulties underlie the series of events that led from the first prokaryotes in the Archean to single-cell eukaryotes in the Proterozoic, and to multicellular organisms in the Phanerozoic.

In agreement with our hypothesis, the genes codifying for the proposed RdRPase in the infected cell must have been transferred in a series of symbioses during evolution:

4.3.1 While algae may have first appeared during the Riphean period late in the Proterozoic eon, the oldest definitive evidence of tracheophytes (except for Ordovician fossilized spore tetrads) is from the Late Silurian, some 410 Mybp (Gray and Shear, 1992).

4.3.2 Progymnosperms (intermediate plants between nonseed-bearing and seed-bearing tracheophytes) were present during the Upper Devonian, and pteridosperms (seed-bearing plants with fern-like foliage) were present in Permian and Carboniferous flora of Gondwanaland.

4.3.3 The angiosperms themselves may have arisen from Bennettitales, an extinct group related to living cycads.

At each of the underlying successive symbiotic events in the evolution of the flowering plants, from its photosynthetic prokaryotic ancestors, plastids must have shared genes with their hosts (in principle even the gene coding for the putative RdRPase).

4.4 ON HORIZONTAL (OR LATERAL) GENE TRANSFER

In fact, it has been argued that factors other than mutation and natural selection may have a role to play in the evolution of new species.

Direct acquisition of the genetic patrimony of foreign species by an organism has been called: *horizontal (or lateral) gene transfer (HGT).*

This postulated mechanism has been discussed extensively (Smith *et al.,* 1987): HGT is a process by means of which genetic information may be implanted into a host species from a donor species, or intracellularly between organelles, or between organelles and the cell nucleus. HGT has been persuasively documented in the prokaryote-eukaryote case and vice versa (cf., Table 4):

TABLE 4: Some cases of unicellular horizontal gene transfer (HGT)

DIRECTION	ORGANISMS	REFERENCES
Prokaryote-unicellular eukaryote	*Escherichia coli-Entamoeba histolytica*	Smith *et al.*, 1987
Prokaryote-unicellular eukaryote	*E. coli-Saccharomyces cerevisiae*	• Smith *et al.*, 1987 • Marsh & Lebherz, 1992 • Heinemann & Sprague, 1989
Prokaryote-multicellular eukaryote	*Agrobacterium tumefaciens-Nicotiana*	Furner *et al.*, 1986
Multicellular eukaryote-prokaryote	Eukaryote-*Bradyrhizobium japonicum*	Carlson *et al.*, 1986
Multicellular eukaryote-prokaryote	*Clarkia ungulata-E. coli*	Smith *et al.*, 1987
Multicellular eukaryote-prokaryote	Eukaryotic donor-*enteric bacteria (E. coli, Salmonella)*	Nelson *et al.*, 1991

HGT in metazoan has also been suggested (cf., Table 5):

TABLE 5: Some examples of putative horizontal gene transfer (HGT) in two cases: intracellularly, and between different species of metazoans.

PUTATIVE HGT/ KINGDOM	ORGANISMS	REFERENCES
Red alga parasite-red alga host PROTOCTISTA	Rhodophyta: *Plocamiocolax pulvinata/ Plocamium cartilagineum*	Goff, 1991
Mitochondria-nucleus FUNGI	Ascomycota: *(Saccharomyces)*	Farrelly & Butow, 1983
Mitochondria-nucleus FUNGI	Ascomycota: *(Neurospora crassa)*	van der Boogaart *et al.*, 1982
Mitochondria-nucleus FUNGI	Ascomycota *(Podospora anserina)*	Wright & Cummings, 1983
Mitochondria-mitochondria ANIMALIA	Amphibians: Anura (*Rana ridibunda/ Rana lessonae)*	Spolsky & Uzzel, 1984
Mitochondria-mitochondria ANIMALIA	Mammals: Rodentia *(Mus domesticus/ M. musculus)*	Ferris *et al,* 1983

HGT in metaphytes has also been suggested (cf., Table 6):

TABLE 6: Some cases of HGT in the Kingdom Plantae.

PUTATIVE HGT IN THE KINGDOM PLANTAE	ORGANISMS	REFERENCES
Mitochondria-nucleus	Monocot: Commelindae *(Zea mays)*	Kemble *et al.*, 1983
Mitochondria-chloroplast	Monocot: Commelindae *(Zea mays)*	Stern & Lonsdale, 1982
Chloroplast-nucleus	Dicots: Caryophyllidae *(Spinacia)*	Timmis & Steele Scott, 1983
Mitochondria-chloroplast	Dicots: Caryophyllidae *(Spinacia oleracea)*	Stern & Palmer, 1984
Mitochondria-chloroplast	Dicots: Rosidae *(Pisum sativum, Vigna radiata)*	Stern & Palmer, 1984
Mitochondria-nucleus	Dicots: Rosidae *(Oenothera)*	Schuster & Brennicke, 1987

The putative gene transfers in terms of the gene products have been well documented in some special cases as we illustrate in Table 7:

TABLE 7: Some putative gene transfers

RECIPIENT ORGANISM	GENE PRODUCT	HOMOLOGY (H)
Entamoeba histolytica	Fe-superoxide dismutase (SOD)	•65% to *E. coli* • 38% to other eukaryotes
Bradyrhizobium japonicum	•Typical bacterial glutamine synthetase: GS I • GS not found in other prokaryotes: GS II	H between the bacterium GS II and other GSs: •*Pisum sativum* 47%•*Nicotiana plumbaginifolia* 42% •*Anabaena 7120* 24%
Escherichia coli	Glucose phosphate isomerase (Gpi)	88% to *Clarkia ungulata*

4.5 SOME COMMENTS ON ASPECTS OF THE EXPERIMENTAL EVIDENCE

4.5.1 HGT between organelles and the nucleus.

(i) Some genes may have been exchanged between the chloroplast and the nucleus. We may understand this argument as follows:

The complete sequence of the 156 kilobase pair (kbp) genome of the chloroplast of a bryophyte (liverwort, *Marchantia polymorpha*) has been established (Ohyama *et al.*, 1986). Its DNA has some genes that are not detected in the larger (121 kbp) dicotyledonean tobacco chloroplast (Gray, 1986).

It may be argued that since these two genomes are remarkably similar in gene content and organization, such nonvascular and vascular plants contain the same number of proteins. Some genes are present in the chloroplast DNA of one species and in the nuclear DNA of the other (Darnell *et al.*, 1990). An illustration of this remark is provided by the *Euglena* chloroplast which codes for the elongation factor Tu, but in both *M. polymorpha* and tobacco it must be codified by the nuclear genome, since its gene does not appear in the chloroplast DNA (Gray, 1986). Thus, some genes have been exchanged between the chloroplast and the nucleus of plants during the wide time span that separates the first appearance of bryophytes, late in the Paleozoic era (Devonian period), from the first appearance of the dicots in the Lower Cretaceous.

(ii) There is some evidence that the nuclei of cells of spinach (*Spinacia*, Chenopodiaceae, Caryophyllales) contain integrated sequences homologous to chloroplast DNA sequences (Timmis and Scott, 1983).

4.5.2 Some genes may have been exchanged between the mitochondrial genome and the nucleus.

There are several experiments that suggest this possibility (cf., Tables 5 and 6). We consider some of them in turn:

(i) The presence of common DNA sequences in the maize nuclear and mitochondrial genomes have been reported (Kemble *et al.*, 1983)

(ii) There is clear evidence of active mobilization of genetic elements from the mitochondrion to the nucleus of the ascomycete fungus *Podospora anserina* (Wright and Cummings, 1983).

4.6 HYPOTHESIS CONCERNING THE MECHANISMS FOR THE PHYSICAL TRANSPORT OF THE GENES PARTICIPATING IN HGT.

In the mitochondria of the North American herb *Oenothera* (Onagraceae, Myrtales), an open reading frame has been described with high homology to reverse transcriptase (Schuster and Brennicke, 1987). In spite of the fact that there is no direct evidence for the mechanism of transfer of the nucleic acid from the organelles to the host nucleus, it has been remarked that in higher plants only transcribed sequences have been found in more than one subcellular compartment. This suggested the Schuster-Brennicke hypothesis that interorganellar transfer of genetic information may occur via RNA. Hence, by means of reverse transcriptase the exogenous RNA may be integrated into the genome.

4.7 RNA REPLICATOR SPECIFICITY SHOULD BE TESTED IN DIFFERENT TAXA THAT ARE IN THE LINEAGE ANCESTRAL TO THE ANGIOSPERMS

One consequence of (4.4.1 - 4.4.2), and the experiments suggested in Secs. 3.2 and 4.2, is that in order to put on solid basis the conjecture that relics of the RNA world may be specific to angiosperms, the following possibility should be considered: The gene coding for the putative RdRPase may have been horizontally transferred in successive evolutionary stages in plant phylogeny from late in the Proterozoic eon (algae) to the Lower Cretaceous (angiosperms), thereby making DNA-independent RNA replicators specific to each of the various tracheophyte taxa in the lineage of the angiosperms. To sum up, RNA replicator specificity should be tested in different plant, protist, and prokaryotic taxa that are in the lineage ancestral to the angiosperms.

5. CONCLUSION

5.1 ON THE PRESERVATION OF RELICS OF THE PRIMORDIAL RNA WORLD

If successful, the proposed extensive experimental work discussed in this paper would remove the difficulty implied by the rather recent time of the first appearance of angiosperms. The preservation of the relics of the primordial RNA world may have followed a possible pathway from cyanobacteria and purple bacteria to charophycean green algae, and, to the tracheophytes. In other words, we cannot exclude the possibility that DNA-independent RNA replicators could have had a continuous pathway from the RNA world to the present.

5.2 MICROFOSSILS OF THE EARLY ARCHEAN

The earliest date assigned to fossils of the Archean is about 3.5 Gybp (Schopf, 1993; Schopf, and Packer, 1987); the micropaleontological work has been done in the Warrawoona Group (part of the Lower Archean Pilbara Supergroup) in northwestern Western Australia. Eleven taxa of filamentous dark brown carbonaceous microfossils were identified in 1993. These microbes resemble cyanobacteria.

This microfossil-containing environment had produced earlier (reported in 1987) fossils in organosedimentary domes known as stromatolites, which we first mentioned in Sec. 4.2. These rocks consist of regular laminated mats built up by microbial communities, which are mainly cyanobacteria. However, the recent discovery of the 11 taxa at the higher sedimentary unit (Apex Basalt) do not contain stromatolite forming cyanobacteria.We assume that stromatolites are fossilized cyanobacteria, due to the fact that extant stromatolites are constructed mainly by cyanobacteria (Walter, 1983).

5.3 RNA REPLICATORS AND THE AGE OF THE EARTH.

If cyanobacteria - the most ancient prokaryotes - may be shown to be capable of being associated with DNA-independent RNA replicators, and this property may be further shown to be possible in various taxa in the lineage of the angiosperms, then we could interpret the data in terms of RNA relics that may have been present during the major part of the duration of life on Earth. However, we should remark that extant sedimentary rocks have been retrieved from the Isua peninsula in western Greenland with dates prior to the earliest cyanobacteria from the Pilbara Supergroup but, unfortunately, these ancient samples are severely altered by metamorphism, thus barring any certain inferences from paleobiological techniques.

To sum up, in spite of the progress in the studies of the carbon isotope record the geochemical evidence for the pathway of life from Isua time (3.8 Gybp) to 3.5 Gybp requires further insights (Schidlowski and Aharon, 1992).

REFERENCES

Carlson,T.A. and Chelm, B.K., 1986: Apparent eukaryotic origin of glutamine synthetase II from the bacterium *Bradyrhizobium japonicum*. Nature **322,** 568-570

Cavalier-Smith, T., 1985: Selfish DNA and the origin of introns. Nature **315**, 283-284.

Chela-Flores, J., 1991: Comments on a novel approach to the role of chirality in the origin of life. Chirality, 3, 389-392

Chela-Flores, J., 1994a: Life in the universe: Towards an understanding of its origin. Invited lecture delivered at the Venice Conference on Cosmology and Philosophy. Ca' Dolfin, Venice, December, 1992. In: **Universe: Origins, Life, Intelligence.** Eds F. Bertola and M. Calvani. Padova: Il Poligrafo: 1994 (in press).

Chela-Flores, J., 1994b: Are viroids molecular fossils of the RNA world?. J. Theor. Biol. **166,** 163-166.

Crossley, R., 1992: The relevance of chirality to the study of biological activity. Tetrahedron 48, 81-55-8178.

Darnell, J., Lodish, H., and Baltimore, D., 1990: Molecular Cell Biology. 2nd Edition. New York W.H. Freeman, Scientific American Books. p. 700.

Diener, T.O., 1989: Circular RNAs: Relics of precellular evolution? Proc. Natl Acad. Sci. USA **86,** 9370-9374.

Diener, T.O., 1991: Subviral pathogens of plants: viroidds ans viroidlike satellite RNAs. FASEB J. **5,** 2808-2813.

Diener, T.O., 1993: Small pathogenic RNAs of plants: Living fossils of the RNA world? In: Ponnamperuma, C. and Chela-Flores, J. (Eds.), 1993: Chemical Evolution Series: I. Origin of Life. A. Deepak Publishing: Hampton, Virginia, USA. pp. 69-83.

Farrelly, F. and Butow, R.A., 1983: Rearranged mitochondrial genes in the yeast nuclear genome. Nature **301** 296-301.

Ferris, S.D. Sage, R.D., Huang, C.-M., Tonnes, J., Ritte, U., and Wilson, A.C., 1983: Flow of mitochondrial DNA across species boundaries. Proc. Natl. Acad. Sci. USA **80**, 2290-2294.

Finnegan, P.M. and Brown, G.G., 1986: Autonomously replicating RNA in mitochondria of maize plants with S-type cytoplasm. Proc. Natl. Acad. Sci. USA **83,** 5175-5179.

Fraenkel-Conrat, H., 1983: RNA-dependent RNA polymerases of plants. Proc. Natl. Acad. Sci. USA **80**, 422-424.

Fraenkel-Conrat, H., 1986: RNA-directed RNA polymerases of plants. CRC Critical Reviews in Plant Sciences **4,** 213-226.

Furner, I.J., Huffman, G., Amasino, R.N., Garfinkel, D.J., Gordon, M.P., and Nester, E.W., 1986: An *Agrobacterium* transformation in the evolution of the genus *Nicotiana*. Nature **319,** 422-427.

Goff, L.J., 1991: Symbiosis, interspecific gene transfer, and the evolution of new species: A case study in a parasitic red algae. In:Symbiosis as a source of evolutionary innovation, speciation and morphogenesis. Eds. L.Margulis and R. Fester. London: The MIT press. pp 341-363

Gray, J., 1986: Wonders of chloroplast DNA. Nature **322,** 501-502.

Gray, J. and Shear,W., 1992: Early life on land. American Scientist **80,** 444-456.

Heinemann, J.A. and Sprague Jr., G.F., 1989: Bacterial conjugative plasmids mobilize DNA transfer between bacteria and yeast. Nature **340,** 205-209.

Ikegami,M. and Fraenkel-Conrat, H., 1979: Characterization of double-stranded ribonucleic acid in tobacco leaves. Proc. Natl. Acad. Sci. USA **76,** 3637-3640.

Ikegami, M. and Fraenkel-Conrat, H., 1978: RNA-dependent RNA polymerase of tobacco plants. Proc. Natl. Acad. Sci. USA **75,** 2122-2124.

Kemble, R.J., Mans, R.J., Gabay-Laughnan, and Laughnan, J.R., 1983: Sequences homologous to episomal mitochondrial DNAs in maize nuclear genome. Nature **304,** 744-747.

Kornberg, A. and Baker, T.A., 1992: DNA Replication. Second edition. New York: W.H. Freeman and Co. pp. 345-346.

Kuntzel, H, Kochel, H., 1981: Evolution of rRNA and origin of mitochondria. Nature **293,** 751-755.

Lambowitz, A.M., 1989: Infectious introns. Cell **56,** 323-326.

Levings III and Brown, G.G., 1989: Molecular Biology of Plant Mitochondria. Cell **56**, 171-179.

Lewis, D., 1991: Mutualistic Symbioses in the Origin and Evolution of Land Plants.In: Symbiosis as a source of evolutionary innovation, speciation and morphogenesis. Eds. L.Margulis and R.Fester. London: The MIT press. pp.288-300.

Lloyd, A.M., Walbot, V., and Davis, R.W., 1992: *Arabidopsis* and *Nicotiana* anthocyanin production activated by maize regulators R and C1. Science **258,** 1773-1775.

Margulis, L., 1991: Symbiogenesis and symbioticism. In: Symbiosis as a source of evolutionary innovation, speciation and morphogenesis. Eds. L. Margulis and R. Fester. London: The MIT press. pp. 1-14.

Margulis, L., 1993: Symbiosis in Cell Evolution. Second Edition. New York: W.H. Freeman and Co. p. 190.

Marsh, J.J. and Lebherz., 1992: Fructose-biphosphate aldolases: an evolutionary history. Trends Biochem. Sci. **17,** 110-113.

Mason, S. F., 1992: Chemical Evolution. Origin of the elements, molecules, and living systems. Oxford: Clarendon Press.

Maynard Smith, J., 1991: A darwinian view of symbiosis. In: Symbiosis as a source of evolutionary innovation, speciation and morphogenesis. Eds. L.Margulis and R. Fester. London: The MIT press. pp. 26-39.

Nelson, K., Whittam, T.S., and Selander, R.K., 1991: Nucleotide polymorphism and evolution in the glyceraldehyde-3-phosphate dehydrogenase gene (gapA) in natural populations of *Salmonella* and *Escherichia coli*. Proc. Natl Acad. Sci USA **88,** 6667-6671.

Ohyama, K., Fukuzawa, H., Kohchi, T., Shirai, H., Sano, T., Sano, S., Umesono, K., Shiki, Y., Takehuchi, M., Chang, Z., Aota, S., Inokuchi, H., and Ozeki, H., 1986: Chloroplast gene organization deduced from complete sequence of liverwort *Marchantia polymorpha* chloroplast DNA. Nature **322,** 572-574.

Perasso, R., Baroin, A., Qu, L.H., Bachellerie, J.P., and Adoutte., 1989: Origin of the algae. Nature **339,** 142-144.

Pring, D.R., Levings,III, C.S., Hu, W.W.L., and Timothy, D.H., 1977: Unique DNA associated with mitochondria in the "S"-type cytoplasm of male sterile maize. Proc. Natl Acad. Sci USA **74,** 2904-2908.

Salam, A., 1991: The role of chirality in the origin of life, J. Mol. Evol., **33,** 105-113.

Saldanha, R., Mohr, G., Belfort, M., and Lambowitz, A.M., 1993: Group I and group II introns. FASEB J. **7,** 15-24.

Schidlowki, M. and Aharon, P., 1992: Carbon cycle and carbon isotope record: Geochemical impact of life over 3.8 Ga of Earth history. M. Schidlowki *et al* (Eds.) Early organic evolution: Implications for mineral and energy resources. Springer-Verlag: Berlin.

Schopf, J.W.(1993). Microfossils of the Early Archean Apex Chert: New Evidence of the Antiquity of Life. Science **260,** 640-646.

Schopf, J.W. and Packer, B.M., 1987: Early Archean (3.3-Billion to 3.5-Billion-Year-Old) Microfossils from Warrawoona Group, Australia. Science **237,** 70-73.

Schuster, W.and Brennicke,A., 1987: Plastid, nuclear and reverse transcriptase sequences in the mitochondrial genome of *Oenothera*:: Is genetic information transferred between organelles via RNA?. The EMBO Journal **6,** 2857-2863.

Smith, M.W., Feng, D-F., and Doolittle, R.F., 1987: Evolution by acquisition: the case for horizontal gene transfers. Trends Biochem. Sci. **17,** 489-493.

Spolsky, C. and Uzzel, T., 1984: Natural interspecies transfer of mitochondrial DNA in amphibians. Proc. Natl. Acad. Sci. USA **84,** 5802-5805.

Stern, D.B. and Lonsdale, D.M., 1982: Mitochondrial and chloroplast genomes of maize have a 12-kilobase DNA sequence in common. Nature **299**, 698-702.

Stern, D.B and Palmer, J.D. (1984). Extensive and widespread homologies between mitochondrial DNA and chloroplast DNA in plants. Proc. Natl. Acad. Sci. USA **81,** 1946-1950.

Swain, T., 1991: Chemical Signals from Plants and Phanerozoic Evolution. In: Environmental Evolution. Eds. L.Margulis and L. Olendzenski. London: The MIT press. pp. 245-263.

Timmis, J.N. and Steele Scott, N., 1983: Sequence homology between spinach nuclear and chloroplast genomes. Nature **305,** 65-67.

van den Boogaart, P., Samallo, J., and Agsteribbe, E., 1982: Similar genes for a mitochondrial ATPase subunit in the nuclear and mitochondrial genomes of *Neurospora crassa*. Nature **298,** 187-189.

Walter, M.R., 1983: Archean stromatolites. Evidence of the earth's earliest benthos. In: Earth's earliest biosphere its origin and evolution. Ed. J.W. Schopf. Princeton: Princeton University Press. pp. 187-213.

Wolfe, K.H., Gouy, M., Yang, Y.-W., Sharp, P.M., and Li, W.-H., 1989: Date of the monocot-dicot divergence estimated from chloroplast DNA sequence data. Proc. Natl. Acad. Sci. USA **86,** 6201-6205.

Wright, R.M. and Cummings, D.J., 1983: Integration of mitochondrial gene sequences within the nuclear genome during senescence in a fungus. Nature **302,** 86-88.

DESIGNING A BIOSPHERE FOR MARS

Robert H. Haynes
York University
Toronto, Ontario M3J 1P3, Canada

Christopher P. McKay
NASA Ames Research Center
Moffett Field, California, U.S.A.

ABSTRACT

Present environmental conditions on Mars are extremely hostile and would be destructive to any organisms that might arrive there unprotected today. However the idea of implanting life on Mars is no longer confined to the realm of science fiction, even though our knowledge of that planet is insufficient to come to any definite conclusion regarding its feasibility. Mars appears to be the only object in the solar system on which such a feat of planetary engineering might be carried out with foreseeable technologies. Its unalterable physical parameters, i.e., its orbit, gravity and possible volatile inventories, are compatible with the existence of a thick, warm, carbon dioxide atmosphere and substantial quantities of liquid water. On the other hand, its surface is cold (-60° C on average in temperate latitudes), desiccated and highly oxidized. Thus a general scenario for implanting life on Mars would include three main phases: (1) Robotic and human exploration to determine whether sufficiently large and accessible volatile and mineral resources do in fact exist; (2) A program of planetary engineering designed to warm the planet, release liquid water and produce a thick carbon dioxide atmosphere; and (3) If no indigenous organisms emerge as liquid water becomes available, a program of biological engineering designed to implant pioneering microbial communities capable of proliferation in the newly clement, though still anaerobic, Martian environment. The word 'ecopoiesis' is used to describe the formation of a primordial ecosystem in a lifeless environment. Ecopoiesis occurred spontaneously on early Earth, and perhaps on early Mars. Even though Mars appears to be a lifeless planet today, it may be possible to bring it to life by planetary engineering and 'directed ecopoiesis'. We have proposed previously, and we reiterate here, that a feasibility study of Martian ecopoiesis should be included in the space exploration programs of the next few decades.

1. INTRODUCTION

Mars holds a special place in the human imagination. It is a popular home for extraterrestrials in science fiction and it is the most probable site for the first permanent human outposts on another planet. Interest in the question of life on Mars and other celestial bodies dates back at least to the seventeenth century (Dick, 1982). To search for signs of life was the major purpose of the Viking missions in 1976. The results of these missions indicate that Mars is presently devoid of life. However, other Viking results also suggest that during the first two billion years of its history, conditions on Mars may have been very different from those of today. There is good geomorphological evidence that large amounts of liquid water flowed freely on the Martian surface during this early period, and perhaps episodically since then (Baker *et al.*, 1991). However, water does not exist in the liquid state on Mars today. For its average surface temperature to rise above the freezing point of water the planet would have to possess a thick carbon dioxide atmosphere (McKay *et al.*, 1991). Thus, early in its history Mars might have enjoyed habitable, though still anaerobic, conditions somewhat similar to those on early Earth.

The early warm, wet period of Martian history did not last long in terms of geological time - perhaps only a few hundred million years (Pollack *et al.*, 1987; McKay and Davis, 1991). It is thought that Mars lost much of its initial atmosphere through reactions of carbon dioxide in the presence of liquid water to form carbonate sediments. Similar reactions occurred on the early Earth. However, tectonic activity on Earth recycles this carbonate back to carbon dioxide, thereby maintaining conditions suitable for life (Brener, 1990). Mars, being a one-plate planet, apparently lacks any long-term recycling mechanism of this kind. Thus, the gradual conversion of atmospheric carbon dioxide to carbonates resulted ultimately in the cold, thin atmosphere that characterizes the planet today (McKay and Stoker, 1989). If this view is correct, most of the carbon dioxide inventory of Mars may be locked up in carbonate rocks, although additional amounts are likely to be present as adsorbed gas in the regolith (about 300 mb, according to Fanale and Cannon, 1979); still smaller amounts exist in the atmosphere-polar cap system (10-100 mb). The initial water budget on Mars - estimated to be as much as a 500-1000 m thick layer - is though to be tied up as ground ice in the permafrost regions poleward of about 40° latitude (Squyres and Carr, 1986).

Liquid water is an essential environmental requirement for life (see McKay, 1991 for a discussion in the context of planetary biology). If Mars once had abundant liquid water it is possible that life may have arisen there during the early period of planetary development. Furthermore, we know from studies of Earth's paleontological record that life came into being and reached a remarkable degree of biochemical development as early as 3.5 billion years ago (Schopf, 1983). Thus, investigation of the Martian geological record for traces of life, or of prebiotic chemical synthesis, has become a major scientific priority for future missions to Mars (McKay, 1986; McKay and Stoker, 1989; Stoker *et al.*, 1990).

While the water, carbon dioxide and nitrogen in the current Martian atmosphere may be only the remnants of a past habitable state they are, nonetheless, potentially valuable resources for supporting human exploration (Meyer and McKay, 1989). The utilization of local materials would make it possible for early explorers to be at least partially self-sufficient in that not all supplies for the missions would have to be brought from Earth.

The previous existence of a climate suitable for life, and the need to construct habitats for humans, perhaps with some food production capacity, suggest that it may be possible, and desirable in the long run, to alter the Martian climate so as to return the planet to a state less hostile to life than it is at present. Even though only microorganisms that are obligate or facultative anaerobes could be introduced initially, the resulting supply of locally generated biomass would be a valuable source of energy, food and perhaps other materials useful for future visitors or colonists.

In discussions of the establishment and development of a biosphere on other planets we distinguish between *terraformation* and *ecopoiesis* (Haynes and McKay, 1992). Terraformation refers to the generation of an Earth-like environment, one suitable for humans by virtue of the presence and maintenance of an oxygen containing atmosphere. Ecopoiesis (from the Greek 'oikos', meaning a 'home' or 'abode', and 'poiesis', meaning 'a making', coined by Haynes, 1990) refers to the establishment of an energetically open ecological system or biosphere (not necessarily one containing *Homo sapiens*) on an initially lifeless planet.

Before the implementation of any program for altering the Martian environment is initiated, it is imperative that the question of the possible existence of indigenous

Martian life be settled beyond reasonable doubt. No signs of life were detected during the Viking missions (Horowitz, 1986; for a contrary view see Levin and Straat, 1981; and Levin, 1988; for recent reviews see McKay and Stoker, 1989; and Thomas, 1991). However, any future discovery of a native Martian biota would force the revision of current proposals for planetary engineering. McKay (1990) has argued that if potentially viable organisms are found on Mars we should consider only those alterations of the environment that would enhance the proliferative capacity and diversity of the indigenous species. However, given the extremely inhospitable conditions on Mars, and the apparent absence of life, we will discuss ecopoiesis on the assumption that there is in fact no life there, nor even any potentially viable microorganisms or spores derived from past life; this, and other major assumptions that underlie current speculations regarding ecopoiesis are discussed in more detail elsewhere (Haynes, 1990).

2. IS IT POSSIBLE?

In recent years several general approaches for altering the climate of Mars have been examined in a preliminary way. Depending on the assumptions made about the nature of the local volatile inventories, these proposals have entailed a range of techniques, ranging from the relatively benign to the highly intrusive; some of the latter have been based on futuristic technologies (Sagan, 1973; Burns and Harwit, 1973; Averner and MacElroy, 1976; Oberg, 1981; McKay, 1982; Lovelock and Allaby, 1984; Fogg, 1989, 1992, 1993a, 1993b; McKay *et al.*, 1991; Pollack and Sagan, 1993). Even these rather crude and highly speculative studies make it evident that we do not have enough information about the planet itself, nor even about the origin and evolution of Earth's biosphere, to assess today the feasibility of ecopoiesis on Mars. However, some of the main problems and uncertainties at least have been identified.

The most serious difficulty with all planetary engineering scenarios for Mars concerns the availability of suitable amounts of the necessary volatiles, in particular, water, carbon dioxide and nitrogen. The quantities present in the atmosphere are woefully inadequate for planetary-scale ecopoiesis and there is as yet no direct proof that subsurface reservoirs of these vital compounds are present in sufficient size to generate a reasonably robust biosphere using foreseeable and acceptable technologies. Even if sufficient volatile inventories do exist they may be chemically combined in mineral forms from which they cannot readily be released.

For example, while it might become feasible to liberate carbon dioxide from the polar caps, and even from the regolith (McKay *et al.*, 1991), it would be much more difficult to secure its release from carbonate rocks except through the use of possibly unacceptable technologies such as the detonation of thermonuclear devices widely distributed over the Martian surface (Fogg, 1989). There also are other potential roadblocks that could forestall or undermine current scenarios for ecopoiesis on Mars (e.g., the amount of nitrogenous compounds in the Martian soil may not be sufficient to support a robust global ecosystem). However, the range of uncertainty concerning the resources of Mars is sufficiently large that the feasibility of ecopoiesis cannot be ruled out.

If ecopoiesis is ultimately judged to be possible, the construction of a biosphere suitable for animals naturally subdivides into two overlapping phases (McKay, 1982; McKay *et al.*, 1991). The main objective of the first phase would be to warm the planet from its currently frigid conditions (averaging about -60° C) to a mean temperature at or above zero. To have average surface temperatures above the freezing point of water on Mars would require a carbon dioxide atmosphere of about 1-3 bars (McKay *et al.*, 1991). Such an atmosphere would be toxic to most organisms but it would be suitable for anaerobic bacteria and perhaps for some forms of plant life (Seckbach *et al.*, 1970). The thermal energy required to produce this amount of carbon dioxide (starting from solid carbon dioxide ice) would require the equivalent of 10 years of the incident sunlight (McKay *et al.*, 1991).

If we assume that the surface can be warmed sufficiently by carbon dioxide release, water from the permafrost would melt and Mars might once again become suitable for life. In fact, its climate might then be similar to what is believed to have existed there some 4 billion years ago. If anaerobic life forms were then introduced, and if they flourished, a minimal form of ecopoiesis would have been achieved. However, to extend the process to ensure the production of an oxygen atmosphere suitable for humans or other animals, then a further phase would become necessary. This would entail the photosynthetic conversion of carbon dioxide to organic compounds and the subsequent sequestering of much of this organic material in order to yield a net accumulation of oxygen in the atmosphere. This process, if effected by plants as it was on Earth, would take a long time, perhaps as long as 100,000 years (McKay *et al.*, 1991). Unfortunately, climate models suggest that a nitrogen/oxygen atmosphere suitable for humans would be much too cold on Mars. A possible mechanism for warming such an atmosphere up to the freezing point of water would involve the addition of broad-band, highly efficient greenhouse gases: however, these gases would have to be even more effective and photochemically stable than currently available chlorofluorocarbons (McKay *et al.*, 1991).

3. IS IT DESIRABLE?

Let us assume that future research reveals that it is feasible to make Mars habitable for at least some forms of anaerobic life, either selected from existing Earth organisms or genetically engineered specifically for survival on a warm, wet Mars. Is such a planetary transformation desirable? This question clearly transcends science. Thus, the societal, ethical and philosophical aspects of the proposal must also be considered. Indeed, one of the more interesting aspects of thinking seriously about ecopoiesis is that several new questions are then placed on the agenda of moral philosophy (Haynes, 1990; McKay, 1990; McKay and Haynes, 1990).

The possible benefits that would accrue from such a program of planetary engineering are difficult to predict. In a recent paper we summarized most of the arguments, both *pro* and *con*, that we and others have brought forward in numerous debates on the merits of ecopoiesis and/or terraformation as a long-term objective for the major space agencies of the world (Haynes and McKay, 1991). It is generally conceded that a greatly increased understanding of planetary science would result from a concerted attempt to design a second biosphere in the solar system, even if the project never came to fruition (McKay and Stoker, 1991). In particular, we would undoubtedly achieve a much deeper understanding of biogeochemical cycles, of the stability and fragility of ecosystems, and of the cogency of certain broad principles that have been put forward regarding the regulation of the planetary scale ecosystems [e.g., Lovelock's (1979, 1987, 1988) *Gaia* hypothesis]. In view of the adverse environmental impact of current human activities here on Earth, such gains in our understanding of 'biospheric physiology' are perhaps the most compelling motivation for undertaking a detailed feasibility study of ecopoiesis and terraformation on Mars. Longer term, more intangible, benefits would arise if it can be shown beyond reasonable doubt that Mars is a potentially habitable planet, a viable and acceptable second home for life in the solar system. The possibility of propagating life on another planet seems certain to have exhilarating implications for humanity's view of its place in the universe (Turner, 1988, 1989).

Perhaps the most interesting philosophical question associated with introducing life to Mars is to ask what is of more value, Mars as it currently exists or a Mars endowed with life? Even to ask this question gives a new perspective on basic issues of environmental ethics (McKay, 1990). Earth's biosphere is inextricably linked with the state of the atmosphere, hydrosphere and lithosphere. The evolution and diversification of life is intrinsic to the geological history of our planet. Thus, a choice between a 'living planet' and 'Nature untouched by human hands' does not exist on Earth. It is inconsequential to ask what has more intrinsic value, Earth with its biosphere as we know it, or Earth as it was before the onset of biological evolution and the appearance of humans (McKay, 1990). Thus, as pointed out by the philosopher Thomas Hurka (1990), the possibility of implanting life on Mars disentangles two moral imperatives that are often conflated in contemporary thinking about the environment. One of these is that humans should respect nature (physical as well as biological) as it is and not interfere with natural processes; the other is that we should value nature for its vitality, diversity and the richness of its living forms as found on Earth. On the basis of the first idea one would be forced to conclude that it would be wrong to implant life on Mars because this would violate the 'natural' integrity of that planet as a dead world. On the other hand, the second claim would seem to imply that if a richly diverse biosphere is good on Earth such a living, ever evolving global community is good wherever it can be established: to bring life to Mars would endow that planet with greater intrinsic, or cosmic, value than it presently possesses. At some future date our descendants may be presented with a choice between one or the other. In thinking about these alternatives, it is a mistake to assume that what *is* also is what *ought* to be (McKay and Haynes, 1990). Clearly, the question whether Mars should be implanted with life has no easy, objective answer. The significance of the question, and the answer itself, depend on where one stands in the solar system: if, in future, human outposts on Mars are established, ecopoiesis may become quite a different priority than it is for Earth-bound thinkers today.

4. HOW CAN IT BE DONE?

NASA's projected program of robotic and human exploration of Mars constitutes the first steps toward an assessment of the feasibility of ecopoiesis. In fact, the identification of Mars as a possible second home for life can be considered to have begun, unwittingly, with the Mariner and Viking missions. Even without an explicit goal to study the feasibility of ecopoiesis, future studies of Mars will yield some of the information necessary for answering the questions that have already arisen in discussions on the habitability of that planet. The only alterations in the near-term program of Mars exploration that would result from a decision to conduct a feasibility study of ecopoiesis might be shifts in the relative priority of the scientific questions asked, and in our responses to the long-term implications of the data returned. For example, the question of the availability of nitrates and phosphates in the Martian soil is scientifically interesting but it becomes crucially important in discussions of ecopoiesis and terraformation.

It is difficult to foresee how ecopoiesis might be implemented if its feasibility is established on the basis of future exploration. However, we would argue that if it is decided to implement such a program of planetary engineering, a slow and conservative approach is essential. Sufficient time must be allowed for a wide range of studies of Mars as it exists at present, and for careful planning, modeling and 'pilot-plant' trials (where possible) of all successive steps in the enterprise.

5. CONCLUSION

It is by no means clear whether ecopoiesis on Mars is scientifically possible or technologically achievable. However, we do believe that a 'green planet' has more intrinsic value than a 'red planet'. Thus we urge that one of the objectives of space science during the 21st century be to assess the feasibility of implanting life on Mars. Such a goal would provide an attractive and coherent focus for robotic and human exploration of that planet. The knowledge gained from studies of the comparative planetology of Earth and Mars would undoubtedly be valuable, and could prove critical, for understanding the origin and development of Earth's biosphere and how best to manage it in future. The construction of a self-sustainable biosphere elsewhere in the solar system, especially if conducted as a cooperative international effort, could provide an extraordinarily exciting challenge, a new vision, and ultimately a new frontier, for humankind.

REFERENCES

Averner, M. M., and R. D. MacElroy, (Eds.), 1976: *On the Habitability of Mars*, NASA Report SP-414, U. S. Government Printing Office, Washington, District of Columbia.

Baker, V. R., R. G. Strom, V. C. Gulick, J. S. Kargel, G. Komatsu and V. S. Kale, 1991: Ancient oceans, ice sheets and the hydrological cycle on Mars, *Nature*, **352**, 589-594.

Brener, R. A., 1990: Atmospheric carbon dioxide levels over phanerozoic time, *Science*, **249**, 382-1386.

Burns, J. A. and M. Harwit, 1973: Towards a more habitable Mars - or - the coming Martian spring, *Icarus*, **19**, 126-130.

Dick, S. J., 1982: *Plurality of Worlds*, Cambridge University Press, Cambridge.

Fanale, F. P. and W. A. Cannon, 1979: Mars: carbon dioxide adsorption and capillary condensation on clays - significance for volatile storage and atmospheric history, *J. Geophys. Res.*, **84**, 8404- 8414.

Fogg, M. J., 1989: The creation of an artificial dense martian atmosphere: A major obstacle to the terraforming of Mars, *J. Brit. Interplanet. Soc.*, **42**, 577-582.

Fogg, M. J., 1992: A synergic approach to terraforming Mars, *J. Brit. Interplanet. Soc.*, **45**, 315-329.

Fogg, M. J., 1993a: Dynamics of a terraformed Martian biosphere, *J. Brit. Interplanet. Soc.*, **46**, 293-304.

Fogg, M. J., 1993b: Terraforming: a review for environmentalists, *The Environmentalist*, **13**, 7-17.

Haynes, R. H., 1990: Ecce Ecopoiesis: Playing God on Mars. In *Moral Expertise*, D. MacNiven (Ed.), Routledge, London and New York, 161-183.

Haynes, R. H. and C. P. McKay, 1992: The implantation of life on Mars: feasibility and motivation, *Adv. Space Res.*, **12(4)**, (4)133-(4)140..

Horowitz, N. H., 1986: *To Utopia and Back: The Search for Life in the Solar System*, W. H. Freeman and Co., New York.

Hurka, T., 1990: Is there life on Mars? Should there be? *The Globe and Mail*, Toronto, December 4, A18.

Levin, G. V. and P. A. Straat, 1981: A search for a nonbiological explanation of the Viking labeled release life detection experiment, *Icarus*, **45**, 494-516.

Levin, G. V., 1988: A reappraisal of life on Mars, Proceedings of the NASA Mars conference. *Amer. Astron. Soc.*, **71**, 187-207.

Lovelock, J. E., 1979: *Gaia*, Oxford University Press, Oxford.

Lovelock, J. E., 1987: The ecopoiesis of Daisyworld. In *Origin and Evolution of the Universe: Evidence for Design?*, J. H. Robson (Ed.), McGill-Queen's University Press, Kingston and Montreal, 153-166.

Lovelock, J. E., 1988: *The Ages of Gaia*, Oxford University Press, Oxford, 252 pp.

Lovelock, J. E. and M. Allaby, 1984: *The Greening of Mars*, Warner Books, Inc., New York.

McKay, C. P., 1982: Terraforming Mars, *J. Brit. Interplanet. Soc.*, **35**, 427-433.

McKay, C. P., 1986: Exobiology and future Mars missions: The search for Mars' earliest biosphere, *Adv. Space Res.*, **6 (12)**, 269-285.

McKay, C. P., 1990: Does Mars have rights? An approach to the environmental ethics of planetary engineering. In *Moral Expertise*, D. MacNiven (Ed.), Routledge, London and New York, 184-197.

McKay, C. P., 1991: Urey prize lecture: Planetary evolution and the origin of life, *Icarus*, **91**, 93-100.

McKay, C. P. and R. H. Haynes, 1990: Should we implant life on Mars, *Scientific American*, **263 (6)**, 144.

McKay, C. P. and C. R. Stoker, 1989: The early environment and its evolution on Mars: Implications for life, *Rev. Geophys.*, **27**, 189-214.

McKay, C. P. and C. R. Stoker, 1991: Gaia and life on Mars. In *Scientists On Gaia,* S. Schneider and P. J. Boston (Eds.), M.I.T. Press, Cambridge, MA, 375-381.

McKay, C. P. and W. L. Davis, 1991: Duration of liquid water habitats on early Mars, *Icarus*, **90**, 214-221.

McKay, C. P., O. B. Toon, and J. F. Kasting, 1991: Making Mars habitable, *Nature*, **352**, 489-496.

Meyer, T. R. and C. P. McKay, 1989: The resources of Mars for human settlement, *J. Brit. Interplanet. Soc.*, **42**, 147-160.

Oberg, J. E., 1981: *New Earths*, Stackpole Books, Harrisburg, Pennsylvania.

Pollack, J. B. and C. Sagan, 1993: Planetary engineering. In *Near Earth Resources*, J. Lewis and M. Mathews (Eds.), University of Arizona Press, Tucson, AZ.

Pollack, J. B., J. F. Kastings, S. M. Richardson, and K. Poliakoff, 1987: The case for a wet, warm climate on early Mars, *Icarus*, **71**, 203-224.

Sagan, C., 1973: Planetary engineering on Mars, *Icarus*, **20**, 513-514.

Schopf, J. W. (Ed.), 1983: *Earth's Earliest Biosphere: Its Origin and Evolution*, Princeton University Press, Princeton, N. J.

Seckbach, J., F. A. Baker, and P. M. Shugarman, 1970: Algae thrive under pure carbon dioxide, *Nature*, **227**, 744-745.

Squyres, S. W. and M. H. Carr, 1986: Geomorphic evidence for the distribution of ground ice on Mars, *Science*, **231**, 249-252.

Stoker, C. R., C. P. McKay, R. M. Haberle, and D. T. Andersen, 1990: Science strategy for human exploration of Mars, *Adv. Space Res.*, **12(4)**, (4)79-(4)-90.

Thomas, D. J., 1991: Mars after the Viking mission: Is life still possible?, *Icarus*, **91**, 199-206.

Turner, F., 1988: *Genesis, an Epic Poem*, Saybrook Publishing Company, Dallas, Texas.

Turner, F., 1989: Life on Mars: cultivating a planet and ourselves, *Harper's Magazine*, **279 (1671)**, 33-40.

UNEXPECTED *IN VITRO* INTRON SPLICING OF COMMON BEAN CHLOROPLAST trnL (UAA) GENE AND PSEUDOGENE BY T7 RNA POLYMERASE

O. Carelse and M.V. Mubumbila
Biochemistry Department, University of Zimbabwe, P.O.Box MP 167, Mount Pleasant, Harare, Zimbabwe.

ABSTRACT

A 600 bp product of a polymerase chain reaction (PCR) amplification, containing the common bean chloroplast trnL (UAA) gene and pseudogene, was transcribed *in vitro* using *E. coli*, T7 and SP6 RNA polymerases. The *in vitro* transcription with *E. coli* RNA polymerase produced two primary RNA transcripts which derived from the gene and pseudogene. The *in vitro* transcription with T7 RNA polymerase resulted in the splicing of the intron and the production of mature RNA transcripts that derived from the 5' and 3' exons and ligated exons. The *in vitro* transcription with SP6 RNA polymerase did not occur. An increase in transcriptional activity from SP6, through *E. coli*, to T7 RNA polymerases, appears to reflect successive intron loss from the SP6, *E. coli* and T7 genomes, through the evolutionary time scale. If we assume that SP6 and T7 genomes evolved from a common ancestor, the SP6 genome may have diverged very early in time through the loss of its introns allowing for the SP6 RNA polymerase to acquire a very stringent specificity for its own promoter. In contrast, the T7 genome may still contain introns, and, as a result, the T7 RNA polymerase has retained its ability to splice an ancient group I intron. On the other hand introns loss from the *E. coli* genome may have altered the ability of the *E. coli* RNA polymerase to splice group I introns. Although of very different phylogenetic origins, both *E. coli* and T7 RNA polymerases identified promoter sequences within the gene's and pseudogene's intron, for transcription to occur.

1. INTRODUCTION

Introns have been found in nuclei, mitochondria and chloroplasts genes of eukaryotes (Cech 1986), cyanobacteria (Xu et al. 1990), archaebacteria (Kaine 1987), and in viral genes of eubacteria (bacteriophages) and those of eukaryotes (Quirk et al. 1989). Absence of introns was observed in eubacteria (Xu et al. 1990). Interest by these introns includes the ability of many of them to catalyse their own splicing (Cech 1990), the activity of the proteins they encode (Perlman and Butow 1989), their properties as recently acquired mobile genetic elements (Copertino et al. 1991), or as primitive elements which arose very early in molecular evolution, in the "progenote" ancestor of all living organisms (Darnell and Doolittle 1986).

Self-splicing of group I introns was first demonstrated for the nuclear large subunit rRNA intron of *Tetrahymena thermophila* (Cech et al. 1981). However, some group I introns are not self-splicing but encode site-specific endonucleases or maturases that are synthesised *in vivo* for the splicing out of these introns (Perlman and Butow 1989). This protein-dependent splicing was shown to occur for chloroplast introns, such as those of the large subunit rRNA of *Chlamydomonas eugamatos* (Gauthier et al. 1991), the trnL (UAA) gene in a number of higher plants (Kuhsel et al. 1990; Xu et al. 1990) and that of *Cyanophora* (Evrard et al. 1988). The presence of a group I intron in the trnL (UAA) gene of all major groups of chloroplasts and in cyanobacteria is strong evidence that it resided in the trnL (UAA) gene of the cyanobacterial ancestor of plastids, and is therefore regarded as being of ancient origin (Kuhsel et al. 1990; Xu et al. 1990).

Chloroplast genes were shown to contain promoters of prokaryotic-type, and some tRNA genes have been transcribed by *E. coli* RNA polymerase (Zech et al., 1981). *In vitro* transcription systems using chloroplast extracts of spinach (Gruissem, 1984) and common bean (manuscript in preparation) have been developed. Cloning vectors containing T7 and SP6 promoters flanking the cloned insert have shown that initiation of transcription with T7 (Dunn and Studier, 1983) and SP6 (Kassavetis et al., 1982) RNA polymerases was highly specific for the T7 and SP6 phage promoters, respectively.

In this paper we are presenting the *in vitro* transcription of a 600 bp Polymerase Chain Reaction (PCR) product, containing the common bean chloroplast trnL (UAA) gene and pseudogene, using *E. coli* , T7 and SP6 RNA polymerases. The evolutionary implication in the transcriptional activity of these RNA polymerases is discussed.

2. MATERIALS AND METHODS

A 600 bp PCR product was prepared as described (Mubumbila et al. 1993) and used as a DNA template for *in vitro* transcription experiments. *E. coli* RNA polymerase holoenzyme, T7 and SP6 RNA polymerases were purchased from Epicentre technologies. *In vitro* transcriptions were performed using reaction mixtures from Pharmacia (1986) with *E. coli* , and New England Biolabs (1991, spermidine omitted), with T7 and SP6 RNA polymerases. Transcription products were phenol extracted, ethanol precipitated, separated on 10% polyacrylamide-8M urea gels and autoradiographed for 1 to 7 days at room temperature.

3. RESULTS AND DISCUSSION

The common bean chloroplast trnL (UAA) gene and pseudogene have a size of 616 and 591 bp, respectively, but the PCR-derived homologues were 611 and 586 bp in size, as delimited by the primers (Mubumbila et al. 1993).

The *in vitro* transcription of the 600 bp PCR product with *E. coli* RNA polymerase yielded two primary RNA transcripts of 611 and 586 nucleotides (nts) in size, which corresponded to the precursors of the tRNA-Leu (UAA) gene and pseudogene (Table 1). These results were identical to those obtained after *in vitro* transcription with a purified chloroplast enzymatic extract of common bean (manuscript in preparation). As neither the gene nor the pseudogene contained promoter regions at the 5' end of the 5' exon, it may be suggested that the trnL (UAA) gene and pseudogene do not require 5' upstream promoter sequences for *in vitro* transcription. It would appear that specific sequences within the intron may have participated in the initiation of this transcription. Although *E. coli* RNA polymerase yielded primary RNA transcripts, they were not processed, by splicing of the intron, into mature tRNA-Leu (UAA) molecules. This strongly suggests that the group I introns of both gene and pseudogene are not self-splicing, and that the purified *E. coli* RNA polymerase holoenzyme did not contain the maturase activity necessary for the production of a functional tRNA.

The expected RNA transcripts deriving from the *in vitro* transcription of the common bean chloroplast trnL (UAA) gene are those of the 5' exon (35 nts), the 3' exon (50 nts) and the ligated exons (85 nts). However, the PCR-derived homologue of the 3' exon was reduced to 45 nts in length (Mubumbila et al., 1993), which in turn produced ligated exons of 80 nts in size. The *in vitro* transcription of the 600 bp PCR product with T7 RNA polymerase yielded such expected RNA transcripts (Table 1). They were identical in size to those obtained from the *in vitro* transcription with a crude chloroplast enzymatic extract of common bean (manuscript in preparation). The 35-nt band derived from the gene's 5' exon, whereas the 26-nt band derived from the pseudo-5' exon. The intensity of the 45-nt band suggested that it derived from both the gene's and the pseudogene's 3' exon. The gene's 5' exon and 3' exon appeared to have been ligated into a tRNA-Leu (UAA) of 80 nts. In contrast, the pseudo-5' exon was not utilised for the synthesis of a pseudo-tRNA-Leu (UAA), as no corresponding band of 71 nts was present.. Other bands, such as the 32-nt band may have derived from a polymorphic form of the pseudo-5' exon. The 14-nt band may represent oligonucleotides that have been synthesised during the initial "abortive" phase of transcription by T7 RNA polymerase (Martin et al. 1988). The 125-nt band may be representing the lariat form of circular introns (Thompson and Herrin 1991). The *in vitro* transcription with SP6 RNA polymerase did not occur. The splicing of a chloroplast group I intron by a viral enzyme was quite unexpected as T7 RNA polymerase has always shown to be very specific for its own promoters. This *in vitro* transcription occurred without the presence of any of the T7 RNA polymerase promoters. The results have also shown that the trnL (UAA) gene could be transcribed without its 5' upstream region. However, an increase in transcriptional activity from the SP6, through E. coli, to T7 RNA polymerases, appear to reflect the successive intron loss from the SP6, *E. coli* and T7 genomes, in that order,

TABLE 1. LIST OF PRIMARY AND MATURE RNA TRANSCRIPTS

Sizes (in nucleotides) of RNA transcripts which were synthesised by *in vitro* transcription of the PCR-derived trnL (UAA) gene and pseudogene with bacterial (E. coli) and phage (T7) RNA polymerases.

RNA transcript number	Expected mature RNA transcript	Expected primary RNA transcripts	RNA transcripts obtained with *E. coli* RNA polymerase	RNA transcripts obtained with T7 RNA polymerase
1		611	611	
2		586	586	
3	80			80
4	45			45
5	35			35
6	26			26
7				32
8				14
9				125

through the evolutionary time scale. If we assume that SP6 and T7 phages evolved from a common ancestral genome (Kassavetis et al. 1982), the SP6 genome may have been the first to diverge from this ancestor, through the loss of its introns. As a result the SP6 RNA polymerase may have developed a very stringent promoter specificity. On the other hand, the T7 genome may still contain introns and, as a result, the T7 RNA polymerase has retained its ability to splice an ancient group I intron. Introns loss from *E. coli* genome may have occured within the time scale that gradually restricted *E. coli* RNA polymerase from splicing group I introns. Although of very different phylogenetic origins, both *E. coli* and T7 RNA polymerases were able to identify promoter sequences within the introns of the chloroplast trnL (UAA) gene and pseudogene, for transcription to occur. Such ability may reflect an ancestral relationship between the two genomes.

Protein-dependent splicing is a mechanism that may have been developed for intron retention, as both introns "early" or introns "late" views (Kuhsel et al. 1990; Coppertino et al. 1991) appear to indicate that introns were either retained, through evolution, or that they were being regained as an evolutionary process. In either ways, the functions of the intron were shown to include the recognition of tRNA precursors (Abelson 1992), its participation in tRNA splicing (Baldi et al. 1992), and in addition, it probably maintains gene stability and may even have some control over gene expression.

ACKNOWLEDGMENTS

We thank Professor Jules Kempf, Institut de Chimie Biologique, 11 rue Humann, 67085 Strasbourg, France, in whose laboratory this work was done.

REFERENCES

Abelson, J., 1992: Recognition of RNA precursors: A role for the intron, *Science,* 255, 1390-1390.

Baldi, M.I., E. Mattoccia, E. Bufardeci, S. Fabbri, and G. P. Tocchini-Valentini, 1992: Participation of the intron in the reaction catalyzed by the *Xenopus* tRNA splicing endonuclease, *Science*, 255, 1404-1408.

Cech, T.R., 1986: The generality of self-splicing RNA: Relationship to nuclear mRNA splicing, *Cell*, 44, 207-210.

Cech, T.R., 1990: Self-splicing of group I introns, *Annu. Rev. Biochem.,* 59, 543-568.

Cech, T.R., A.J. Zaug, and P.J. Grabowski, 1981: *In vitro* splicing of the ribosomal RNA precursor of *Tetrahymena*: Involvement of a guanosine nucleotide in the excision of the intervening sequence, *Cell*, 27, 487-496.

Copertino, D.W., D.A. Christopher, and R.B. Hallick, 1991: A mixed group II/group III twintron in the *Euglena gracilis* chloroplast ribosomal protein S3 gene: evidence for intron insertion during gene evolution, *Nucl. Acids Res.*, 19, 6491-6497.

Darnell, J.E., and W.F. Doolittle, 1986: Speculations on the early course of evolution, *Proc. Natl. Acad. Sci. U.S.A.*, 83, 1271-1275.

Dunn, J.J., and F.W. Studier, 1983: Complete nucleotide sequence of bacteriophage T7 DNA and the locations of T7 genetic elements, *J. Mol. Biol.*, 166, 477-535.

Evrard, J.L., M. Kuntz, N.A. Strauss, and J.H. Weil, 1988: A class I intron in a cyanelle tRNA gene from *Cyanophora paradoxa*: Phylogenetic relationship between cyanelles and plant chloroplasts, *Gene*, 71, 115-122.

Gauthier, A., H. Turmel, and C. Lemieux, 1991: A group I intron in the chloroplast large subunit rRNA gene of *Chlamydomonas eugamatos* encodes a double stranded endonuclease that cleaves the homing site of this intron, *Curr. Genet.*, 19, 43-47.

Gruissem, W., 1984: A chloroplast transcription system from higher plants, *Plant Mol. Biol. Rep.*, 2, 15-23.

Kaine, B.P., 1987: Intron-containing tRNA genes of *Sulfolobus solfataricus*, *J. Mol. Evol.*, 25, 248-254.

Kassavetis, G.A., E.T. Butler, D. Roulland, and M.J. Chamberlin, 1982: : Bacteriophage SP6-specific RNA polymerase. II. Mapping of the SP6 DNA and selective *in vitro* transciption, *J. Biol. Chem.*, 257, 5779-5788.

Kuhsel, M.G., R. Strickland, and J.D. Palmer, 1990: An ancient group I intron shared by eubacteria and chloroplasts, *Science*, 250, 1570-1573.

Martin, C.T., D.K. Muller, and J.E. Coleman, 1988: Processivity in the early stages of transcription by T7 RNA polymerase, *Biochemistry*, 27, 5755-5762.

Mubumbila, M.V., O. Carelse, and J. Kempf, 1993: Isolation by asymmetric PCR and sequencing of the common bean chloroplast trnL (UAA) gene and pseudogene, *Phytochem. Anal.*, 4, 145-148.

Perlman, P.S., and R.A. Butow, 1989: Mobile introns and intron-encoded proteins, *Science*, 246, 1106-1109.

Quirk, S.M., D. Bell-Pedersen, and M. Belfort, 1989: Intron mobility in the T-even phages: High frequency inheritance of group I introns promoted by intron open reading frames, *Cell*, 56, 455-465.

Thompson. A.J., and D.L. Herrin, 1991: *In vitro* self-splicing reactions of the chloroplast group I intron Cr.LSU from *Chlamydomonas reinhardtii* and in vivo manipulation via gene-replacement, *Nucl. Acids Res.*, 19, 6611-6618.

Xu, M.Q., S.D. Kathe, H. Goodrich-Blair, S.A. Nierzwicki-Bauer, and D.A. Shub, 1990: Bacterial origin of a chloroplast intron: Conserved self-splicing group I introns in *cyanobacteria*, *Science*, 250, 1566-1569.

Zech, M., M.R. Hartley, and H.J. Bohnert, 1981: Binding sites of *E. coli* DNA-dependent RNA polymerase on spinach chloroplast DNA, *Curr. Genet.*, 4, 37-46.

ON DYSON'S MODEL OF THE ORIGINS OF LIFE AND POSSIBLE EXPERIMENTAL VERIFICATION

J.N. Islam
Research Centre for Mathemetical and Physical Sciences,
Chittagong University, Chittagong 4331, Bangladesh

M.K. Pasha
Department of Botany, Chittagong University,
Chittagong 4331, Bangladesh

ABSTRACT

Some years ago F.J. Dyson put forward an interesting 'double-origin' hypothesis concerning the origin of life on the earth. His essential contention was that protein-based life capable of metabolism originated first, about three billion years ago, and that much later on RNA-based replication evolved in existing cells. Dyson based his hypothesis on the experimental evidence that amino-acids, which require the basis of proteins, are much easier to synthesize in prebiotic primordial conditions than nucleotides, the constituents of RNA and DNA. The purpose of this review paper is to examine critically Dyson's theory and consider recent experimental evidence for and against the theory.

1. INTRODUCTION

Some important organic molecules such as amino acids and nucleotides can associate to form long variable length polymers. One nucleotide can join with another by a phosphodiester bond and one amino acid can join with another by forming a peptide bond. The repetitious units lead to linear chains known as polynucleotides and polypeptides, respectively. Once a polymer has formed it can influence the formation of other polymers. But the nucleic acids, in particular, have the ability to specify the sequence of nucleotides by acting as templates for the polymerization reaction. The proteins and nucleic acids are viewed as the most important constituents of life. We can not say clearly how the primary structure of a protein is transferred to protein function though a great deal is known about how primary structure is built into secondary and tertiary structures. The number of amino acid, sequences of amino acids and possible folding (intermoleccular, non-covalent forces and by disulfide bridges) arrangement in a protein is astronomical. Most proteins are composed of between 50 and 750 residues. The number of possible sequences even in proteins with as few as 50 residues, is again astronomical (20^{50}), a number much higher than the number of metabolic and physical functions needed to sustain contemporary life process. Biochemical reactions associated with metabolism, growth and differentiation are catalyzed by enzymes (proteins) which are specific to a reaction or to a family of related reactions.From both a chemical and a structural standpoint, proteins are among the most sophisticated molecules known to us.

The nucleic acid is a long series of polymeric molecules comprised of few monomers. Out of the three different components **of monomers: nucleotide,** sugar, phosphate and base, the real difference lies only in the bases of just four types. The backbone of polynucleotide is linked by phosphodiester bond and free rotation is not permissible around the axis of the primary chain. Base composition and its sequence in a polymeric chain is a crude measure of differences and similarities between nucleotides. The earliest polymer that formed as RNA chain showed some unique versatility in many ways. The primary structure of RNA is acting and transferred to RNA function. The number of bases, sequence in a chain and their possible folding by its complementary base pairing is also astronomical. Unlike proteins the base composition is not limited. **Probably, a functional** RNA is about 75 to several thousand **base pairs. The number of** sequences in small RNA molecule with as few as 75 nucleotide monomers is again astronomical (4^{75}), a number sufficient to serve a variety of functions that are required to control metabolic and physical function for metabolism, growth and differentiation of life. The ability of small RNA molecules to act as a protein like catalyst has been widely interpreted (Cech, 1986; Altman et al 1986).

In the present day **organisms, a** collaborative system of RNA molecules (t RNA, r RNA and mRNA) plays a central role in directing the synthesis of polypeptides. It seems likely that RNA, acting as enzyme, guided the primordial synthesis of protein in a very simple way and primitive fashion. Subsequently RNA was able to create mechanisms for more efficient biosynthesis of both proteins and nucleic acids. Some of these molecules could have been put to use in the replication of RNA. The polypeptide formation under the guidance of RNA required the evolution of code by which the polynucleotide sequence specifies the amino acid sequence. After acquiring the efficient genetic mechanism for protein synthesis, the less stable RNA molecule created more stable DNA molecule and then took over the primary genetic function in a more **efficient** way and became the permanent genetic repository of genetic information in living

systems. The proteins became the **catalysts** and RNA remained primarily as the intermediary **connection** between the DNA and protein. Crick (1958) postulated that the biological information contained in the DNA, which constitute the gene, is transfered first to RNA and then to protein, in a unidirectional manner.

2. DYSON'S MODEL FOR THE ORIGIN OF METABOLISM

According to the Dyson's hypothesis (1982, 1985) metabolism originated first in protein based cells which produced suitable enzymes, and that genes appeared through nucleotides much later. Accordingly, Dyson gave a simple model for the origin of metabolism which essentially explains quantitavely how a population of molecules in the primordial 'soup' could have made a transition from a 'disordered' state to an 'ordered' state. The model itself attempts to define 'metabolism' and how it can be made to correspond to an 'ordered' state and the manner in which a transition to such a state may occur. As mentioned by Dyson himself, his model is simple and crude and it can be generalized in several directions, although it is claimed that the essential features that he is studying are incorporated in the model. We will give here a brief explanation of the model, and indicate one or two directions in which we propose to generalize the model. Although these generalizations may not add anything new qualitatively, we believe these may give some minor insights and help to understand the model better. The full consequences of these generalizations has still to be worked out.

Dyson makes ten assumptions to define his model which briefly are as follows:

1. Cell came first, then enzymes; genes appeared much later.

2. A 'cell' is a droplet with polymer molecules, made of monomers of a certain number in N sites. The monomers can enter or leave the cell, and there is enough energy available, causing chemical reactions between monomers and polymers.

3. Evolution of the population of the molecules in a cell proceeds by random drift.

4. Changes in the polymers occur by a single substitution (point mutation) illustrated by the following (A, B, C, ... are monomers)

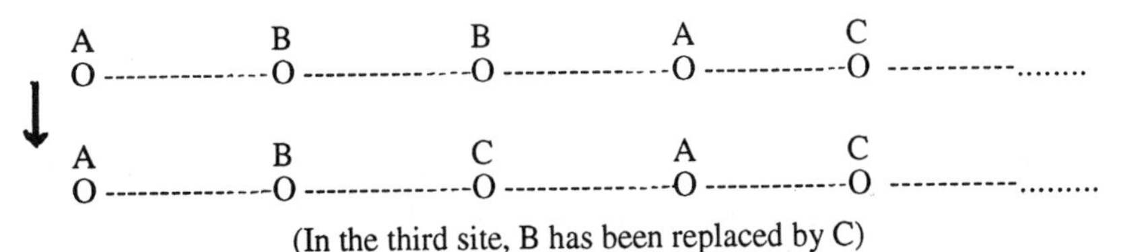

(In the third site, B has been replaced by C)

5. At each step, the N sites in a polymer mutate with equal probability (I/N).

6. The bound monomers are either "active" or 'inactive"

7. The "active" monomer enhances the ability of a polymer to act as an enzyme. To "act as an enzyme" means to assist the mutation of other polymers so that the correct species of monomer moves into an active site. "Metabolism" here is defined as the cyclic shuffling around of monomer units; the population is metabolically active if the proportion of active monomers is maintained above a certain level.

8. In a cell if x: 1 is the ratio of active monomers to the total, the probability that a new insersion of monomer will be active is ϕ (x), (In analogy with magnetism, this may be called the "mean-field approximation").

9. The curve y = ϕ (x) is S-shaped as in Fig. 1.

10. $\phi(x) = (1 + ab^{-x})^{-1}$. Here (1 + a) = number of species of monomers, and b is another constant related to the "degree of discrimination" of the catalyst. This function is obtained from thermodynamic considerations.

3. A GENERALIZATION OF DYSON'S MODEL

We will indicate briefly two directions in which generalization of the above model can be considered.

(A) Instead of two kinds of sites "active" and "inactive", we consider three kinds, "active", "semi-active" and "inactive", or "hyperactive", "active" and "inactive". In this case one has a function ϕ (x, y) of the variables x, y which are the ratios of (if we consider the first labels) "active" and "semi-active" monomers. We will not consider this further in this paper, but hope to do so in a future work.

(B) In Fig. 1, x= α, β, γ are equilibrium points of which x = α, γ are stable but x = β is unstable. The stable points are those for which the slope of y = ϕ (x) is less than unity. The point x = α is metabolically "inactive' while x = γ is "active" (death and life). This figure pertains to Dyson's model. One could generalize this to the following function ϕ' (x) parametrically (for the present this function is chosen purely "phenomenologically") through the parameter ξ :

$$\phi'(X) = A\ \{\ \xi + \cos\ (\ \xi o + \omega\xi\ )\ \}$$
$$X = A\ \{\ \xi - c'os\ (\ \xi o + \omega\xi\ )\ \}.$$

This function is illustrated in Fig 2. By chossing the constants A, C, C′, ξo and ω suitably, one can have any number of equilibrium points (depending on the constant ω) and make the curve y = ϕ' (x) as close as one pleases to y = x. One such curve is illustrated in Fig 2. The equlibrium points are x = α, β, γ, δ, ξ of which x = α, γ, ξ are

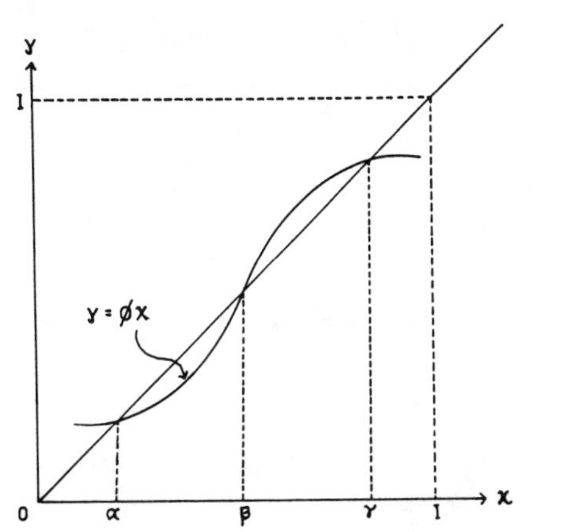

Fig. 1. The function ϕ (x) = $(1+abx^{-1})^{-1}$ used by Dyson.

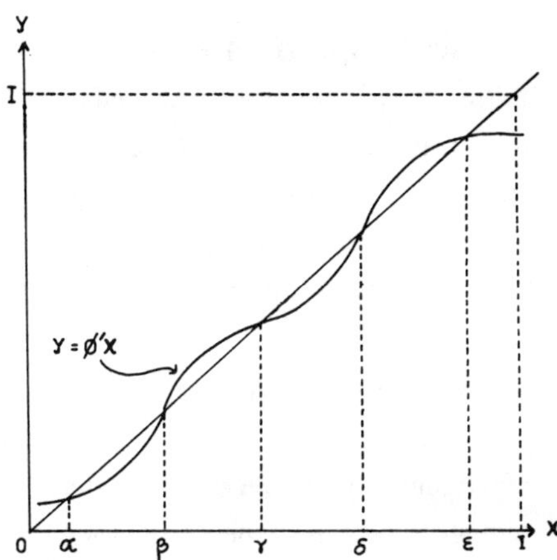

Fig. 2. A generalization ϕ' (x) of Dyson's function.

stable and x = β, δ are unstable. The points x = γ, ξ have different degrees of metabolic activity. Working out fully the consequences of the above modifications may give useful insights in this approach to the definition and the origin of metabolism.

4. DISCUSSION

Several experiments were performed by nature for polymer synthesis. Most surprising illustrations of early events are still performed by primitive organisms. A definite sequence of amino acids is maintained without involvement of RNA (tyrocidine, a circular polypeptide antibiotic). The SH group of pentothenic acid residues of the enzyme complex selectively activates a particular amino acid followed by peptide linkage resulting in the formation of polypeptide chain The maintenance of the amino acid sequence of the small peptide chains formed in bacterial cell wall is also independent of the conventional translational mechanism. The "central dogma" thus, although useful, is not necessarily the only method by which genetic information can be transferred (Sen, 1985). At the same time Dyson implies that the dogma was not true for the earliest forms of life. There is no logical reason, as stated Dyson, why a population of enzymes mutually catalysing each others' synthesis should not serve as carrier of genetic information.

Altman et al (1986) and some other experiments clearly indicated that we can no longer assume that biological catalysts are exclusively the realm of proteins. It has been demonstrated now that RNA acte as enzyme. So, we now get the dual properties of RNA-that RNA itself has templating properties essential for replication and potential to foldup to form complex surfaces that can also catalyse specific reactions. Thus special versatility of RNA molecules is likely to have provided the basis for evolution of the first living systems. Eigen et al (1981) and Orgel's (1981) experiments clearly indicated that RNA can be created using neither a template nor an enzyme which is the material to raise the hypothesis exploring the evolution of RNA, according to Dyson, under conditions appropriate to the second origin of life. It is a challange to chemists and biologists to devise experiments which will refute or vindicate Dyson's hypothesis.

REFERENCES

Altman, S., M. Baer, C. Guerrier-Takada, and A. Viogue, 1986: Enzymatic Cleavage of RNA by RNA, Trends Biochrm. Sci., 11, 515-518.

Cech, T., 1986: RNA as an Enzyme, Sci. Amer., 255 (4), 76-84.

Crick, F.H.C., **1958:** On Protein Synthesis, Symp. Sci. Expt. Biol, 12, 138-163.

Dyson, F.J., 1982: A Model for the Origins of Life, J. Mol. Biol., 18, 344-350.

Dyson, F.J., 1985: Origins of Life, Cambridge University Press, Cambridge.

Eigen, M., W. Gardiner, P. Schuster, and Winkler Oswatitsch, 1981: The Origin of Genetic Information, Sci. Amer., 244 (4), 78-118.

Orgel, L.E. 1986: RNA Catalysis and the Origin of Life. J. Theo. Biol., 123, 127-149.

Sen, S.P., 1985: The Origin and Evolution of Life: Some Unanswered Questions, Presidential Address, 72nd Indian Science Congress, Lacknow.

THE ROLE OF INFORMATION PROCESSING IN THE EVOLUTION OF COMPLEX LIFE FORMS

K. Tahir Shah
International Centre for Theoretical Physics
P.O.Box 586, 34100 Trieste, Italy

ABSTRACT

Biological evidence suggests that non-coding regions of genome are involved in information processing. Our theoretical analysis shows that there are some highly conserved motifs in virtually all species from viroids to human that are found near important sites. These motifs have their analog in number-theoretic sequences constructed from a finite set of symbols, called automatic sequences. A theorem of Cobham relates these sequences to information-processing automata. Based on this analysis we put forward the following hypothesis.
1. Formation of these motifs is the result of prebiotic chemistry. Information-processing sequences, i.e., ribozymes, were formed before the genetic code and protein enzymes. 2. Introns, satellite DNA, telomers, highly conserved TATA and ALU sequences, as well as viroids and transposons, are all involved in some aspect of information processing. Editing, proof reading, error detection and correction are examples of this capability. As an application, a plausible explanation of the C-value paradox is suggested. Moreover, using this approach it would be possible to predict critical determinants of specific genes. This may help design effective antisense and ribozyme-directed strategies for gene therapy.

1. INTRODUCTION

The existence of non-coding regions (e.g., introns, ALU sequences, satellite DNA) is a mystery. Why nature has kept them during the long evolutionary period if they played no role in the development of complex life forms? More precisely, what kind of capability is encoded within linear nucleotide sequences, which is necessary (but not sufficient) condition for the evolution of complex life forms, keeping in mind the C-value paradox and the omnipresence of non-coding segments. The role of ribozymes is now well established in many enormously difficult tasks such as splicing, editing, regulation and control (Cech, 1987), (Joyce, 1989), (Barinaga, 1993). All this is encoded and achieved within the genome, a linear sequence. It is natural to think, therefore, that non-coding regions are involved in something essential. It seems that there is no other explanation except that non-coding regions contain algorithms for processing information in both the coding and non-coding regions. It is true that protein enzymes are also a major player in information processing tasks, but they are evolutionary late comers, only after nature invented the genetic code. It is more likely that only ribozymes were present in the RNA world. We argue that this capability played a major role in evolution and without which we would not have been here today! Moreover, information-processing segments were formed early in the RNA world along with informational macromolecules. Introns, viroids, telomers and other noncoding sequences are possibly more specialized versions of such segments.

In the physico-chemical description of enzymatic/ribozymatic processes the algorithmic aspect remains hidden. The description is hardware oriented and does not highlight algorithmic or information processing aspect. However, the information processing capability depends on the abstract combinatorial nature of enzymatic/ribozymatic sequences. To understand how this capability played an essential role in the development of complex life forms, it is necessary that we give some precision to the notion of complexity. The genotype complexity is not simply the size of DNA (C-value paradox). What makes a genome complex is its internal structure that encodes not only various functions but also intricate mechanism to replicate and translate DNA in a energy cost-effective and reliable manner. An efficient and reliable way to encode a large number of functions is to distribute them in smaller and independent chunks. In our information processing approach, the size of a linear sequence does not reflect on its information content or computational capability. It is the *structural complexity* of the information-carrying symbolic sequence that determines its computational capability. As a possible solution to C-value paradox. we suggest is that it is the structural complexity of a genotype, not its size, that determines the complexity of its phenotype. Thus, the human DNA need not to be the longest one, only rich in its structure.

We shall show that the formation of nucleotide segments described later are the result of prebiotic chemistry and combinatorial laws. We shall call these segments our *fundamental motifs*. Larger segments constructed out of fundamental motifs may or may not fold to form any specific ribozymes. It depends on the presence of inverted palindromes and appropriate correlations. Many sequences constructed out of these motifs are same as those generated by automata. They are called automatic sequences for obvious reason.

Our line of argument is as follows. After explaining the computational paradigm, we show why one should use it in Genetics. The abstract computational model we use is the rewrite rules system. Such a representation is most suitable for linear sequences Next, we search fundamental motifs in various species, especially in intronic and other noncoding segments. A comparison between well-known automatic sequences and sequences constructed out of fundamental motifs proves the point: noncoding regions are a sort of "hardwired" programs, i.e., they are computational devices after appropriate folding. Thus in our model, a noncoding region, e.g., an intron contains a program while an exon contains its corresponding acceptable data. After establishing the model, we show its usefulness in ribozyme-directed and antisense gene therapy.

2. A NEW PARADIGM TAKES ROOTS IN GENETICS

In the sequel we discuss how and why of a new paradigm to describe physico-chemical processes in molecular biology.

2.1 THE COMPUTATIONAL PARADIGM

The word paradigm means a model, a pattern, or a description that is deep, but close to reality. Normally, one entity or concept is used in the description process. For instance, all physical and chemical processes are described using energy or energy related concepts. The dynamics of a system is derived from the temporal description in the sense that dynamical equations such as, Hamiltonian or Lagrangian equations are obtained by minimization of action, an energy related concept. The energy paradigm dominates the physical and engineering sciences. However, the algorithmic aspect of any process remains hidden in this paradigm since it is hardware oriented. To understand the abstract algorithmic nature of physico-chemical processes, it is useful to use the computational paradigm, in which the information (not in the Shannon's sense) is the basic entity and computation is performed on an input leading to an output. This suggests that dynamics and computation are somehow equivalent. A strong version of Church-Turing thesis suggests exactly this. Thus, in this equivalent paradigm basic processes are computational and act on information encoded in a symbolic form. It is useful in Genetics because ribozymes, enzymes and other regulatory and control genes are involved in complex tasks that can best be explained in computational terms. Church-Turing Thesis: Any number-theoretic function is computable by an effective procedure if and only if it is Turing computable Any effective procedure in which a finite set of symbol is manipulated can be translated into a number-theoretic function by a suitable encoding of the symbols as non-negative integers. Strong versions of the thesis relates to physico-chemical processes (Conrad, 1985), or to quantum mechanical processes (Deutsche 1985). In both dynamics and computation, it is the change of states of a system that determines their direction.

Markov algorithms, Post production system, rewrite rules, lambda calculus, and infinite register machines are all equivalent to Turing machine (Salomaa, 1985). We use rewrite rules since a theorem of Cobham (1974) and Christol et al (1980) relates automata with linear sequences. The theorem states that a sequence is recognizable by an automaton if and only if it is generated by a fixed length rewrite rule. In our model, C, G, A, U (or T) is a set of symbols and we compare nucleotide sequences with those generated symbolically through applications of rewrite rules. Since not all sequences are generated by fixed length rewrite rules, they may not be recognized by equivalent automata. They may, however, be acceptable to other automata.

Rewrite systems generate sequences, while Markov systems accept what can be called automatic sequences. Both have production rules. One may envisage three types of situations in Genetics:

a. Generate and compare situation by ribozymes or enzymes.

b. Simple acceptor and actuator situation where messages are accepted and tasks are actuated.

c. Simple acceptor situation: proof-reading to see if the language is correct.

In linear nucleotide sequences, both software and hardware aspects merge together. Their structure encodes the algorithm or data while the chemistry is their hardware implementation. After an appropriate folding into secondary and tertiary structures, the program is "hardwired" and a ribozyme (or enzyme) becomes a dedicated finite state machine/automata.

2.2 WHY USE THE COMPUTATIONAL PARADIGM?

We are motivated by a large number of facts that can best be explained only in computational paradigm. However, we illustrate our viewpoint using RNA editing as an example, which is a highly complex task. Its complexity suggests that only an "intelligent" system can achieve it. The RNA editing might be prebiotically related mechanism to RNA splicing and in its primitive form was perhaps a simple error detection and correction mechanism.

Consider an specific case of RNA editing in plant mitochondria and chloroplasts (Gray & Covello 1993). This example shows clearly how word structure is changed by editing of wheat mitochondrial *cox2* mRNA. Fourteen out of sixteen edited words are of the type XYYZ , XYYYZ, and 1-overlapping XYYXXY, a subset of our fundamental motifs. In the table below, words (4), (9), and (10) are 1-overlapping words. In other cases (not included in Table 1) of editing, 1-overlapping words are created or deleted. The computing significance of such changes is that words are modified either to change the instructions or some data is transformed into a new data or instructions. In this example, given the type of edited words, the data or instructions are transformed into instructions of specific type. Nature has devised such a system possibly for their dual or multiple purpose use to optimize energy cost of creating new programs and data, especially if they all belong to the same class of tasks. We may call such editing as the dynamic or adaptive change of algorithm or data. This is a remarkable case of "intelligent" information processing.

Table 1. An Example of RNA Editing

Original Word	Edited Word
1. AUUCC	AUUUC
2. AUCUC	AUUUC
3 AUCA	AUUA
4. AUUCGGA	AUUUGGA
5. AUCCA	AUUUA
6 AUCA	AUUA
7. CCUC	CUUC
8. CUCA	CUUA
9. GUUCAAAU	GUUUAAAU
10. CUCGGU	CUUGGU
11. AUUCG	AUUUG
12. AUCGGA	AUUGGA
13. ACCCG	A.UCCG
14. AUUCG	AUUUG

3. THE MODEL

Our model is based on a comparison between our fundamental motifs and number-theoretic automatic sequences. In the sequel we discuss their formation in the RNA world and propose an interpretation in the computational paradigm.

3.1. FORMATION OF NUCLEOTIDE SEQUENCES IN THE RNA WORLD

Very often it is stated that random sequences were formed in the early RNA world. We think it is quite the contrary, the formation of nucleotide sequences was governed by chemical and combinatorial laws. In the prebiotic soup, the only possible and simplest chemical constraint that can influence the formation of nucleotide sequences was Crick-Watson base-pairing/affinity. Complementary nucleotides were attracted towards each other and concatenated to form longer and longer sequences. Moreover, due to energy related stability properties, initially only C-G pairs and later only C-G rich sequences were formed. The C-G rich structure of viroids (Diener 1991) confirms our assumption.

Van der Waerdon's theorem (Lothaire, 1983) in combinatorics asserts that sequences formed by a finite set of symbols will eventually have subsequences with regularity properties. This means, whether or not there are chemical constraints, nucleotide sequences with regularity are sure to be formed. Random sequences simply cannot exist if finite (only four in this case) symbols are used to form them.

Our combinatorial analysis shows that Morse-Thue, Fibonacci as well as other number-theoretic sequences are highly probable. For instance, starting from a single nucleotide C, the binary Morse-Thue sequence can be formed as follows:

C CG CGGC CGGC.GCCG CGGC.GCCG.GCCG.CGGC etc.

Similarly, the binary Fibonacci sequence can be formed as:

C GC CGGC GCCG.C CGGC.G.CGGC GCCG.C.GCCGGCCG.C etc.

Both type have palindromes as their subsequences. Let F be a Fibonacci sequence and k is the iteration number. Then, for all k equal or greater than 3, one has F = P.d, where P is a palindrome and d = CG, if k even and d = GC if k is odd (de Luca, 1981).

Both these sequences can also be generated by rewrite rules as follows:

C --> CG and G --> GC generates Morse-Thue or its complement sequence and C --> CG and G --> C generates Fibonacci sequence or its complement. Fundamental motifs CGGC, GCCG, GCCGGC ,and CGGCCG are present in both Morse-Thue and Fibonacci sequences.

3.2 SOME AUTOMATIC SEQUENCES

Most of the following sequences are well-known in mathematical literature (Dekking, 1982a and 1982b). The set of four nucleotides is used here as a set of four abstract symbols.

1. *Morse-Thue* (Non overlapping words): CGGCGCCGGCCGCGGC...
2. *Fibonacci* (1-overlapping words): GCGGCGCGGCGGC...
3. *Fredholm* (1-overlapping words): CGGCGCCCGCCCCCCG...
4. *Kolakoski* (1- and 2-overlappping words): G.GCCG.CGGCGGCCGCCGGCGCCG...
5 *Paperfolding* (1- and 2-overlapping words): GGCGGCCGGGCCG...
6. *Periodic sequences* can also be generated by rewrite rules.

6a. Let (C, G, A, U) be the symbol set, and C--> CG; G-->AG; A --> AU; and U --> CG, be the set of rewrite rules. It generates a sequence consisting of the following motifs: (1) CGAGAUAG, (2) AUCGAUAG, (3) AUCGCGAG.

6b. Let (C, G, A, U) be the set of symbols and A --> AU; U --> GU; G --> GC; C-->AC, be the set of rewrite rules. It generates a sequence consisting of motifs (1) AUGUGCGU, (2)GCACGCGU, (3) GCACAUAC, (4) AUGUAUAC.

7. *Milnor-Thurston* sequence (1988): This is also known as the kneading sequence in symbolic dynamics and is of the form CGGGC.GC.GCCG....It can fold itself into a hair pin configuration.

8 *Telomeric sequences* contain segments of the type TTGGGG and TTAGGG. They can be generated by the following rewrite rules.

8a. Rewrite rules TT -->TTG, G --> GGG; GG --> TTG generates TTGGGG repetitively. However, rules are to be applied to 2-symbol word and 1-symbol word alternatively. The initial sequence is TTG. Such rewrite rule systems are not known in the theory of automatic sequences.

8b. Rewrite rules TT --> TTA; A --> GGG; GG --> TTA; and G -->GGG, generate TTAGGG repetitively in the same manner as in (a) with the initial sequence TTA.

Although rules are applied in parallel, like in a L-system, however, both (8a) and (8b) do not belong to an L-system.

3.3 FUNDAMENTAL MOTIFS IN RNA AND DNA

In biology, the importance of our fundamental motifs is derived from the following facts:

Highly conserved nature;
Found at or near important sites;
Play a direct role in most enzymatic processes;
Found in introns, viroids, transposons and other ancient systems;
Found in other non-coding regions such as telomers, repetitive;
DNA, ALU sequence etc.

Using these motifs as the building blocks, we can construct the following types of automatic sequences:

1. Automatic sequences of DOL (deterministic, context-free, parallel, rewrite) systems.
2. Automatic sequence of L-type (simply parallel rewrite).
3. Automatic sequences generated by only sequential application of rewrite rules.
4. Data sequences (may or may not any of the type (1) to (3).

3.4. EXAMPLES OF FUNDAMENTAL MOTIFS.

Although we collected a large amount of data, we give here only a brief list of such motifs found in species ranging from viroids to human -- from the simplest and most ancient to most complex of all and latest in evolution.

1. T-cell receptor gamma-gene junctional sequences,
GCCT.GC.TGGGA (Asarnow et al., 1993).
2. Dopamine D4 receptor in human and rat where D4 repeat-consensus sequence contains, GCCG, ACCA, ACCG etc (Makoff 1992).
3. The recognition sequence of restriction endonulease of E. Coli, EcoRI contains, GAAT.TC, where G is the splice site (Brock, 1991).
A completely conserved leader sequence ribozyme that inhibit HIV-1 strain (Mang et al., 1993) contains, T.GCCCG.TCT.GTTG.GT.
4. The HIV-1, *NC7* nucleocapsid protein has the viral splice site exactly in the motif UGGU. This protein plays a critical regulatory role in the dimeraztion/packaging process (Sakaguchi et al., 1993).
5. The human chromosome 16 contains the sequence,
TCCT.X.TCCT.CTTC.C.ACCCT.CAG.TGGA.ATGA. Here X is 2-9 nucleotides.

This is a mini-satellite type repeat present only in primates. It structure suggests that it was possibly acquired through a retrovirus (Stalling and Doggett, 1993).
6. A 16 nucleotides group I consensus sequence is present in all of the viroids and virosoids and contains motifs, UCCU, AGGA, GAAG, and CUUC (Diener, 1993).
7. An ALU sequence contains a non-overlapping motif CTTC and an overlapping motif CTTCTTG (Goldberg, 1993).

3.5 THE HYPOTHESIS

From the above discussion it is clear that there are two aspects of our model.

a. Evolutionary aspect: Automatic sequences were formed in the RNA world alongside information carrying nucleotide sequences. The existence of fundamental motifs in viroids, introns and transposons, and other noncoding segments and the highly conserved nature of such motifs confirms the above statement.

b. Correlation between coding and non-coding regions: A complete genome (otherwise it needs a host to replicate) contains informational as well as information processing segments. Typically, a genome consists of the "program-data" (e.g., intron-exon) segments, each for some specific function. The task of a program is to make sure that data is replicated faithfully and at a proper section of the genome. Nature has designed such a system of "product verification and quality control" so as minimize errors in reproduction of a species. Other salient points of our aproach are:

1. Intron-exon correlation suggests that an intron structure is exon (data)-dependent. This suggests that introns are instructions or programs acting as special purpose computational devices

2. The secondary structure of a sequence depends on the rewrite rule system that generates it (more precisely, an equivalent symbolic structure). Not all rewrite systems produce foldable sequences, only some do it.

3. Fundamental motifs are critical parts of either a program (intron), data (exon), or other non-coding sequences. Their mutation can cause partial or complete malfunction of the respective processes. This description is obviously valid only for what we call the first level, theRNA and pre-translation DNA worlds. The second level description is the post genetic code DNA world in which enzymes are involved in information processing.

4. EXPERIMENTAL VERIFICATION

Since information processing ribozymes are built out of fundamental motifs and that these motifs are their critical determinants, any change or damage to critical segments will lead to suppression or malfunction of its activity. To block unwanted mechanisms or correct damaged mechanisms, either ribozyme-directed or antisense therapies can be utilized. Endogenous regulatory mechanisms based on antisense RNA exist in eukaryotes. A wide range of natural processes use antisense strategies (Nellen and Lichtenstein, 1993). It is also possible to restore correct splicing of pre-mRNA using this strategy (Dominski and Kole, 1993). Such a wide range of antisense phenomenon can only be explained in computational terms as the program disabling and enabling process known to computer scientists. We suggest, therefore, that segments built out of these motifs can be used in both kinds of therapies. A few examples are given below . Their successful testing confirms our hypothesis.

Antisense oligodeoxynucleotides corresponding to *c-myc, c-myp*, and *cdc 2*, have all been shown to interfere with mitogenic activation of T-lymphocytes (Green et al., 1992). The gene sequences are:

c-myc GAAGTTC.AC.GTTG.AGGGGC.AT
c-fos GTTGAAACCCG.A.GAAC.ATCAT
c-able CTTCAAAC.AGATC.TCCAAC.AT
bcl-2 T.GTTC.TCCCGGCTTG.C.GCCA.T
c-myb GTGT.CGGGGT.C.TCCG.GCCAT

Their corresponding antisense sequences are:

(AS)-*c-myc* CAC.GTTG.AGGGGC.AT
(AS).*c-fos* ACCCG.A.GAAC.ATCAT

Fundamental motifs of the type XYYYZ, XYYZ and XYYX, where X,Y, and Z are C,G,A, and T, in the antisense oligonucleotides control gene function in T-lymphocytes. Similar data suggests (Zaia et al., 1992) that these motifs occur at splice sites of HIV-1 RNA, rbzA, rbzB, and rbzC and the antisense sequences constructed out of such motifs can be used for anti-HIV-1 gene therapy. Pagano et al., (1992) used sense,

5' - AUG.TCTGACG.AGGGGCCAGGT.AC.ACCACCTGGA,
and anti sense,
3' - AGACTGC.TCCCGGTCCA.TG.TCCTGGACCTTT.ACCG

oligomers for the treatment of EBV-infected Raji-Burkitt lymphoma cells. In this case, 1- and 2-overlapping sequences are used to construct sense and antisense segments.

5. CONCLUSION

We have investigated the computational nature of non-coding regions and ribozymatic processes. We discovered that some highly conserved fundamental motifs are their critical determinants. An information processing approach is proposed that can enhance our understanding of genetic processes in ribozyme-directed and antisense gene therapy.

ACKNOWLEDGMENTS

Thanks are due to Professors C. Ponnamperuma and J. Chela-Flores for their kind invitation, Professors R.H. Haynes and J. Chela-Flores for many interesting discussions, Professor Abdus Salam for hospitality at ICTP, and Professor A. Zichichi, Director, World Laboratory, for the grant of a fellowship.

REFERENCES

Altman, S., 1993: RNA enzyme-directed gene therapy, *Proc. Natl. Acad.Sci.* (USA), **90**, 10989-10900.

Asarnow, D.M., 1993: Selection is not Required to Produce Invariant T-cell Receptor Gamma-gene Junctional Sequences, *Nature*, **362**, 158-160.

Barinaga, M., 1993: Ribozymes: Killing the messenger, *Science*, **262,** 1512-1514.

Brock, T.D. and Madigan, M.T., 1991: *Biology of Microorganisms*, Prentice-Hall, NJ. USA.

Cech, T.T., 1987: The chemistry of self-splicing RNA and RNA enzymes, *Science*, **236**, 1532-1539.

Christol, G., Kamae, T., Mendez/France, M., and Rauzy, G.,1980: Suites algebriques, automates et substitutions, *Bull. Soc. Math. France*, **108**, 401-419.

Cobham, A., 1974: Uniform tag sequences, *Math. Systems Theory*, **6**, 164-192.

Conrad, M., 1985: On the design principles for a molecular computer, *Comm. ACM*, **28**, 464-480.

Dekking, F.M., 1982a: Recurrent sets, *Advances in Math.*, **44**, 78-104.

Dekking et al., 1982b: Folds!, *Math. Intelligencer*, **4,** 130-138, 173-181, and 190-195.

deLuca, A., 1981: A Combinatorial Property of the Fibonacci Words, *Information Processing Letters,* **12**, 193-195.

Deutsch, D., 1985: Quantum Theory, the Church-Turing Priciple and the universal computer, *Proc. R. Soc. Lond.*, **A400**, 97-117.

Diener, T.O., 1991: Subviral pathogens of plants: Viroids and Viroidlike Satellite RNAs, F*ASEB J.*, **5**, 2808-2813.

Dominski, Z., and Kole, R., 1993: Restoration of Correct Splicing in Thalassemic pre-RNA by Antisense Oligonucleotides, *Proc. Natl. Acad. Sci.(USA)*, **90**, 8673-8677.

Goldberg, Y.P., 1993: Identification of an Alu Retrotransposition Event in close Proximity to a Strong Candidate Gene for Huntington's Disease, *Nature*, **362**, 370-373

Gray, M.W., and Covello, 1993: P.S., RNA Editing in plant Mitochondria and Chloroplasts, *FASEB J.*, **7**, 64-71.

Green, D.R., Zheng, H., and Shi, Y., 1992: Antisense Oligonucleotides as Probes of T-lymphocytes Gene Function, in *Antisense Strategies*, eds: R. Barsega and D.T. Denhardt, *Ann of NY Acad. Sci.*, **660**, 193-203.

Jelinek, W.R., and Schmid, C.W., 1982: Repetitive Sequences in Eukaryotic DNA and their Expression, *Ann. Rev. Biochem*, **51**, 813-844.

Joyce, G.F., 1989: RNA Evolution and the Origins of Life, *Nature*, **338**, 217-224.

Lothaire, M., 1983: *Combinatorics on Words*, Addison-Wesley, Reading (MA), USA..

Makoff, A.J., 1992: Echoes of D4 receptor repeats, *Nature*, **360**, 424.

Mang, Yu, Ojwang, J., Yamada, O., Hampel, A., Rapapport, J., Looney, D., and Wong-Staal,., 1993: The Leader Sequence Ribozyme Inhibits Diverse HIV-1 Strains, *Proc. Natl. Acad. Sci. (USA)*, **90,** 6340-6344.

Milnor, J. and Thurston, W., 1988: *Lecture Notes in Math.* no **1365**, Springer-Verlag, Berlin, p.465.

Moyzis, R.K., 1991: The Human Telomer, *Scientific American*, August, 34-41.

Nellen, W., and Lichtenstein, C., 1993: What makes an mRNA anti-sense-itive?, *TIBS*, **18,** 419-423.

Pagano, J.P., Jimenez, G., Sung, N.S., Raab-Traub, N., and Lin, J-C., 1992: Epstein-Barr Viral Latency and Cell Immortalization as targets for Antisense Oligomers, in *Antisense Strategies*, eds: R. Barsega and D.T. Denhardt, *Ann of NY Acad. Sci.*, **660**, 107-116.

Sakaguchi, K., Zambrano, N., Baldwin, E.T., Shapiro, B.A., Erickson, J.W., Omichinski, J.G., Clore, M.G., Gronenborn, A.M., and Appella, E., 1993: Identification of a Binding Site for the Human Immunodeficiency Virus Type 1 Nucleocapsid Protein, *Proc. Natl. Acad. Sci. (USA)*, **90**, 5219-5223.

Salomaa, A. 1985: *Computation and Automata*, Cambridge University Press, Cambridge, U.K.

Stalling, R.L., and Doggett, N.A., 1993: *Los Alamos Science*, USA, p.211.

Zaia, J.A., Chatterjee, S., Wong, K.K., Elkins, D., Taylor, N.R., and Rossi, J., 1992: Status of Ribozymes and Antisense Based Developmental Approaches for Anti-HIV-1 Therapy, in *Antisense Strategies*, eds: R. Barsega and D.T. Denhardt, *Ann of NY Acad. Sci.*, **660**, 95-106.

PART V
CHIRALITY AND SELF-ORGANIZATION

THE WEAK FORCE

AND THE ORIGIN OF LIFE AND SELF-ORGANIZATION

Alexandra J. MacDermott

School of Biological and Molecular Sciences,
Oxford Brookes University, Oxford OX3 0BP, England
and
Physical Chemistry Laboratory, South Parks Road,
Oxford University, Oxford OX1 3QZ, England.

ABSTRACT

All biomolecules are of one hand - but what determines which hand? Why is life based on L-amino acids and D-sugars rather than D-amino acids and L-sugars? We believe the symmetry-breaker could be the weak force, which causes enantiomers to differ very slightly in energy. In this paper we present our calculations of this parity-violating energy difference (PVED) for a range of important biomolecules, and in nearly all cases it is indeed the "natural " enantiomer which is the more stable.

1. HOMOCHIRALITY: A NECESSARY PRE-CONDITION FOR LIFE?

It has been clear since Fischer (1894) that all biomolecules have to be of one hand: this "homochirality" is a hallmark of life, being essential for an efficient biochemistry, just like the universal adoption of right-hand screws in engineering. It has only recently become recognized, however, that homochirality is not just a consequence of life, but may also be a pre-condition for life. This is because life and self-organization seems to require polymers and polymerizations don't go in racemic solution. It is found that polymerization to give stereoregular biopolymers - in particular poly-D-ribonucleotides (Joyce et al., 1984) and poly-L-peptides (Blair et al., 1981) - proceeds efficiently only in almost homochiral monomer solutions: in racemic solution, addition of the "wrong" hand to the growing chain tends to terminate the polymerization.

A pre-biotic symmetry-breaker would thus appear necessary to get life started, unless the need for homochirality could be circumvented with a "prochiral" ancestral replicator such as the glycerol-based polymer (essentially RNA with C2 removed), proposed by Bada and Miller (1987) and Schwartz and Orgel (1985), which is based on achiral monomers and owes its chirality purely to its helical conformation. However the choice must still be made between left and right-hand helices. Some clays are totally achiral, so an ancestral clay replicator would completely circumvent the chiral start-up problem, although symmetry-breaking would still be necessary later in the transition to the present polymer-based biochemistry.

Initial selection of one hand in ancestral biomolecules fixes the handedness of the rest of biochemistry through diastereomeric interactions (Fischer, 1894); we therefore need only account for the handedness of ancestral biomolecules. The diastereomeric connection between the L-amino acids and the D-sugars (Wolfrom et al., 1949; and Melcher, 1974) probably precludes an L-amino acid/L-sugar or D-amino acid/D-sugar biochemistry. However D-amino acid/L-sugar "mirror life " should be just as viable as terrestrial L-amino acid/D-sugar life, so we may ask whether selection of the latter was a "frozen accident" (Miller and Orgel, 1974) or whether it was determinately selected by a chiral influence.

Pasteur (1894, 1922) believed in a universal chiral influence, and we now know that there is one. Elementary particles themselves have a handedness, felt only by the weak force, which is one of the four forces of nature:

Force	Carried by	Strength
Electromagnetic	photon	1
Weak	W^{+}, Z°	10^{-11}
Strong	mesons, gluons	100
Gravity	gravitons	10^{-38}

The electromagnetic and weak forces were unified in 1968 by Salam and Weinberg, and the strong force is being brought in with "grand unification" theories, although a superunification theory which encompasses gravity is proving more elusive. The four forces are quite clearly one, because they all operate by the same mechanism: the world is made of fermions, which interact by exchange of virtual bosons. For example, the electostatic repulsion between two electrons is mediated by exchange of virtual photons, while the weak interaction between an electron and a neutron is mediated by virtual Z° bosons.

The weak force is the only one of the four that is chiral: it can feel the difference between left and right. Fermions exist in two states of opposite helicity, corresponding to spin and momentum vectors parallel (right-handed) or anti-parallel (left-handed). The two helicity states participate equally in the electromagnetic, strong and gravitational interactions, but they do not participate equally in the weak interaction, which is therefore said to violate parity. Left-handed electrons participate in the weak interaction preferentially compared with right-handed electrons; similarly right-handed positrons are preferred over left-handed positrons in the weak interaction. So although there are normally equal numbers of left and right-handed electrons, the weak force can feel the left-hand ones better, and so electrons can be viewed as left-handed as far as the weak force is concerned; similarly positrons can be viewed as right-handed. The universe therefore has an intrinsic "left-handedness" because it is made of matter and not anti-matter, and we believe this all-pervading chiral influence could determine biomolecular chirality. It should be emphasized, however, that electrons are not "really" left-handed: there are normally equal numbers of left and right-handed electrons. But the weak force can feel the left-hand ones better, so it is convenient to regard electrons as left-handed as far as the weak force is concerned.

The handedness of elementary particles means that L and D molecules are really diastereoisomers, not enantiomers: the true enantiomer of an L-amino acid is the D-amino acid made of anti-matter. Left and right-handed molecules should therefore differ very slightly in many properties, e.g. NMR chemical shifts (Barra et al, 1987), and most importantly, in energy. This parity-violating energy difference between enantiomers - the "PVED" or ΔE_{pv} - arises from weak neutral current interactions, mediated by the Z° boson, between electrons and neutrons. These interactions impart a parity-violating energy shift (PVES), E_{pv}, to the energy of a chiral molecule, and an equal and opposite shift, $-E_{pv}$, to that of its enantiomer, giving a parity-violating energy difference (PVED) of $E_{pv} = 2E_{pv}$. The PVED produces a very slight excess of the more stable enantiomer, which can be amplified to produce the observed homochirality.

2. CALCULATION OF THE PVED

To calculate the PVED (Mason and Tranter, 1985; and MacDermott and Tranter, 1990) we start from the parity-violating Hamiltonian density for the weak neutral current interaction between electrons and nuclei. This is obtained from the Feynman diagram

by associating a current J^μ with each fermion involved, a coupling constant g_N with each vertex, and a propagator (here $1/M^2_{Z^\circ}$) with each virtual boson exchanged, giving

$$\mathcal{H}_{pv} = J_\mu(e)\; g_N \; \frac{1}{M^2_{Z^\circ}} \; g_N \; J^\mu(n)$$

The weak coupling constant is not small ($g_N \approx e$), but the mass M_{Z° of the Z° boson is large, which is why the weak force is weak.

The neutron current $J^{\mu}(n)$ is broken down into currents of the constituent quarks, and the current for each fermion - electron or quark - is broken down into sums of left and right-handed fermion currents. The handedness of the weak force enters through the unequalness of the coefficients of left and right-handed fermion currents, e.g. for electrons $c_L = -1/2 + \sin^2\Theta_W$, $c_R = \sin^2\Theta_W$. Using the theoretical value of 30° for the Weinberg angle Θ_W the Hamiltonian becomes

$$\mathcal{H}_{pv} = -(G_F/2\sqrt{2})\ J_{\mu}(e)\,J^{\mu}(n)$$

where $J_{\mu}(e) = \bar{e}\gamma_{\mu}\gamma_5 e$, $J^{\mu}(n) = \bar{n}\gamma^{\mu}n$. On reduction to non-relativistic quantum mechanics by explicitly multiplying out the γ-matrices, the Hamiltonian density $\mathcal{H}_{pv}$ becomes the parity-violating Hamiltonian

$$\hat{H}_{pv} = -(\Gamma/2) \sum_a \sum_i N_a \left\{\vec{s}_i.\vec{p}_i\ ,\ \delta^3(\vec{r}_i - \vec{r}_a)\right\}_+$$

The sums over i and a are over all electrons and nuclei respectively in the molecule, and N_a is the neutron number of nucleus a. This beautifully elegant expression summarizes the physical origin of the PVES. The term in s.p represents the projection of the spin onto the direction of momentum, thus touching directly on the left-handedness of the electron. $\hat{H}_{pv}$ is of opposite sign for enantiomers: p changes sign under parity, being a polar vector, while s remains the same, being an angular momentum and therefore an axial vector. The delta function expresses the contact nature of the weak force. The smallness of the PVES arises from the smallness of the constant Γ, which contains the very small weak coupling constant G_F.

One evaluates the PVES E_{pv} by taking matrix elements of $\hat{H}_{pv}$ over the ground state wave functions. In the absence of spin-orbit coupling one can assume separability of the wavefunction into spin and orbital parts, which results in E_{pv} being identically zero because $\hat{p}$ is imaginary. If the ground state wavefunction is corrected for the effect of spin-orbit coupling, the PVES no longer vanishes, and we obtain

$$\Delta E_{pv} = 2 \sum_j^{o} \sum_k^{u} P_{jk}\ (\varepsilon_j - \varepsilon_k)^{-1}$$

where $$P_{jk} = \mathrm{Re}\ \langle\psi_j|\hat{V}_{pv}|\psi_k\rangle.\langle\psi_k|\hat{V}_{so}|\psi_j\rangle$$

is the "parity-violating strength". The sums are over all occupied MOs j and all unoccupied MOs k, ψ_j and ψ_k are the energies of the MOs, $\hat{V}_{pv}$ is a one-electron version of the parity-violating Hamiltonian, and $\hat{V}_{so}$ is a spin-orbit coupling operator.

The parity-violating strength P_{jk} is closely analogous to the rotational strength

$$R_{OA} = \text{Im} \langle \psi_o | \vec{\mu} | \psi_A \rangle . \langle \psi_A | \vec{m} | \psi_o \rangle$$

in optical activity. $\hat{V}_{pv}$ and $\vec{\mu}$ are both parity-odd (being polar vectors), while $\hat{V}_{so}$ and $\vec{m}$ are both parity-even (being axial vectors), with the result that both P_{jk} and R_{OA} are oppositely signed for enantiomers.

The PVED can be calculated by ab initio methods (Mason and Tranter, 1985; MacDermott and Tranter 1989 and 1990), using the LCAO approximation

$$|\psi_j\rangle = \sum_c \sum_\gamma C^j_{c\gamma} |\phi_{c\gamma}\rangle$$

to express the molecular orbitals $|\psi_j\rangle$ as linear combinations of atomic orbitals $|\phi_{c\gamma}\rangle$ of type γ on nucleus c. To evaluate the PVED, we need the MO coefficients $C^j_{c\gamma}$ and the MO energies ε_j (the matrix elements of $\hat{V}_{pv}$ and $\hat{V}_{so}$ over the atomic orbitals are pre-evaluated for the required elements using a STO-6-31G basis set). The computation is therefore divided into two stages. First we obtain the MO coefficients and energies from GAUSSIAN 92. Second, we use our own program PVED 84 to combine the MO coefficients and matrix elements to give ΔE_{pv}.

3. AMPLIFICATION OF THE PVED

The calculated PVEDs are of the order of 10^{-20} - 10^{-17} hartree, about 10^{-17} - 10^{-14} kT at room temperature, for typical biomolecules, giving, from the Boltzmann distribution, an enantiomeric excess $|L-D|/L+D = \Delta E_{pv}/kT$ of 10^{-17} - 10^{-14}. These small excesses need amplifying to produce the observed homochirality, and the possible mechanisms fall into three classes: (a) the <u>Yamagata</u> cumulative mechanism, applicable to crystallizations or polymerizations of optically <u>labile</u> or achiral monomers, and soundly based in equilibrium thermodynamics; and (b) the <u>Kondepudi</u> catastrophic mechanism, applicable to optically <u>non-labile</u> molecules, and rather more speculatively based on non-equilibruim statistical mechanics; and (c) the <u>Salam</u> mechanism, even more speculatively based on quantum mechanical tunnelling.

An example of the Yamagata cumulative mechanism (Yamagata, 1966) is the crystallization of quartz, which consists of helical crystals made of achiral silica units. During crystal growth an achiral unit A may add on to a growing crystal of either hand:

$$L_{n-1} + A \rightleftharpoons L_n \qquad \Delta G_L \neq \Delta G_D$$
$$D_{n-1} + A \rightleftharpoons D_n$$

But owing to the PVED the corresponding free energy changes are unequal, which means that A will add preferentially to one hand of crystal rather than the other, resulting in a fractional excess

$$f_L = \frac{L_n/L_{n-1}}{D_n/D_{n-1}} = e^{-(\Delta G_L - \Delta G_D)/RT}$$

which is operative at each of N stages of crystallization and leads cumulatively to an excess of one hand, given by

$$\frac{L_n}{D_N} = \frac{L_1}{D_1} f_L^{N-1} = (1 + (\Delta G_L - \Delta G_D)/RT)^N$$

$$\approx 1 + N(\Delta G_L - \Delta G_D)/RT$$

This is equivalent to the PVED of the crystal being N times the PVED of the individual units within the crystal:

$$\Delta E_{pv} \text{ (N-unit crystal)} = N\Delta E_{pv} \text{ (one unit)}$$

The Kondepudi catastrophic mechanism (Kondepudi, 1987) is based on a kinetic scheme involving autocatalysis and enantiomeric antagonism, i.e. the presence of one enantiomer encourages production of itself but inhibits production of its mirror image. Many polymerization reactions essential to life have precisely these characteristics, and some enantiomeric enrichment can be demonstrated in the laboratory (Brack and Spach, 1980). Kondepudi envisages an open-flow reactor system, such as a lake, fed by an input of achiral substrates A and B, which react both directly (k_1) and autocatalytically (k_2) to form chiral products X_L and X_D:

$$A + B \underset{k_{-1}}{\overset{k_1}{\rightleftharpoons}} X_{L(D)}$$

$$X_{L(D)} + A + B \underset{k_{-2}}{\overset{k_2}{\rightleftharpoons}} 2X_{L(D)} \quad \text{autocatalysis}$$

$$X_L + X_D \xrightarrow{k_3} P \quad \text{enantiomeric antagonism}$$

The reaction scheme can accommodate unequal reaction rates for enantiomers, and can be extended to include racemization, thermal fluctuations, etc.

Using the methods of non-equilibrium statistical thermodynamics, Kondepudi showed that for $\Delta E_{pv} > 10^{-17}$ kT, amplification to homochirality will eventually occur, but for smaller ΔE_{pv} the amplification effect will be overcome by thermal fluctuations. The amplification time for a PVED of 10^{-17} kT is 10^4 years, a very short time on an evolutionary timescale. This figure is based on a small lake of 4×10^9 dm^3 (1km x 1km x 4m), with modest concentrations of order 10^{-3} M and a realistic reaction rate of 10^{-10} mol dm^{-3} s^{-1}. The amplification can withstand a racemization half-life as low as 10^2 - 10^3 years (cf. typical values of 10^5 - 10^6 years for most amino acids).

The amplification time is very sensitive to the magnitude of the PVED and the lake size. If E_{pv} increases by one order of magnitude, the amplification time decreases by four orders of magnitude, so that whereas a PVED of 10^{-17} kT takes 10^4 years, a PVED of 10^{-16} kT takes only 1 year. Alternatively, if ΔE_{pv} increases by one order of magnitude, one can decrease the lake volume by two orders of magnitude for the same amplification time. So whereas the smaller PVEDs of 10^{-17} kT take 10^4 years in 4×10^9 dm^3, the larger PVEDs of 10^{-14} take 1 year in only 4×10^5 dm^3.

The Salam mechanism (Salam 1991) is based on the idea that below a certain critical temperature tunnelling to the most stable form should occur in a cooperative phase transition effect. Thus, if a racemic mixture were cooled to very low temperatures, it should become homochiral and optical rotation should be detected. Figureau has attempted to verify this experimentally (see this volume) by cooling racemic cystine crystals, but so far no optical rotation has been detected. Many chemists might be sceptical because the barrier to L-D inversion is so high that tunnelling times might be of the order of thousands or millions of years. However, there is plenty of time available on an evolutionary timescale, and the advantage of the Salam mechanism is that unlike the Kondepudi mechanism it works at low temperatures, and so could be effective for an extraterrestrial origin of life.

4. THE PVEDS OF IMPORTANT BIOMOLECULES

Since only the PVED of ancestral biomolecules is relevant (the handedness of modern biomolecules is fixed not by their own PVED but by diastereomeric connection with their ancestors), we have to consider what came first. Scenarios for the origin of life fall into three main categories: (1) nucleic acids first, (2) proteins first, (3) clays first. We have calculated the PVEDs of molecules from all three classes of possible ancestral replicators.

The natural L-amino acids L-alanine, L-valine, L-serine and L-aspartate were all found to be more stable than their D-enantiomers, in both the solution and α-helix and β-sheet conformations, by 10^{-17} kT (Mason and Tranter, 1984 and 1985), just enough for Kondepudi amplification. The aldose sugar series $(H_2CO)_n$ was investigated for n=3 to 5, but the results were not so clear-cut as for the amino acids. For glyceraldehyde (n=3), the parent of the higher sugars, the D form is PVED-stabilized by about 10^{-17} kT (Tranter, 1986); D-deoxyribose is also PVED-stabilized in the C2-endo form found in DNA; but D-ribose in the C3-endo form found in RNA is <u>less</u> stable than its enantiomer (Tranter and MacDermott, 1992). These latter results are explained by the PVED of the basic furanose skeleton, tetrahydrofuran (THF): the C2-endo ring conformation of THF is more stable than the enantiomeric C3-endo conformation by about 10^{-17} kT (Tranter and MacDermott, 1986). So it would appear that D-ribose cannot have been selected by its own PVED (since L-ribose is more stable). But the D-sugar series may have been selected either by the PVED of its parent D-glyceraldehyde, or alternatively by the diastereomeric connection with the PVED-stabilized L-amino acids. Ribose is in fact the only molecule so far studied in which the "natural" enantiomer is the less stable, and this may reflect the fact that ribose, being difficult to synthesize, is probably not pre-biotic anyway but a later biological evolution after the first self-replicating systems appeared, in which case its own PVED is irrelevant and one should look to more primitive replicators. Furthermore, we had examined β-D-ribose, the form found in RNA. But more recently we learnt from Benoit Prieur (private communication 1993) that the pre-cursor of the β-D-nucleotides is not β-D-ribose but α-D-ribose, which undergoes an S_N2 reaction with a Walden inversion at the C1 chiral centre. Moreover, the α and β forms of ribose have opposite optical rotation, reflecting their opposite configuration at C1. Would they also have opposite PVEDs? Our new calculations show that they do indeed : whereas for β-ribose the L form is more stable, we found that for α-ribose the D form is more stable. This is extremely satisfactory, showing that β-D-ribose could indeed be PVED-selected, but by the PVED of its α-D-ribose precursor rather than by its own PVED. This result also opens the way to further studies to see if the signs of the PVED and the optical rotation show more widespread correlations.

Thus the most important biomolecules - amino acids and sugars - have PVEDs of 10^{-17} kT, which is only just amplifiable by the Kondepudi mechanism. If the PVED were a little larger, Kondepudi amplification would be very much easier. The PVED is proportional to the sixth power of the atomic number (largely due to the importance of spin-orbit coupling), so attention was turned next to molecules incorporating second-row heavy atoms such as P, Si, S, in the hope of finding larger PVEDs (MacDermott and Tranter, 1989b).

We studied fragments of the sugar-phosphate backbone of DNA without the bases (which are anyway planar and in themselves achiral), and the right-hand B double helix showed a PVED-stabilization of about 10^{-17} kT per sugar-phosphate unit. Right-hand A RNA was not PVED-stabilized, however: the negative contribution of the phosphate group in the right hand helix was cancelled by the positive contribution of the D-ribose moiety. For the more primitive glycerol-based polymer (in which the momomer units are achiral and the chirality comes only from the helical conformation) the right-hand helix was again found to be PVED-stabilized, in both B and A conformations, by 10^{-17} kT per sugar-phosphate unit. It had been hoped that larger PVEDs would be obtained with the heavy phosphorus atom. The fact that the phosphate PVED was still only 10^{-17} kT was traced (MacDermott and Tranter, 1989b) to the phosphorus atom being very electropositive in phosphates due to the electron-withdrawing effect of the four oxygens, resulting in too little electron density on the phosphorus to feel the potentially larger parity-violating effect.

Clearly more electronegative heavy elements such as sulphur would be better candidates for a large PVED, and more recently (MacDermott et al, 1992) we have studied DNA fragments where the normal $-O-PO_2^--O-$ linkages are modified by thiosubstitution. Thiosubstitution of the side oxygens to give $-O-POS^--O-$ and $-O-PS_2^--O-$ produced for the first time PVEDs of 10^{-16} kT, while thiosubstitution in the helix itself to give $-S-S-CH_2-$ produced the enormous PVED of 10^{-14} kT, with the right-hand helix the more stable in all cases. This is our most exciting result yet: we saw that a PVED of 10^{-17} kT is only just amplifiable by the Kondepudi mechanism, taking 10^4 years in 4×10^9 dm^3 lake, while our new PVEDs of 10^{-14} kT are amplifiable in 1 year in a small 4×10^5 dm^3 pond.

Thiosubstituted links may have large PVEDs, but can they actually form a viable double helix, and are they relevant to the origin of life? They certainly do form a base-paired double helix, as shown by their recent use as "antisense inhibitors" (Cohen, 1989). If an oncogene, for example, results in an undesirable gene product, one can use a piece of DNA of complementary base sequence ("antisense") to bind to the corresponding m-RNA transcript to form a DNA-RNA hybrid double helix; this prevents that piece of m-RNA from being translated into undesirable proteins. But if one tries to introduce a normal antisense oligonucleotide into a cell, nucleases break it down, so it is necessary to alter the DNA backbone to make it nuclease-resistant. A thiosubstituted backbone is found to be ideal because it forms a double helix with little change of geometry. Thiosubstituted DNA analogues could also be relevant to the origin of life because one of the problems is that the Miller-Urey experiment assumed a reducing atmosphere, while the primitive Earth probably had a neutral atmosphere; but deep-sea volcanic vents do provide a reducing environment, and the sulphurous gases present could make a thiosubstituted replicator highly plausible. Moreover, incorporation of phosphate is also a problem in the origin of life, making a purely sulphur-linked ancestral replicator even more likely.

We had previously neglected the bases when considering DNA and its analogues because of CPU time limitations, and we had assumed they would contribute little to the PVED because they are planar and therefore achiral in themselves. Direct SCF methods have now made it possible to confirm this, and preliminary results suggest that the bases do have a small positive PVED, but it is not large enough to affect the conclusion that right-hand helical DNA is more stable.

Another sulphur-based system which we have very recently examined is the amino acid L-cysteine, $CH_2(SH)C^*H(NH_2)COOH$, and its dimer L-cystine, which have been used in Figureau's experiments to verify the Salam mechanism. Figureau chose a sulphur-containing molecule because it would be expected to have a large PVED, especially in the disulphide-linked cystine dimer. Our calculations show that it does indeed have a large PVED, but surprisingly, it is the D form that is the more stable! In retrospect, this might have been anticipated from the earlier results for serine, the oxygen containing analogue, $CH_2(OH)C^*H(NH_2)COOH$. Here the asymmetric carbon centre has a negative PVED but the beta-hydroxyl gives a small positive contribution, although this is not sufficient to affect the overall PVED-stabilization of L-serine. In cysteine however, the beta-sulphur gives a much larger positive contribution, making L-cysteine less stable overall than D-cysteine. In the dimer cystine, a huge positive PVED of 0.6×10^{-14} kT is obtained, due mainly to the right-hand helical arrangement of the $-CH_2-S-S-CH_2-$ link in hexagonal L-cystine crystals. This is surprising, because a right-hand helix has normally been found to give a negative PVED; however the sign may depend on the radius and pitch of the helix in a manner which needs further investigation. None of this affects the validity of Figureau's experiments: it is simply that if the tunnelling effect does occur at low temperature it will be to the D form and not the L, and so any detected optical rotation will be of the opposite sign.

Turning now from polymers to crystals, L-quartz consists of right-hand 3-fold helices of silica tetrahedra. As a chiral mineral made of achiral units it can undergo Yamagata amplification, and indeed a 1.4% excess of l-quartz has been reported in a large collection of crystals from all over the world (Palache et al, 1962). Our calculations (MacDermott et al, 1992) show that l-quartz is PVED-stabilized by 10^{-17} kT per SiO_2 unit. This was again disappointingly small for the heavy Si atom: but silicates have the same problem as phosphates in that the electron-withdrawing effect of the oxygens leaves the Si atoms very electropositive. However the PVED does not need to be larger than 10^{-17} kT to account for the 1.4% excess of l-quartz: according to the Yamagata mechanism,

$$\Delta E_{pv}(\text{crystal}) = N\Delta E_{pv}(\text{unit})$$

so $\triangle E_{pv}$(crystal) = 10^{-2} kT (corresponding to a 1% enantiomeric excess) can be obtained from
$\triangle E_{pv}$(unit) = 10^{-17} kT if N = 10^{15}, which corresponds to a realistic small crystal of side 0.1 mm (Tranter, 1985). These results for quartz thus predict almost exactly the observed 1% excess of l-quartz, which can now for the first time be regarded as evidence for the global symmetry-breaking effects of the weak force: whereas biomolecular chirality <u>could</u> be the result of a chance effect in a single ancestor, chiral bias in unconnected mineral deposits <u>must</u> be the result of a global influence. Even if quartz-like clay minerals were not ancestral replicators, they could have acted as a chiral surface for pre-biotic catalysis. L-quartz adsorbs L-amino acids preferentially from a racemic mixture, with a 1% enantioselectivity (Kavasmaneck and Bonner, 1977); combining this asymmetry of 10^{-2} in adsorption with the 10^{-2} asymmetry in the quartz crystals themselves gives an overall electroweak enantioselectivity of 10^{-4}, which is much greater than that from the PVED of individual molecules, and could be amplified to homochirality much more easily. Realistic chiral mineral catalysts would not be quartz itself, but aluminosilicate quartz analogues, possibly containing heavy metal cations which might produce larger PVEDs.

The example of quartz shows that symmetry-breaking is altogether easier if a chiral surface is involved in transferring the weak chiral bias to biology. Another example of a chiral surface could be afforded by Buckminsterfullerenes, or "buckyballs": these are closed balls of graphite-like sheets with some hexagons substituted by pentagons to provide the necessary positive curvature. The first buckyball was the highly symmetrical C_{60} (Kroto, 1991), but many are chiral, e.g. C_{28}, the smallest chiral buckyball, and C_{76} (Ettl et al, 1991), which contains helical ribbons of six hexagons and has recently been resolved. If heptagons are used instead of pentagons, negative curvature is introduced and we obtain a "buckygym" (Vanderbilt and Tersoff, 1992) - a diamond-type lattice of chiral tetrahedral units which have chiral propellor-like structures on each face of the tetrahedron. Buckyballs would be expected to show a large PVED because they are giant chiral aromatic chromophores (cf. the huge optical rotation of hexahelicene). It is believed (Kroto, 1991) that fullerenes could form in large quantities in outer space in the outflow from carbon stars. There they could act as surface catalysts for prebiotic reactions. The expected large PVED would lead to a substantial excess of one hand of fullerenes and hence enantioselectivity in the reactions their surfaces catalyzed. The effect is likely to be best for buckygyms, which form a huge chiral surface with molecules trapped in the cavities, etc..

At the last conference in this series we reported some initial trial PVED calculations using approximate coordinates supplied by Manolopoulos of Nottingham University. The results were disappointingly small: 10^{-19} kT for C_{28} and 4 x 10^{-17} kT for C_{76}. A possible reason is that Buckyballs are not all that aromatic afterall: there has been some controversy about their aromaticity because although their NMR chemical shifts indicate aromaticity, they do in fact undergo addition reactions quite easily. Our PVED results showed a lot of cancellation of contributions from different atoms, so we thought the overall result might be extra senstiive to the exact coordinates of the atoms. We therefore did a proper optimization calculation, using the STO-6-31G basis set in GAUSSIAN92, and with the new optimized coordinates we obtained a PVED of 10^{-17} kT for C_{28}, which is quite impressive for a small ball. One might therefore hope that properly optimized coordinates would produce much larger PVEDs for bigger buckyballs such as C_{76}, because the reduced curvature would improve the pi-overlap and thus increase aromaticity.

In conclusion, the weak force appears to predict the correct hand whatever came first. For nucleic acids first, D-glyceraldehyde and α-D-ribose are more stable, so that the D-sugars would be selected, and the right-hand helical backbone of DNA is also PVED-stabilized. For proteins first, the L-amino acids are more stable in most cases studied. For clays first we find that the weak force correctly predicts the reported 1% excess of l-quartz (which preferentially adsorbs L-amino acids) - but it remains to be seen whether this natural bias towards the PVED-stabilized form extends to the aluminosilicates which represent more realistic candidates for Cairns-Smith's clay replicators. So, if homochirality is indeed a pre-condition for life, we could well owe our existence to the Z^{o} boson!

References

Bada, J.L., and Miller, S.L., 1987: BioSystems 20 21.

Barra, A.L., Robert, J.B., and Wiesenfeld, L., 1987: BioSystems 20 57.

Blair, N.E., Dirbas, F.M., and Bonner, W.A., 1981: Tetrahedron 37 27.

Brack, A., and Spach, G., 1980: J. Mol. Evol. 15 231.

Cohen, J.S. (ed), 1989: "Oligodeoxynucleotides: antisense inhibitors of gene expression", Topics in Mol. and Struct. Bio. 12.

Ettl, R., Chao, I., Diederich, F., and Whetten, R.L., 1991: Nature 353 149-153.

Fischer, E., 1894: Chem. Ber. 27 2985 and 3189.

Joyce, G.F., Visser, G.M., van Boeckel, C.A.A., van Boom, J.H., Orgel, L.E., and van Westresen, J., 1984: Nature 310 602.

Kavasmaneck, P.R., and Bonner, W.A., 1977: J. Amer. Chem. Soc. 99 44.

Kondepudi, D.K., 1987: BioSystems 20 75.

Kroto, H.W., 1991: Chem. Rev. 91 1213-1235.

MacDermott, A.J., and Tranter, G.E., 1989a: Croatica Chemica Acta 62 165.
MacDermott, A.J., and Tranter, G.E., 1989b: Chem. Phys. Lett. 163 1-4.
MacDermott, A.J., and Tranter, G.E., 1990: Symmetries in Science IV (B.Gruber and J.H.Yopp, eds), Plenum, NY, p.67.
MacDermott, A.J., Tranter, G.E., and Trainor, S.J., 1992: Chem. Phys. Lett. 194 152-156.
Mason, S.F., and Tranter, G.E., 1984: Mol. Phys. 53 1091.
Mason, S.F., and Tranter, G.E., 1985: Proc. R. Soc. Lond. A 397 45.
Melcher, G., 1974: J. Mol. Evol. 3 121.
Miller, S.L., and Orgel, L.E., 1974: "The Origins of Life on the Earth", Prentice Hall, p.171.
Palache, C., Erman, G.B., and Frondel, C., 1962: Dana's System of Mineralogy 7th ed. Vol III, Wiley, New York, p.16.
Pasteur, L., 1894: Rev. Scientifique 7 1.
Pasteur, L., 1922: Oeuvres de Pasteur I (Pasteur Vallery-Radot, ed), Masson et Cie, Paris, p.369.
Salam, A., 1991: J. Mol. Evol. 33 105.
Schwartz, A.W., and Orgel, L.E., 1985: Science 228 585.
Tranter, G.E., 1985: Nature 318 172.
Tranter, G.E., 1986: J.Chem. Soc. Chem. Commun. p.60.
Tranter, G.E., and MacDermott, A.J., 1986: Chem. Phys. Lett. 130 120.
Tranter, G.E., MacDermott, A.J., Overill, R.E., and Speers, P.J., 1992: Proc. R. Soc. Lond. A 436 603-615.
Vanderbilt, D., and Tersoff, J., 1992: Phys. Rev. Lett. 68 511-513.
Wolfrom, J.L., Lumieux, R.U., and Olin, S.M., 1949: J. Am. Chem. Soc. 71 2870.
Yamagata, Y., 1966: J. Theor. Biol. 11 495.

TUNNELING, CHIRALITY AND COLD PREBIOTIC EVOLUTION

Vitalii I Goldanskii
N.N.Semenov Institute of Chemical Physics
Russian Academy of Sciences
Ul.Kosygina 4, 117334 Moscow, Russia

ABSTRACT

The extra-terrestrial scenario of the origin of life suggested by Svante Arrhenius (1908) as the 'panspermia' hypothesis was revived by the discovery of a low-temperature quantum limit of a chemical reaction rate caused by the molecular tunneling (Goldanskii et al., 1973). Entropy factors play no role near absolute zero, and slow molecular tunneling can lead to the exothermic formation of quite complex molecules.

Interstellar grains or particles of cometary tails could serve as possible cold seeds of life, with acetic acid, urea and products of their polycondensation as quasi-equilibrium intermediates. Very cold solid environment hinders racemization and stabilizes optical activity under conditions typical for outer space.

Neither 'advantage factors' can secure the evolutionary formation of chiral purity of initial prebiotic monomeric medium - even being temporarily achieved it cannot be maintained at subsequent stages of prebiotic evolution because of counteraction of 'enantioselective pressure'.

Only bifurcational mechanism of the formation of prebiotic homochiral - monomeric and afterwards polymeric - medium and its subsequent transformation in 'homochiral chemical automata' ('biological big bang' - passage from 'stochastic' to 'algorithmic' chemistry) is possible and can be realized.

Extra-terrestrial (cold, solid phase) scenarios of the origin of life seem to be more promising from that point of view than terrestrial (warm) scenarios. Within a scheme of five main stages of prebiological evolution some problems important for further investigation are briefly discussed.

1. INTRODUCTION

Life can be defined as the form of existence of complex polymeric systems (proteins, polynucleotides etc) able to self-replicate under the conditions of permanent exchange of energy and substance with the surrounding medium (and in accordance with the universal genetic code). Age of life on the earth is restricted at one end by the age of our planet - ca. 4.5 Bln. years - and even more rigidly by the age of the earth's solid crust, ca. 4 Bln. years. Limitations from the other end is determined by the age of cell-like fossils (Schopf et al., 1983; Schidlowski, 1988) and the genetic code (Eigen et al., 1989) as ca. 3.8 Bln. years. Thus, the duration of chemical (prebiological) evolution on the earth could not exceed 200 million years and was possibly even much shorter.

Two main classes of scenarios of the origin of life on the earth have been suggested.

The first of these, based on the pioneering ideas of Oparin (1924) and Haldane (1929) and classical experiments of Urey (1952), Miller (1974) and their followers is the so-called terrestrial or warm scenario. This 'standard model' of the origin of life deals with chemical conversion in primitive atmosphere and/or 'primordial soup' initiated by various physical agents, e.g. light, ionizing radiation, electric discharges, shock waves etc and resulted in the formation of basic building blocks of biopolymers - amino acids, carbohydrates, purines, pyrimidines etc. The hypothesis of the extra-terrestrial origin of life dates back to the 'panspermia' hypothesis of Arrhenius (1908). Its modification by Crick (1981) as a 'directed panspermia' variant is based on idea of transportation of the genetic material from a certain source populated by a supposed supercivilization.

Discovery of the quantum low-temperature limit of a chemical reaction rate and its explanation as the manifestation of molecular quantum mechanical tunneling (Goldanskii et al., 1973; Goldanskii, 1979) lead us to the idea of a 'cold prehistory of life' which also belongs to the class of extra-terrestrial scenarios.

However, neither the terrestrial nor extraterrestrial scenarios of the origin of life, in the early stages of development, included any attempts to combine the explanation for the existence of two main properties of living species which are unique from the standpoint of physics - namely, the functional property of ability for self-replication and structural property of their homochirality; that is, the chiral purity of amino acids in all proteins (L-enantiomers) and of sugars - ribose and deoxyribose - in RNA and DNA (D-enantiomers).

The absence of such attempts was particularly surprising in view of the firm conviction already expressed in 1860 by the discoverer of dissymmetry, Pasteur - 'Homochirality is the demarcation line between living and non-living matter'.

Now, it seems obvious that just the coexistence of these two (and only these two) above properties may serve as Ariadne's thread in the labyrinth of hypotheses of the origin of life, and predetermines the path of prebiological evolution (Goldanskii and Kuz'min, 1989, 1991; Avetisov et al., 1991; Bonner, 1991 and references therein). Detailed combined analysis of two properties of living matter lead us to the following three conclusions:

(1) Chiral purity, typical for the stage of prebiotic evolution was a necessary condition for the subsequent development of self-replication.

(2) Chiral purity of the bio-organic world was achieved neither by the continuous (gradual, evolutionary) accumulation of fluctuations nor as a result of the systematic production of enantiomeric excess by some global external 'advantage factor' (AF) but by a bifurcational type breaking of the mirror symmetry in nature.

(3) The sign of chiral purity in the earth's bio-organic world (L-amino acids and D-sugars) is random rather than predetermined by some global AF (e.g. by the non-conservation of parity in weak interactions - particularly weak neutral currents, (WNC).

This paper will not give a detailed description of the arguments which lead us to the above conclusions, nor do we treat here the fundamentals of stereochemistry of optical antipodes (enantiomers), nor are we going to expound the origin and physical sense (and particularly any mathematical formula) of such widely known phenomenon of quantum mechanics as tunneling of particles through potential barriers. The article is devoted mainly to the connections between the peculiarities of chemical reactions at low temperatures (in particular, in space conditions) and the origin of life on earth. It touches also briefly some topics which seem to be among the most interesting subjects for further investigations.

2. TUNNELING PHENOMENA IN CHEMICAL CONVERSIONS

One of the most important consequences of the wave properties of matter is tunneling - the ability of particles to penetrate the potential barriers whose height exceeds the particles' kinetic energy: classically, the sub-barrier region is forbidden for such particles.

Tunneling starts to become important when the so-called de Broglie wavelength of a particle becomes comparable with the barrier width in such a way that the probability of tunneling decreases with an increase of the width, the height of the potential barrier, and the mass of the tunneling particle.

The tunneling concept had its first successful applications in nuclear physics (alpha-decay, spontaneous fission, thermonuclear reactions) and later greatly contributed to solid-state physics, electronics and even to cosmology.

Since the late 1920s and early 1930s numerous articles devoted to the role of tunneling in kinetics of chemical reactions have appeared (Bell, 1973, 1980). However, for several decades the experimental search for chemical tunneling was restricted to the reactions in solutions, i.e. at comparatively high temperatures. Whereas, the role of tunneling is particularly significant for cryochemical reactions - below the so-called 'crossover temperature' T_c (Goldanskii, 1959, 1959a) which corresponds to equal contributions of classical 'over-barrier' transitions described by the Arrhenius law and quantum tunneling, 'under-barrier' transitions. For $T < T_c$ tunneling starts to dominate the over-barrier transitions, and the rate of exothermic reactions in many typical cases gradually reaches its low-temperature plateau. Thus, one could expect the existence of significant temperature-independent chemical reactivity of various substances even in the vicinity of absolute zero.

The low-temperature limit of the rate of chemical reactions in the full sense (which implies the rearrangement of atoms and/or molecules, changes of nature, lengths and angles of the valence bonds) was predicted (Bell, 1973, 1980; Goldanskii, 1959, 1959a) and then discovered in 1973 (Goldanskii et al., 1973) in the case of radiation-induced polymerization of formaldehyde and explained as molecular tunneling (Goldanskii et al., 1973; Goldanskii, 1979).

At present, scores of examples of low-temperature kinetical plateaus in processes of various classes are known: intramolecular and intermolecular transfer of hydrogen, transfer of heavy particles or groups in monomolecular (e.g. isomerization) and bimolecular reactions (e.g. numerous chain reactions), rotational tunneling and quantum diffusion. The theory of tunneling in chemical conversion is also well developed. However, these problems lay outside the framework of this article, and it would be better to address the readers to various reviews (Goldanskii, 1978, 1979; Goldanskii et al., 1987, 1989; Benderskii et al., 1992,1993) and monograph (Goldanskii et al., 1989a).

3. COLD PREHISTORY OF LIFE: INTERSTELLAR POLYMERIZATION

The last paragraph of the article (Goldanskii et al., 1973) on the discovery of a quantum low-temperature limit of a chemical reaction rate reads: 'Near absolute zero, entropy factors play no role, and all equilibria are displaced to the exothermic side, even for the formation of highly ordered systems. Therefore, it would be of interest to establish the role of slow chemical reactions at low and ultralow temperatures in chemical and biological evolution (cold prehistory of life?)'.

Indeed, although polymerization or polycondensation processes with the formation of such complex products as polypeptides and polynucleotides should be thermodynamically profitable when chemical equilibria are determined exclusively by changes of enthalpy, i.e. by the thermal effects of chemical reactions, the very problem of approaching chemical equilibrium near absolute zero seemed to have no sense in the framework of classical chemical kinetics. In contrast, tunneling phenomena open up broad possibilities for exothermic cryochemical reactions; in particular, chain reactions triggered by light or ionizing radiation.

The above mentioned hypothesis of cold prehistory of life, i.e. synthesizing rather complex molecules under the combination of space cold and various radiations of cosmic origin postulated in 1973 (Goldanskii et al., 1973) found the support of Wickramasinghe and Hoyle (Wickramasinghe, 1974, 1975; Hoyle and Wickramasinghe, 1977,1983) whose interpretations were widely disputed. Starting with claims that formaldehyde undergoes polymerization in interstellar space with the formation of polyoxymethylene and even of polysaccharides, these authors soon came to the hypothesis of 'living interstellar clouds' (Hoyle and Wickramasinghe, 1978) and even of the extra-terrestrial origin of some viruses (Hoyle and Wickramasinghe, 1979) invading the Earth, e.g. the virus of the influenza pandemia in 1919. These claims met various reasonable objections. Convincing evidence of the existence of polyoxymethylene in space should be dated much later (Mitchell et al., 1987; Huebner, 1987; Moore and Tanabe, 1990) and referred to the comparison of data obtained for the coma of comet Halley by heavy-ion analyzers aboard the Giotto spacecraft, laboratory mass spectra and infrared absorption spectra of formaldehyde polymerized at the surface of silicate grains under the irradiation by protons at 20 K . Molecular tunneling has been regarded as the most likely mechanism (Goldanskii, 1977, 1977a; Mitchell et al., 1987; Huebner, 1987; Moore and Tanabe, 1990; Goldanskii and Kuz'min, 1990). Thus, the main stage of the formation of prebiotic polymers (cold seeds of life) is hypothetically represented by the grains of dust of dense interstellar clouds and more explicitly, by the dirty ice mantles surrounding the cores of these grains. The temperature of these grains is estimated as $T_g \approx$ 10-20 K . Polymerization can be triggered by ultra-violet radiation - in the outer region of dark clouds - and by long-range cosmic protons - in their depths. The main components of dirty ice mantles, listed in order of their decreasing volatility are: ethane, ammonia, formaldehyde, hydrogencyanide, hydrogen isocyanide, water and, finally, polyformaldehyde. The rate of condensation does not depend on volatility while the rate of sublimation (both spontaneous and radiation - induced) decreases rapidly with diminishing volatility. Therefore, the above sequence corresponds also to the increasing steady-state enrichment of dirty ice by various components compared with the gas phase. While the relative abundance of formaldehyde in the gas phase is not larger than several tenths of a per cent of sum of other listed compounds, its abundance in the solid phase - if in the polymer form - could be much higher. Moreover, one should take into account that the cosmic abundance of various elements provides for water and formaldehyde only, the possibility to form the whole mass of dust grains while the maximum total mass of interstellar ethane is half, that of HCN and HNC a quarter, and ammonia one sixth.

Estimates of the rate of gas ⟷ solid processes in both directions (condensation and sublimation) lead to the conclusion that the lifetime of clouds is sufficient to ensure 'shuffling of the deck of cards' - a kind of numerous repetitions of condensation and evaporation of the molecules physically absorbed (adsorption heat D $\approx$0.05-0.1 eV) which leads to the enrichment of the solid phase by less volatile molecules. The evaporation of the mantle would be strongly hindered by its conversion to a crystal with larger binding energy. There will be no evaporation at all of molecules incorporated into polymer chain. Therefore, one can expect to find a particularly strong increase of formaldehyde abundance caused by its polymerization in the outer layers of interstellar grains. Criteria of the possibility (or impossibility) of the effective accumulation of polymers in the interstellar dust grains include the interplay of three characteristic times - the time of the addition of one new link to the growing polymer chain - τ_0 ; the average time of the absorption of an ultra-violet quantum by each molecule of the outer region of the dark cloud ($\tau_{uv} \approx$ 100 years) because such absorption can lead not only to the initiation of the polymer chains but also to their rupture, and the life-time of the clouds determined by their collisions or gravitational collapse ($\tau_{cloud} \approx 10^5$-$10^7$ years).

Extrapolation to T_g = 20 K of our kinetic data on the polymerization of solid formaldehyde at higher temperature (up to 140 K) by an Arrhenius plot gives the duration of the addition of several hundred links to the growing polymer chain which exceeds, by many orders of magnitude, not only the τ_{uv}, but also the τ_{cloud} values. The situation is the complete opposite for the experimental data on the rate of polymerization of formaldehyde at the low-temperature kinetic plateau caused by molecular tunneling - τ_o is shorter than τ_{uv} by about ten orders of magnitude. Thus, tunneling could significantly increase the number of possible low-temperature reactions in dense clouds. One case would be, for example, the possibility of tunneling polymerization at the surface of dust grains with the formation of a very thin (several molecular layers) polymer film around the inner region of dirty ice. Such a film could protect the surface of this inner region from both condensation and sublimation.

For chemical and prebiotic evolution, the reactions of polycondensation in dirty ice mantles with the participation of CH_2O , HCN, HNC, NH_3 and H_2O are of interest. Such reactions could lead to the formation of amino acids, polypeptides, sugars and nucleotide bases (purines and pyrimidines), they are exothermic, but not chain-type.

There are no reasons why there should be a 'pure' molecular tunneling mechanism of such reactions - the rate of tunneling falls steeply to a vanishingly small limit with increasing barrier widths and tunneling masses. However, each single step of chemical conversion which represents an elementary gas phase process (such as the reaction $H_2C = O + NH_3 \longrightarrow H_2C = NH + H_2O$), proceeds in the solid as a sequence of many individual or collective conformational rearrangements of molecules, complexes or indeed whole regions of molecular crystals. The collision of a dust grain with a cosmic proton or ultra-violet quantum, or the release of recombination energy at the grain surface can induce the transfer of the 'driving' particle, such as the electron, which determines the number of conformational rearrangements. As long as quantum effects open the possibilities of various low temperature chemical conversions it seemed to be of interest to calculate the equilibrium composition of cold interstellar dust grains. In fact, there was no need to take into account the entropy in such calculations, they were based exclusively on enthalpies and the cosmic abundance of H, C, N and O atoms. The maximum release of heat was found to correspond to the formation of acetic acid, urea and certainly also the products of their exothermic polycondensation (Goldanskii et al., 1981). As gravitational instability develops in the dark dust-gas cloud, a differentiation of matter occurs, and protostar forms. Planetesimals accrete from the dust-gas disc forming around the star, and enlarge to planets together with the formation of meteorites and comets. Consequently, the organic compounds which had formed in the dust-gas cloud can reach the planet by two processes: first, during the accretion of the planet; second, after the planet had formed, through the adsorption of these compounds on the surface of the planet from the surrounding space. The organic compounds which reached the planet in this fashion might then have served as the raw materials for the formation of the 'primordial soup'.

The formation of even the most complex biopolymers in interstellar dust grains does not guarantee their conservation during the later formation of new stars and planetary systems, when the dense clouds collapse. One cannot exclude, for example, the strong heating at the surface of planet at certain stages in their evolution. Nevertheless, one should also keep in mind the variety of possible chemical reactants and reactions in 'warm' systems depending on their history. For example, polymers created and stabilized by endcapping at low temperatures could survive the consequent warming, but they cannot be formed directly at higher temperatures. Meanwhile the existence of such polymers opens additional possibilities of chemical conversions not only in cold, but also in warm systems, such as the integration of molecules in shock-wave induced solid-phase reactions (Barkalov et al., 1967; Goldanskii et al., 1972) and in the vicinity of phase transitions, which proceed during the alternative heating and cooling of reactants (Kargin et al., 1964). Favourable conditions for the latter type would be provided, for example, by multiple transportation of stable organic compounds between circumstellar discs and interstellar clouds (Sagan and Khare, 1979) and in comets with extended orbits which alternately suffer short heating when approaching the Sun, and prolonged deep cooling, away from the Sun.

4. COLD PREHISTORY OF LIFE: FACTORS FAVOURING AN ENANTIOMERIC EXCESS

The introduction emphasized the inseparable ties between the appearance of chiral purity of the predecessors of the bio-organic world and the subsequent

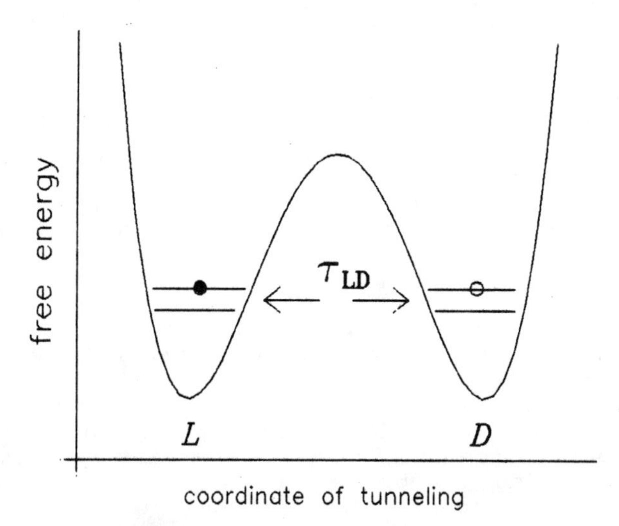

Figure 1. Tunneling between L and D states of chiral molecule: τ_{LD} is the average lifetime of the chiral state.

origin of life. Although there is no evidence (and no definite conjecture) of the achievement of chiral purity under extra-terrestrial conditions it should be noted that some of the above-mentioned factors typical for such conditions (in particular, the combination of deep space cold and various radiations) work in favour of the achievement of some enantiomeric excess during the cold prehistory of life.

For the better understanding of the differences between the problems of the breaking of mirror symmetry in warm and cold scenarios it would be worthwhile considering the picture of two equal potential wells separated by a barrier (Fig. 1) - one of them is the mirror reflection of the other, so we can call one of them the L-well and the other the D-well and treat them as analogous of enantiomers with L and D chirality. Neither of these two wells (or two molecules with opposite chirality) represents the stationary state of the system, because tunneling leads to spontaneous L $\longleftrightarrow$ D conversion, i.e. the system is delocalized between two equal potential wells (the so-called Hund's paradox). In a warm scenario, chemical processes go quite rapidly on the evolutionary time scale and the sign of the chirality is conserved (i.e. LD-delocalization can be neglected) during the chemical transformations of the enantiomer molecules. In a 'cold' scenario the rates of chemical transformations are exceedingly low, and molecules may repeatedly undergo L ⟷ D conversions - either via tunneling or as radiation-induced processes, for which the time scale are comparable with the rates of the chemical processes themself. Accordingly, the LD-delocalization in a 'cold' scenario may lead to a situation in which the concept of the certain sign of the chirality of an isomer molecule 'gets lost' over the time scales of the physico-chemical processes. The first problem associated with the breaking of mirror symmetry in a 'cold prehistory of life' is thus the problem of the stabilization of chirality of the molecules of two mirror isomers-enantiomers.

It has been shown (Simonius, 1978; Harris and Stodolsky, 1978, 1982, 1983) that if chiral molecules interact with the optically inactive medium consisting of a strongly cooled gas of low density, stabilization of the chirality of isomer molecules would be possible over times much longer than the tunneling oscillation time τ_{LD}

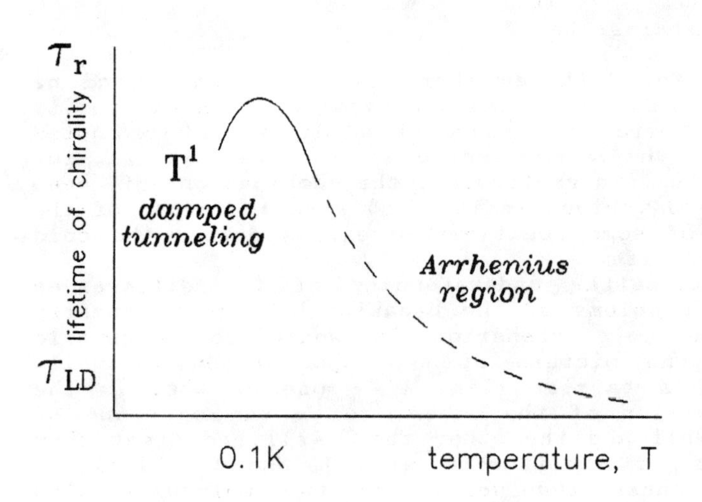

Figure 2. Stabilization of chirality in low-temperature gases (racemization time $\tau_r > \tau_{LD}$).

(Fig.2). Consequently, although racemization does occur, the time scale for the processes increases sharply compared with the racemization time for an isolated particle.

The very fact that the chirality of molecule is stabilized at low temperatures by the interaction with the medium is, understandably, an extremely attractive aspect of a cold scenario but one should also keep in mind that the problem of the deracemization of the medium as a whole arises here. Specifically, each of the isomer molecules is initially in a state with a definite chirality, i.e. in either the L-state or the D-state. The ensemble of such molecules, however, is probably in racemic state. Therefore, it is quite important to analyze the problem of the stabilization of the optical activity of the ensemble of molecules incorporated in solid low-temperature matrices taking into account the contributions of various types of relaxation processes.

Such analysis shows (Berlin et al., 1989) that at

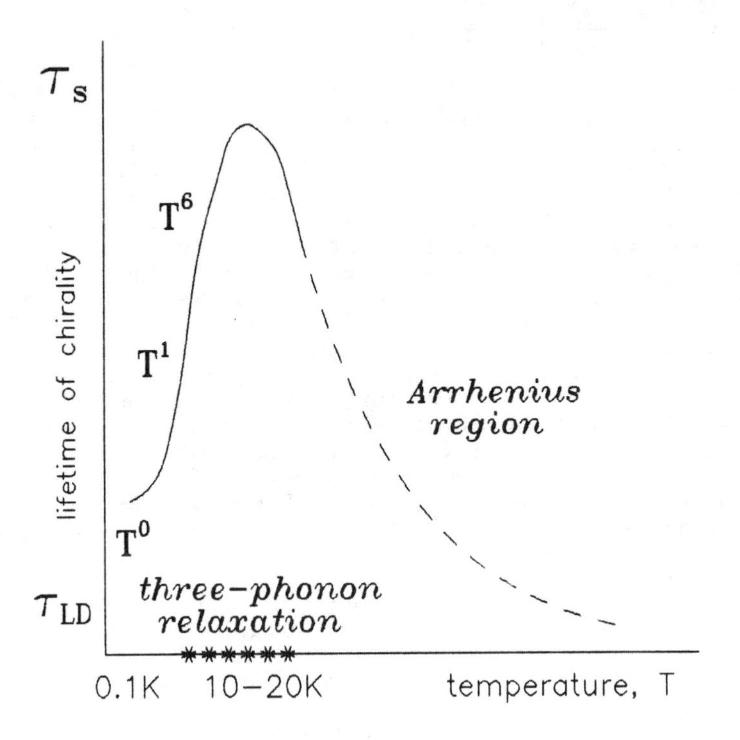

Figure 3. Stabilization of chirality in low-temperature solids ($\tau_s > \tau_{LD}$).

very low temperatures the time τ_S of L $\longleftrightarrow$ D transitions in solid strongly exceeds the time τ_{LD} of 'free' tunneling oscillations, and moreover, τ_S rises in the vicinity of absolute zero with the increase of temperature, and the most effective suppression of racemizing processes in molecular ensembles should be observed at T ≈ 10 - 20 K, i.e. just under the conditions typical for dirty-ice mantles or interstellar dust grains (Fig.3).

One of the most widely discussed problems of the origin of life is the search for various local or global advantage factors (AF) (Goldanskii and Kuz'min, 1989, 1991; Bonner, 1991) which can work in favour of preferential accumulation of one of two enantiomers. By definition, the advantage factors (AF) characterize the relative difference in the rate constants K^L and K^D for mirror conjugated reactions of two enantiomers:

$$g = \left| \frac{K^L - K^D}{K^L + K^D} \right| , \qquad 0 \leq g \leq 1 .$$

Several authors have made suggestions concerning AFs which are specific just for the extra-terrestrial (cold) scenario of the origin of life. Gladyshev *et al.* (1978, 1981, 1990) invoked the role of such factors as magnetic, electric and gravitational field of various cosmic objects and the orientation of reacting molecules absorbed at the surface of dust grains. They also mentioned that chiral molecules might persist unaltered because of low temperature and low frequency of intermolecular (or grain) collisions in space. However, no numerical estimations were given and, moreover, none of these hypothetical extra-terrestrial AFs, belong to the range of 'true' physical AFs as defined by Barron (1986, 1986a, 1987) (i.e. such physical fields, radiations etc which possess the property of 'helicity', which can exist in two enantiomeric forms and transform into its antipode under the effect of spatial inversion, but do not change upon time inversion in combination with any spatial rotation).

Demands for such properties of 'true' AF are satisfied for circularly polarized light as well, of course, as for circularly polarized ultra-violet synchrotron radiation emitted by the neutron star remnants of supernovae. According to the hypothesis of Bonner *et al.* (Bonner, 1991; Rubenstein et al., 1983; Bonner and Rubenstein, 1987) this radiation has produced chirally enriched organic mantles of interstellar grains in the asymmetric photolysis of 'dirty ice'.

As mentioned above there is still no evidence of chiral purity of any space object. However, at least some of the 74 amino acids identified in the widely described Murchison meteorite (which fell in Australia in 1969) and just those which belong to amino acids typical for proteins of the earth's biosphere (e.g. alanine and glutamic acid) were found to be not racemic but slightly enriched by L-enantiomers (Engel and Nagy, 1982).

Suspicion that such enrichment was caused by contamination of terrestrial origin seem to be excluded by the recent observation of ^{13}C abundance in these amino acids (Engel et al., 1990), i.e. the L-enantiomeric excess was accompanied by a ^{13}C enrichment typical of extra-terrestrial organic materials (up to 3%). This result was reasonably interpreted as an argument in favour of the presence of deracemized compounds in the early Solar System (Engel et al., 1990; Chyba, 1990). Moreover, if the partially deracemized state represents an intermediate stage in the transition to a chirally pure state, the abovementioned results can provide an indication of the timescale required for the process of spontaneous breaking of mirror symmetry (Goldanskii and Kuz'min, 1989): according to our estimations (Morozov et al., 1984, 1984a) it could be as short as 10^6 - 10^7 years, whereas the whole prebiotic stage of the earth's history lasted several hundreds of millions of years.

5. COLD PREHISTORY OF LIFE: THE BIOLOGICAL BIG BANG

Our detailed analysis of the connections between chirality, the origin of life and evolution (Goldanskii and Kuz'min, 1989, 1991; Avetisov and Goldanskii, 1991, 1992; Avetisov et al., 1991) is, to a large extent, based on the use of two main parameters - chiral polarization $\eta=(X_L - X_D)/(X_L + X_D)$ (X_L and X_D are the concentrations of L and D enantiomers, and thus η is the dimensionless normalized enantiomeric excess) and enantioselectivity γ of the incorporation of optical antipodes into the growing polymeric chains:

$$\gamma = \omega_{LL} - \omega_{LD} = \omega_{DD} - \omega_{DL}$$

(the first of the two indexes below ω designates the end link of the chain; the second the added link).

The treatment of enantioselectivity - as a parameter additional to chiral polarization - is quite important for the analysis of the possibility of a specific (enzyme-like) activity not only for homochiral, but also for regular heterochiral polymers.

It was concluded that heterochiral polymers cannot

have any specific activity, either because of strong structural limitations or because of strong kinetic limitation (for a certain 'unique' sequence of chiral fragments). It is even more important that the takeover of organic medium by homochiral chains which possess biochemical functions involves overcoming two successive critical points (Avetisov and Goldanskii, 1991; Avetisov et al., 1991). The first is the formation of chirally pure organic medium ($\eta_{pur} \cong 1$) for providing of a selection of any homochiral informational chains in abiogenic condition, while the second point connects to the appearance of chemical functions with very high specifity, in particular, absolute enantioselective polymers assemblage ($\gamma_{abs} \cong 1$) . Overcoming both critical points in some local "seat of life" is necessary for the origin of self-replicating macromolecular patterns. The exponentially fast propagation of such patterns looking like an explosive chain reaction (so-called "Biological big bang") might be a reason for the formation of the early biosphere.

Taking into account all the above circumstances one can represent the main stages of prebiological evolution by the scheme (Fig. 4).

Two features are fundamentally important: first, a strong mirror symmetry breaking in the organic medium preceded the polymeric takeover and predetermined the formation of homochiral polymers. Second, the chiral purity of the medium had to be maintained not only at the stages of primary homochiral polymers but also subsequently, during the formation of structures and functions possessing the biochemical level of complexity. Only after the appearance of structures having enantiospecific activity the requirement of a chirally pure medium can be dropped.

Just the need to reach the second critical point is the main cause which makes the maintaining of chiral purity of a medium impossible, due to any AFs, however strong, because while the AF enriches the monomeric environment by some enantiomer, the enantioselective polymeric takeover has the opposite effect (the so-called stereoselective or enantioselective pressure (Avetisov and Goldanskii, 1991)) that is, the impoverishment of a monomeric system by that enantiomer, inhibition and finally complete blocking of the effect of AF.

Thus, processes in which symmetry breaking depends exclusively on the action of an AF and occur by gradual accumulation of asymmetry, could not be crucial for early steps of prebiotic evolution. One needs a fundamentally different type of processes that can effect a strong mirror symmetry breaking even without an AF and can withstand the enantioselective pressure throughout the

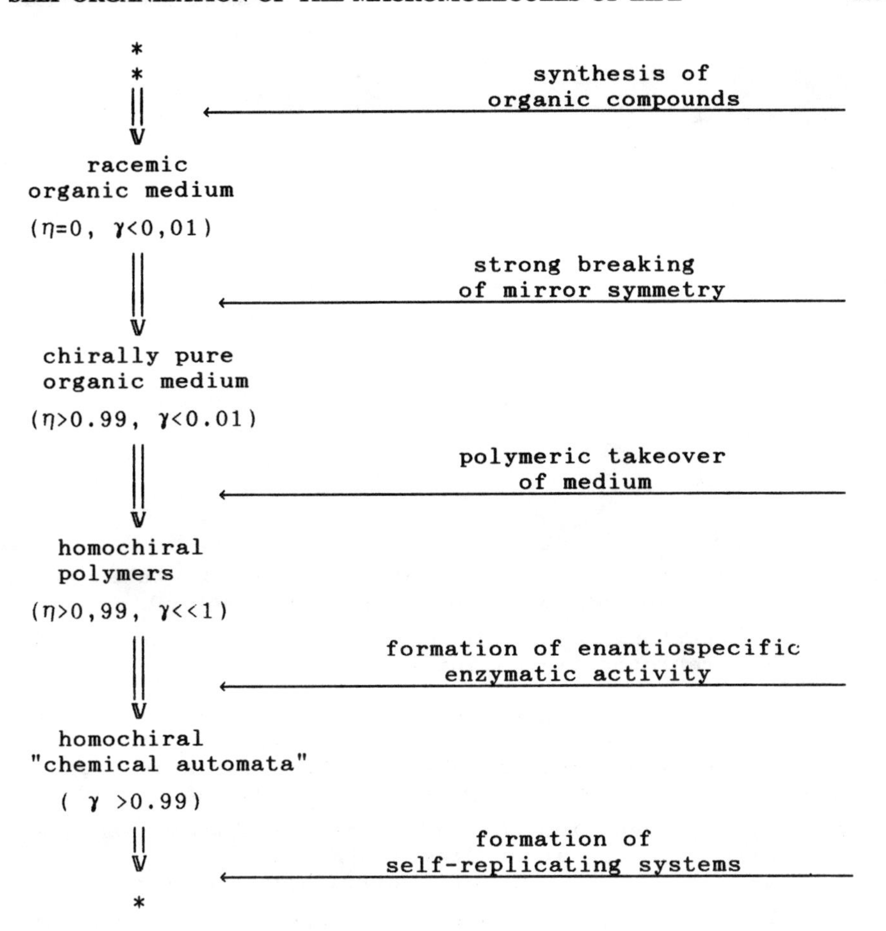

Figure 4. Five main stages of prebiological evolution.

stages of formation of enantiospecific activity.

Processes of the 'bifurcation' type, well known from the theory of non-equilibrium phase transition, possess the required properties. In this case the connection between chiral polarization η of monomeric medium and enantioselectivity γ of the polymeric takeover is described by universal equation of bifurcation type (Avetisov et al., 1991):

$$-\eta^3 + (1-\rho)\eta - \gamma(1-\eta^2) = 0 \ ,$$

where $\rho \geq 0$ depends on kinetics of transformations of L- and D-monomers in some organic area.

When $\gamma=0$, the first two terms in the equation

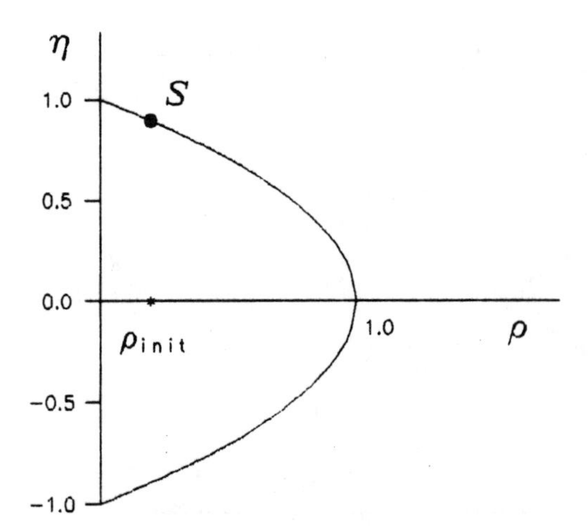

Figure 5. Bifurcation diagram before polymeric takeover. The point S illustrates the set ρ_{init} of starting conditions for the beginning of prebiotic transition.

describe the symmetry breaking of monomeric medium before its polymeric takeover. Truly, the racemic state of chemical system becomes unstable at the bifurcation point $\rho_c=1$ and for $\rho<1$, there are two stable states with broken symmetry. Such behavior of a chemical system can be illustrated by bifurcation diagram (Fig.5).

Note, that the value of controlling parameter ρ_{init} should be much less than unity for strong mirror symmetry breaking connected to overcoming of the first critical point of prebiotic evolution (point S on Fig.5). Since the parameter ρ describes the degree of deviation of chemical transformation from thermodynamic equilibrium, by this is meant that the starting conditions for primary polymeric takeover of organic medium should be strongly non-equilibrium.

The last term in our bifurcational equation corresponds to enantioselective pressure which looks as some compensating "chiral field" (Fig.6). As the result, the enantioselective pressure shifts the position of the bifurcation point ρ_c towards ρ_{init} . The shift depends on the selectivity of a polymeric takeover and when γ trends to absolute enantioselectivity $\gamma_{abs}\cong1$, the

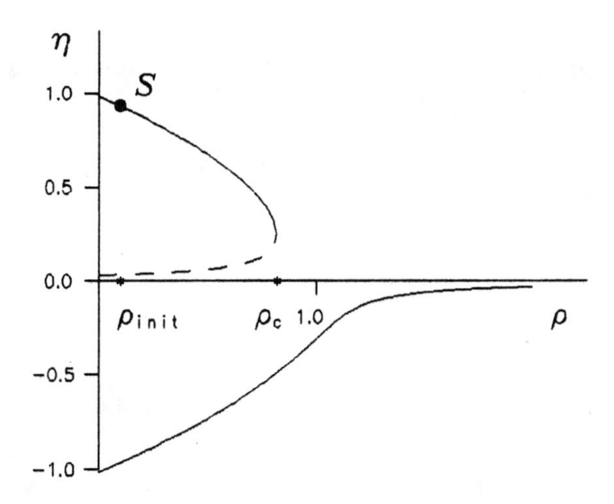

Figure 6. Bifurcation diagram under evolution of specific catalytic functions.

bifurcation point $\rho_c(\gamma)$ shifts to 0.

Therefore for any $\rho_{init} << 1$ such a value of γ_{cr} exists when the bifurcation point ρ_c reaches ρ_{init}. This is most critical situation for prebiotic transition, because the chirally pure state, which had been formed at the beginning of the prebiotic transition, becomes here unstable and the monomeric environment is racemized. Could an enantiospecific function be formed before this critical moment? This is the major question now.

The analysis shows (Avetisov and Goldanskii, 1991; Avetisov and Goldanskii, 1993) that if the condition for overcoming the first critical point is satisfied, the chirally pure state loses its stability just at the moment when enantioselectivity of polymeric takeover reaches the value γ_{abs}.

It is not less important that biological big bang is accompanied by the racemization of the monomeric environment while the broken mirror symmetry for the living matter is preserved in the form of homochirality of macromolecules possessing an enantiospecific activity. Just this specifity of Biosphere is observed on the earth.

Thus we come to the conclusion that the concept of prebiotic evolution based on the bifurcation with symmetry breaking far from the thermodynamic equilibrium satisfies both important features of the prebiotic transition.

The appearance of strongly non-equilibrium conditions resulting in the bifurcation with symmetry breaking in some "warm" or "cold" organic area seem to be the main physical demand for biological big bang. However, if such demand is satisfied the early biosphere can be formed and can develop without any principle restrictions.

6. COLD PREHISTORY OF LIFE: PUZZLES AND CONSIDERATIONS FOR THE FUTURE

Here we come to the first of the still unsolved key problems of any scenario of the origin of life - what was the mechanism of bifurcation with mirror symmetry breaking in the racemic organic medium? This needs a deracemizing process based on the co-operative (non-linear) interactions of enantiomers which can lead to the self-organization of chirality in the system; mirror-symmetry breaking occurs spontaneously as soon as critical conditions are reached. Although many schemes of interactions have been suggested (i.e. the pioneering scheme of Frank (1953) and the generalized treatment of this problem by Morozov (1979)) which fulfil such requirements, neither of these schemes have suggested any concrete, specific chemical reactions for a liquid phase. The situation looks more promising in the solid phase since cooperative interactions in solids are more pronounced than in liquids. Let us refer here to the most detailed example of spontaneous generation of optical activity in 1,1'-binaphthyl (Pincock and Wilson, 1971, 1975). This chiral hydrocarbon exists in two crystalline forms, one of them an eutectic mixture of individual optically pure crystals, the other a racemic compound. The generation of optical activity was found to be a strongly non-equilibrium process. Recently it has been proposed that a phenomenological description of these results might be caused by the non-linear kinetics of the inversion of mirror isomers (Avetisov et al., 1991a), and the experimental data were reproduced by computer simulation. Kondepudi (Kondepudi et al.,1990) have observed and investigated a peculiar example of a total spontaneous resolution of L-and D-crystals of sodium chlorate when their solutions were stirred during crystallization.

Thus, the strong contribution of co-operative interactions to various chemical and/or physico-chemical conversions in solids (particularly, at low temperatures - contrary to warm liquids) speaks in favour of the extra-terrestrial (cold) scenario of the origin of life. One should note here that the crucial role of the driving force of causality in co-operative stochastic systems can be played by correlated fluctuations treated in detail in recent publications (Berlin et al., 1992, 1992a, 1992b; Goldanskii and Mikhailov, 1993). However, in view of the above scheme of prebiological evolution there remains an important intriguing problem. It is not sufficient just to find the mechanisms of spontaneous breaking of mirror symmetry in any scenario of the origin of life, it is also necessary to explain the way of formation of enantiospecific enzymatic activity of homochiral polymers. Indeed, even if the homochiral polymers of extra-terrestrial origin ('cold seeds of life' (Goldanskii, 1977)) arrived at the racemic earth before acquiring enzyme-type activity (biological AF), they would have been racemized in the same way as amino acids from proteins after the death of their possessors (Bada, 1971, 1985). Therefore, any comprehensive hypotheses of 'panspermia' should include some ideas about the mechanism of extra-terrestrial formation of structures having enantiospecific activity and their subsequent delivery to the earth (whose prebiotic chemical medium could, at that stage, be racemic). Two more points need mentioning. What were the chronologically first prebiotic compounds which did acquire chiral purity and, afterwards, enzymatic activity? Were they amino acids or carbohydrates (sugars)? Priority of chiral purity of amino acids would mean the priority of proteins as enzymes while the priority of chiral purity of sugars (e.g.ribose) would serve as a serious argument in favour of a RNA-world, with the ribozyme as an initial enzyme.

From my point of view the priority of chiral purity of carbohydrates seems the more probable. Indeed, among the 74 amino acids found in the Murchison meteorite (i.e. of natural origin) only 20 are found on the earth as components of proteins and thus chirally pure (in fact 19 of them because glycine is optically inactive). On the contrary, practically all natural carbohydrates possess chiral purity and, moreover, they have several asymmetric carbon atoms per molecule, which could widen the possibilities of co-operative interactions in the solid phase.

Another point is connected with the problem of the sign of chirality of the earth's biosphere. As was briefly mentioned in the introduction, we assert that the choice of this sign was random rather than predetermined by some global AF (e.g. by weak neutral currents). However, one can put a question, why - in spite of the accidental choice of the sign of chirality in the course of prebiotic spontaneous breaking of mirror symmetry - this sign turns to be the same all over the earth.

This question has been discussed in detail (Goldanskii and Kuz'min, 1989) and the answer is very simple, it is determined by the interplay of two characteristic times ,- the time required for the takeover of the entire biosphere by the very first domain of life and the expectation time τ_{ex} for the formation of the embryo of the new phase with the opposite symmetry.

The expectation time is a criterion which expresses the possibility that life arises under certain conditions or others, on some certain celestial objects or others. Specifically, if we know the parameters of the medium which are characteristic of the given celestial object (e.g. a planet, a dust-gas cloud, etc) we can estimate the expectation time for the beginning of an irreversible deracemization of the medium. If we find that τ_{ex} exceeds the age of celestial object, we should acknowledge that the appearance of life is impossible in this case, since a deracemization of the medium - a necessary prerequisite for the appearance of living structures - does not occur.

The concept of an expectation time for the beginning of a breaking of mirror symmetry makes it possible to move on the solution of yet another problem which is being widely debated; was the appearance of life on the earth a consequence of single event or the result of competition among several prebiospheres which arose independently? We can approach the resolution of this question along the path of a deracemization of the prebiosphere. The appearance and co-existence of natural habitats within which the chirality of the organic matter has different signs are equivalent to the appearance of a set of competing prebiospheres, i.e. to a multiplicity of nucleation events. If, on the other hand, the deracemization process generated by a critical fluctuation spanned the entire planet, the 'act of organization' was unique.

Let us consider a gedanken experiment: we assume that the racemic 'primeval soup' has reached a critical state required for a transition to chiral order and occupies two habitats which communicate with each other (e.g. northern and southern hemispheres). We also assume that an advantage factor is operating in each habitat and that nature and measure of the advantage factors are identical in two regions, but the signs of the advantage factors are opposite. The action of the advantage factor leads to an excess of one of the antipodes - but different ones in the two habitats. If the mixing of matter is sufficiently intense, however, the medium as a whole will remain racemic. Since such a medium is in a state which is unstable with respect to fluctuations of the chiral polarization, the very first critical fluctuation which appears after a time τ_{ex}, marks the beginning of a deracemization process.

Since the appearance of a critical fluctuation is equally probable in each of the two habitats, we should recognize that the sign of the chirality of the prebiosphere will be determined with equal probabilities by the sign of the advantage factor of each of the habitats. In other words, it will be random for the prebiosphere as a whole. It is not difficult to see that the result of this gedanken experiment does not depend on the number of habitats which we consider or on the number of the local advantage factor. Since the 'colonization' of a medium by a critical fluctuation occurs in a time of 10^2 - 10^4 years (mixing due to flows, etc) and since this time is substantially shorter than the expectation time for the critical fluctuation, which is the next to follow the first, 1-10 million years (Morozov et al., 1984; Goldanskii and Kuz'min, 1989), we can confidently say that the deracemization of the prebiosphere was a result of a single event, rather than a consequence of a set of local deracemization events (this is true regardless of whether there is an advantage factor).

ACKNOWLEDGMENT

The author is greatly indebted to his colleagues and friends Drs Vladik A.Avetisov and Vladimir V Kuz'min for valuable remarks and advices.

REFERENCES

Arrhenius S., 1908: *World in the Making*, Harper and Row, NY.

Avetisov V.A., V.I. Goldanskii, 1991: *BioSystems*, 25, 141.

Avetisov V.A., V.I. Goldanskii, V.V. Kuz'min, 1991: *Phys. Today*, 44(7), 33.

Avetisov V.A., V.I. Goldanskii, S.N. Grechukha and V.V. Kuz'min, 1991: *Chem. Phys. Lett.*, 184, 526.

Avetisov V.A., V.I. Goldanskii, 1993: *Phys. Lett. A*, 172, 407.

Bada J.L., 1971: *Adv. Chem. Ser.*, 106, 309.

Bada J.L. and R. Schroeder, 1975: *Naturwissenschaften*, 62, 71.

Bada J.L., 1985: *In Chemistry and Biochemistry of the Amino Acids*, Chapman and Hall, London.

Barkalov I.M., G.A. Adadurov, A.N. Dremin, V.I. Goldanskii, T.N. Ignatovich, A.N. Mikhailov, V.L. Talrose and P.A. Yampolskii, 1967: *J. Polym. Soc. C*, 16, 2597.

Barron L.D., 1986: *J. Am. Chem. Soc.*, 108, 5539.

Barron L.D., 1986a: *Chem. Phys. Lett.*, 123, 423.

Barron L.D., 1987: *Chem. Phys. Lett.*, 135, 1.

Barron L.D., 1987: *BioSystems*, 20, 7.

Bell R.P., 1973: *The Proton in Chemistry*, (2nd Edn) Cornell University Press, Ithaca, NY.

Bell R.P., 1980: *The Tunnel Effect in Chemistry*, Chapman and Hall, London.

Benderskii V.A. and V.I. Goldanskii, 1992: *Int. Rev. Phys. Chem.* 11, 1.

Benderskii V.A., V.I. Goldanskii and D.E.Makarov, 1993: *Physics Reports*, 233, 195.

Berlin Yu.A., S.O. Gladkov, V.I. Goldanskii and V.V. Kuz'min, 1989: *Sov. Physics - Doklady*, 306, 844.

Berlin Yu.A., D.O. Drobnitsky, V.I. Goldanskii and V.V. Kuz'min, 1992: *Phys. Rev. A*, 45, 3547.

Berlin Yu.A., D.O. Drobnitsky, V.I. Goldanskii and V.V. Kuz'min, 1992a: *BioSystems*, 26, 165.

Berlin Yu.A., D.O. Drobnitsky, V.I. Goldanskii and V.V. Kuz'min, 1992b: *Chem. Phys. Lett.*, 189, 316.

Bonner W.A. and E. Rubenstein, 1987: *BioSystems* 20, 99.

Bonner W.A. and E. Rubenstein, 1990: *In Prebiological Self Organization of Matter*, C. Ponnamperuma and F. R. Eirich (Eds.), A. Deepak Publishing, Hampton, Virginia, USA.

Bonner W.A., 1991: *Orig. Life Evol. Biosphere*, 21, 59.

Chyba C.F., 1990: *Nature*, 348, 113.

Crick F., 1981: *Life.Itself: Its Origin and Nature*, Simon and Schuster, NY.

Eigen M., B.F. Lindemann, M. Tietze, R. Winkler-Oswatitsch, A. Dress and A.von Haeseler, 1989: *Science*, **244**, 673.

Engel M.H. and B. Nagy, 1982: *Nature*, **296**, 837.

Engel M.H., S.A. Macko and J.A. Silfer, 1990: *Nature*, **348**, 47.

Frank C.F., 1953: *Biochem. Biophys. Acta.*, **11**, 459.

Gladyshev G.P., 1978: *The Moon and the Planets*, **19**, 89.

Gladyshev G.P. and M.M. Khasanov, 1981: *J. Theor. Bio.*, **90**, 191.

Gladyshev G.P. and D.Kh. Kitaeva, 1990: *Zh. Vses. Khim. Obshch. im. Mendeleeva*, No 5, 625.

Goldanskii V.I., 1959: *Doklady AN SSSR*, **124**, 1261.

Goldanskii V.I., 1959a: *Doklady AN SSSR*, **127**, 1037.

Goldanskii V.I., T.N. Ignatovich, M.Yu. Kosygin and P.A. Yampolskii, 1972: *Doklady Biochemistry* (Proc. Acad. Sci. USSR),**297**, 218.

Goldanskii V.I., M.D. Frank-Kamenetskii and I.M. Barkalov, 1973: *Science*, **182**, 1344.

Goldanskii V.I., 1977: *Nature*, **258**, 612.

Goldanskii V.I., 1977a: *Nature*, **269**, 583.

Goldanskii V.I., 1978: *Chem. Scripta* **13**, 1.

Goldanskii V.I., 1979: *Nature*, **279**, 109.

Goldanskii V.I., L.V. Gurvich, V.V. Muzylev and V.S. Strelnitskii, 1981: Scientific Informations of the Astronomical Council of the Academy of Sciences of the USSR, No 47, 3.

Goldanskii V.I., L.I. Trakhtenberg and V.N. Fleurov, 1987: *Sov. Sci. Rev. B (Chem.)*, **9**, 59.

Goldanskii V.I., V.A. Benderskii and L.I. Trakhtenberg, 1989: *Adv. Chem. Phys.*, **75**, 349.

Goldanskii V.I., L.I. Trakhtenberg and V.N. Fleurov, 1989a: *Tunneling Phenomena in Chemical Physica*, Gordon and Breach.

Goldanskii V.I., V.V. Kuz'min, 1989: *Sov. Phys. Uspekhi*, 32, 1.

Goldanskii V.I., V.V. Kuz'min, 1990: *J. Brit. Interplanet Soc.*, 43, 31.

Goldanskii V.I., V.V. Kuz'min, 1991: *Nature*, 356, 114.

Goldanskii V.I. and A.S. Mikhailov, 1993: *Phys. Lett. A*, 176, 6.

Haldane J.B.S., 1929: *Ration Ann.*, 148, 3.

Harris R.A. and L. Stodolsky, 1978: *Phys. Lett. B*, 78, 313.

Harris R.A. and L. Stodolsky, 1982: *Phys. Lett. B*, 116, 464.

Harris R.A. and L. Stodolsky, 1983: *J. Chem. Phys.*, 78, 7330.

Hoyle F. and N.C. Wickramasinghe, 1977: *Nature*, 258, 610.

Hoyle F. and N.C. Wickramasinghe, 1978: *Lifecloud*, Dent and Sons, London.

Hoyle F. and N.C. Wickramasinghe, 1979: *Diseases from Space*, Harper and Row, NY.

Hoyle F. and N.C. Wickramasinghe, 1983: *Nature*, 306, 420.

Huebner W.F., 1987: *Science*, 237, 628.

Kargin V.A. , V.A. Kabanov and I.M. Papisov, 1964: *J. Polym. Sci.*, 4, 767.

Kondepudi D.K., R.J. Kaufman and N. Singh, 1990: *Science*, 250, 975.

Miller S.L. and L.E. Orgel 1974: *The Origins of Life on the Earth*, Prentice-Hall, Englewood Cliffs, NY.

Mitchell D.L., R.P. Lin, K.A. Anderson, C.W. Carlson, D.W Curtis, A. Korth, H. Reme, J.A. Sauvaud, C. d'Uston and D.A. Mendis, 1987: *Science* 237, 626.

Moore M.H. and A.T. Tanabe, 1990: *Astrophys. J.*, 365, L, 39.

Morozov L.L., 1979: *Origin of Life*, **9**, 187.

Morozov L.L., V.V. Kuz'min and V.I. Goldanskii, 1984: *Doklady-Biophys.* (Proc. Acad. Sci. USSR), **274**, 55.

Morozov L.L., V.V. Kuz'min and V.I. Goldanskii, 1984a: *Doklady-Biophys.* (Proc. Acad. Sci. USSR), **275**, 71.

Oparin A.I., 1957: *The Origins of Life on the Earth*, Oliver and Boyd, Edinburgh.

Pincock R.E. and K.R. Wilson, 1971: *J. Am. Chem. Soc.*, **93**, 1291.

Pincock R. E. and K. R. Wilson, 1975: *J. Am. Chem. Soc.*, **97**, 1474.

Rubenstein E., W.A. Bonner, H.Y. Noyes and G.S. Brown, 1983: *Nature*, **306**, 118.

Sagan C. and B.N. Khare, 1979: *Nature*, **277**, 102.

Schidlowski M., 1988: *Nature*, **333**, 313.

Schopf J.W. (Ed), 1983: *Earth's Biosphere. Its Origin and Evolution*, Princeton University Press, Princeton, NY.

Simonius M., 1978: *Phys. Rev. Lett.*, **40**, 980.

Urey H.C., 1952: *Proc. Nat. Acad. Sci.*, **38**, 351.

Wickramasinghe N.C., 1974: *Nature*, **254**, 452.

Wickramasinghe N.C., 1975: *Mon. Not. R. Astr. Soc.*, 170, 111.

FINDING THE NECESSARY CONDITION AND SCOPE OF L-AMINO ACID SURVIVING UNDER BETA ELECTRON IRRADIATION

W.Q.Wang, J.L.Wu, X. M. Pan
Peking University
Beijing 100871 , P.R. of China
L.F.Luo
Inner Mongolia University
Huhehot 010021 , P.R.of China

ABSTRACT

According to the difference of inelastic scattering oss sections of D and L type molecules under irradiation β electron, the necessary condition of L-amino acid rviving is: When $R^{+}_{n} > 0$, $\sigma_{D} > \sigma_{L}$. A new classification of amino acids, based on the rotatory strength sign of L-ino acid and experimental results of electric discharge ve been presented. In order to test the above mentioned ea one's effort must be focused on finding the surviving ndition of first kind L- amino acids.

INTRODUCTION

The origin of homochirality of biological molecules has en a constant problem since the discovery of parity non-nservation by Lee and Yang; a series of experiments were rried out to investigate the differential interaction of larized electrons emitted in β decay with enantiomers of ino acids. Until now, no proposed mechanism has been able explain how the completely asymmetric roles of mirror-age biological molecules could arise.

The weak interaction may influence chiral molecules in o ways: changing chemical reaction rate of chiral mole-les under the irradiation of polarized electrons emitted β decay and shifting the atomic energy asymmetrically by utral current. The effect of the latter is very small, out 10^{-19} eV, which is smaller than the value of atomic ergy by 20 orders. Recently Abdus Salam (1991) proposed that irality among the twenty amino acids, which make up the oteins may be a consequence of a phase transition. This is alogous to that due to BCS superconductivity and explore the ideas that a crucial form for the transition temperature Tc involves dynamical symmetry breaking, which could eventually lead to enantiomeric purity. Many scientists including our group tried to confirm the theoretical speculation but this is still in the process. As for the former, the interaction effect between polarized electrons and chiral molecules was believed to be very small; though a thorough analysis has not been given and the experimental results were quite uncertain. It seems that more theoretical work should be done.

THE DIFFERENCE OF INELASTIC SCATTERING SECTIONS OF D- AND L-TYPE MOLECULES UNDER IRRADIATION OF β ELECTRONS

In 1985 J.Y.Wang and L.F.Luo analysed the problem theoretically. The electrons emitted from β decay are all left-polarized with polarization ξ =-v/c (v is the speed of electron). The momentum transfer $\vec{q} = \vec{k} - \vec{k}_0$, after the inelastic collision the quantum state of the chiral molecule changes from ground state 0 to excited state n. Here $\vec{k}_0$ and $\vec{k}$ represent the momentum of incident and scattered electron respectively. The inelastic cross section σ (0 → n) is dependent on the dipole strength D_n and rotatory strength R_n of the molecule,

$$\sigma(0\rightarrow n) = \frac{k}{k_0}\left(\frac{m}{2\pi h^2}\right)^2 \frac{16}{3}\frac{\pi^2 e^2}{q^2}\left(D_n + \frac{2}{3}\xi\frac{hqk_0}{mck_0}R_n\right) \tag{1}$$

The cross section σ (0 n) consists of two parts. The first term is proportional to dipole strength D_n and independent of electron polarization ξ . The second term is proportional to the rotatory strength R_n and electron polarization ξ . For enantiomer amino acids, R_n are equal in absolute value and opposite in sign. Denoting the rotatory strength of L (D-) type molecule by R_n^+, (R_n^-), then $R_n^- = -R_n^+$. The asymmetry of cross section F is defined as

$$F = \frac{\sigma_L(0\rightarrow n) - \sigma_D(0\rightarrow n)}{\sigma_L(0\rightarrow n) + \sigma_D(0\rightarrow n)} = \frac{2}{3}\xi\frac{hqk_0}{mck_0}\cdot\frac{R_n^+}{D_n} \tag{2}$$

Since $h\vec{q}.\vec{k}_0 / 2\pi mck_0 \approx (E_n - E_0)/mc^2 \approx 10^{-4}$ and $R_n / D_n \approx 10^{-2}$, therefore $F \approx 10^{-6}$ and its sign depends on the sign of R_n^+.

The transition of a molecule from ground state to excited state would change its activity and reaction rate. So the asymmetry of cross sections would cause an asymmetry of reaction rates of chiral molecules under the irradiation of β electrons,

$$(K_L - K_D) / (K_L + K_D) \approx 10^{-6} \quad (3)$$

The left - right asymmetry is in the order of one millionth which is much greater than other weak interaction processes, such as the asymmetry of cross section and energy shift due to neutral current and the asymmetrical absorption of bremsstrahlung of β electron by chiral molecule etc.

THE NECESSARY CONDITION OF L-AMINO ACID SURVIVING UNDER IRRADIATION OF BETA ELECTRON

According to the theoretical study, Equation (3) gives the asymmetry of decomposition rate of enantiomers of amino acids under irradiation of β electron (left helicity). When $R_n^+ > 0$, $\sigma_D > \sigma_L$, it favours the L-amino acid surviving. When $R_n^+ < 0$, $\sigma_L > \sigma_D$, it favours the D-amino acid surviving.

In previous studies many scientists have paid attention to the helicity of the polarized β electrons (left or right) which may stereoselectively decompose one of the two configurations (D or L form), but not sufficient attention has been given in the rotatory strength sign of different L amino acid.

Considering the above mentioned equation (1)-(3), a new classification of 20 amino acids is proposed. The first kind L-amino acids are Ala, Val, Ile, Asp, Glu, Gln, Arg and Lys. As their rotatory strength sign are positive, they could survive under the action of β electron on the racemic form. The second kind L-amino acids are Leu, Ser, Asn, His, Cys, Met and Trp, their rotatory strength sign would be changed by acidity. It is possible that the second kind amino acids can survive at acidic medium. L types of Thr, Phe,

Table 1 The rotatory strength sign of L amino acids (298K)

Amino acid	$[\alpha]_D$ (H_2O)	$[\alpha]_D$ (5N HCl)	Amino acid	$[\alpha]_D$ (H_2O)	$[\alpha]_D$ (5N HCl)
Ala	+1.8	+14.6	Leu	-11.0	+16.0
Val	+5.6	+28.3	Ser	-7.5	+15.1
Ile	+12.4	+39.5	Asn	-5.3	+33.2
Asp	+5.0	+25.4	His	-38.5	+33.2 *
Glu	+12.0	+31.8	Cys	-16.5	+6.5
Gln	+6.3	+31.8 **			
			Met	-10.0	+23.2
			Trp	-33.7	+2.8 **
Arg	+12.5	+27.6	Thr	-28.5	-15.0
Lys	+13.5	+26.0	Phe	-34.5	-4.5
			Pro	-86.2	-60.4
			Tyr	----	-10.0

* in 3N HCl; ** in 1N HCl

Tyr and Pro are the third kind of amino acid. When methyl group is introduced to L-Ser and phenyl group to Ala, L-form of Thr, Phe are formed respectively. Their R_n signs are negative. The necessary condition of L-amino acid surviving under irradiation of β electron is $R^+_n > 0$. Therefore the L-type surviving cannot be formed from weak interaction of β particle but can originate either from biosynthesis or by biosuppression.

Table 2 shows that all the three kinds of amino acids can be prebiotic synthesized by electric discharge of different systems in weak acidic condition. In neutral system, the R_n sign of second kind is negative. It reveals that under neutral condition the second kind L-amino acids cannot have survived under the β irradiation on their racemic form synthesized by cosmic electric discharge. In this case, L-amino acid of second sort only can be survived by bioseparation or biosuppression. It must be mentioned that in basic condition, only first kind of amino acids were formed.

We predict that the research has to be concentrated on study-ing the first kind of amino acids ---L type of Ala, Val, Asp, Glu and Gln, Arg and Lys surviving under β irradiation.

Table 2 Amino acid products of $HCN-H_2O$(I), $HCN-CH_4-H_2O$(II), $HCN-CH_4-H_2S-H_2O$(III), $CH_4-N_2-NH_3-H_2O$(IV) and $PH_3-CH_4-N_2-NH_3-H_2O$(V) system by Electric Discharge

Amino acid	I		II		III		IV	V	
	n	h	n	h	n	h*		P-1	P-2**
Gly	5.80	101	7.56	111	61.8	127	5.80	0.97	6.70
Ala			0.88	22.9	3.36	29.6	18.0	0.057	0.54
Val			0.38	1.99	0.52	3.24	34.0	0.044	
Ile	--	0.28			--	0.104		0.073	0.28
Leu					0.076	0.190		0.059	0.055
Ser	3.73	0.21	0.13	--	--	6.58		1.00	0.063
Thr			1.02	3.42	9.16	4.51		0.11	0.079
Asp	--	1.06	--	2.32	1.34	4.90	11.0	0.16	0.28
Glu			0.72	25.2	2.16	18.5	0.10		
Phe			0.17	1.22	0.41	2.79		0.046	0.160
Tyr					1.58	0.406		0.005	---
Arg					1.74	---			
Lys			0.39	0.37	0.17	0.60		0.031	0.110
His			1.05	0.63	10.2	0.52			
Cys					0.841	8.36			
Met					0.17	1.01			
Pro								0.032	

(Wang W. Q., Kobayashi K, Ponnamperuma C. 1984. Qi, S. C. Wang W. Q. et al. 1991)

* n=nonhydrolysis; h=hydrolysis product (μ mole)

** P-1: PH_3 $4x10^3$ Pa; P-2 PH_3 $9.2x10^3$ Pa.

In fact, among 20 amino acids, Gly, Ala, Ser, Asp and Glu are early members in the primitive Earth and others can be biosynthetically produced from these early amino acids. Gly has no handedness. Ala, Val, Asp and Glu all have positive rotatory strength for their L-isomers, L-Ser has positive rotatory strength in HCl. So the theoretical prediction of the handedness seems to be in agreement with observation data.

According to the theoretical prediction, the disputation and suspicion about the controversial experimental results between Garay (1968) and Darge (1976) could be explained, though they have not considered the radiolysis of solvent H_2O under β irradiation. In Darge case, the L-trp was decomposed to a great extent than the D-trp, whereas Garay found that D-tyr was preferentially destroyed, It may be attributed to the fact that the rotatory strength sign of L-tyrosine ($R_n^+ > 0$) and L-tryptophan ($R_n^+ < 0$) are of opposite sign at experimental condition.

ACKNOWLEDGMENT

This paper was supported by Grants of State Science Commission and Doctoral Program Foundation of Institution of Higher Education.

REFERENCES

Darge,W., 1974: Stereoselectivity of β irradiation of D,L-trypotophan in aqueous solution, Nature, 261, 522-524.

Garay,A.S., 1968: Origin and Role of Optical Isomery in Life, Nature 219, 338-339.

Garay,A.S., 1974: Origin of Asymmetry in Biomolecules, Nature, 250, 332-333.

Qi S. C., Wang W. Q., Wang Y. G., and Wu J. L., 1991: Electric Discharge of the Gas Mixture of HCN, H_2S, PH_3, N_2-NH_3 and CH_4-H_2O Stimmlating Prebiotic Synthesis, Chemical Journal of Chinese Universities 12, (1), 114-116

Salam, A., 1991: The role of chirality in the origin of life. J. Mol. Evol. 33, 105-113.

Wang W. Q., Kobayashi K., Ponnamperuma C., 1984: Prebiotic Synthesis in a Mixture of Methane, Nitrogen, Water and Phosphine, Acta Scientiarum Naturalum Universitatis Pekinensis, 6, 36-44.

Wang,J.Y. and Luo,L.F., 1985: The Origin of Biological Chirality and Weak Interaction, Scientia Sinica (Series B), Vol. XXVIII No.12, 1265-1277.

CHIRAL INTERACTION IN MOLECULAR SYSTEMS

G. Gilat
Technion-Israel Institute of Technology
Haifa 32000, Israel

ABSTRACT

Chiral interaction presents a new approach that may have certain dynamical advantages due to chirality. A few macroscopic examples, such as windmills, Crookes' radiometers and electric cells are discussed. A distinction between geometric and physical chirality is made. A simple model for chiral interaction between soluble proteins and ionic solvents is described. It is shown that two major structural features are required for chiral interaction, namely, chiral structure and an interface separating between two media. It is well known that all soluble proteins become globular before they can function as enzymes and it is the globular shape which provides for an interface separating between the solvent and the molecular interior. Chiral interaction generates an intrinsic perturbation or current, flowing in a single preferred direction out of two possible ones. This mode of perturbation is time-irreversible but obeys PT-invariance and it is non-ergodic as well. Such features may be related to a beginning of organization which, presumably, could be associated with a control mechanism of enzymatic activity. Nonergodicity may also be regarded as a necessary condition for molecular evolutionary capacity. Two different experiments aimed to verify the validity of chiral interaction are proposed. One is a direct experiment which involves the effect of a magnetic field at the solvent-air interface, whereas the other is indirect and involves the effect of ionic strength of the solvent. Chiral interaction in conjunction with terrestrial magnetic field might have also played a significant role in the evolutionary selection of the L-enantiomer of amino acids.

1. INTRODUCTION

It is well known that basic molecules of life, such as amino-acids and nucleic acids possess left (L) - right (D) asymmetry, known as chirality. The definition of a chiral body is that it cannot be made to overlap precisely its mirror-image. One of the main and most intriguing questions asked by researchers in moleuclar biology is how did nature select one enantiomeric species out of the two possible ones in order to construct complex molecules such as RNA, DNA and proteins. It is not the purpose of the present article to discuss this specific problem but rather to refer to an alternative question such as: "Why chiral? What is it good for?" or more specifically, "Is there any biological advantage to chiral over achiral molecular structures?"

As will be shown in the present article the answer to this question seems to be *positive* and this is strongly associated with certain phenomenological interactions to be named "chiral interaction" (CI) for which chirality is a *necessary* condition. Such modes of interaction have been introduced recently (Gilat 1985, 1986, 1991; Gilat and Schulman 1985). Before describing the concept of CI it may be of interest to make a significant distinction between, so-called, geometric and various physical chiralities. Recently these concepts have been clarified, as well as quantified (Gilat 1989, 1990, 1994). Geometric chirality concerns the shape of any geometric body whereas physical chirality is related to the spatial distribution of physical properties such as mass, or charge, or wave-function that can be distributed in space as chiral or achiral sets. It is important to realize that different physical properties of the same body or molecule can be distributed in space so that any one of them may yield a different degree of chirality that represents separately each mode of distribution. It is proposed hereby that the relevant degree of molecular chirality may be associated with that of its bonding electronic wave-function distribution. It ought to be also noted that the distinction between physical and geometric chiralities is often being confused in scientific literature. A simple example that stresses such a distinction is a perfectly symmetric tetrahedron which is geometrically *achiral*. By atttaching 4 different point masses to its corners this body becomes *chiral* with respect to its mass distribution. Last but not least it is important to emphasize that the idea and model of CI is largely based on symmetry arguments which cannot be readily refuted and this contributes to the CI hypothesis a certain degree of validity and viability in the absence of a direct experimental support.

2. CHIRAL INTERACTIONS

The concept of chiral interaction (CI) is to be related to any interaction between a system or a device and certain media such as electrolytes, EM radiation, flow of air or liquid, etc. for which the structural *chirality* of the device is a *necessary* condition. In order to describe better such a phenomenon, well recognized macroscopic devices are to be treated first.

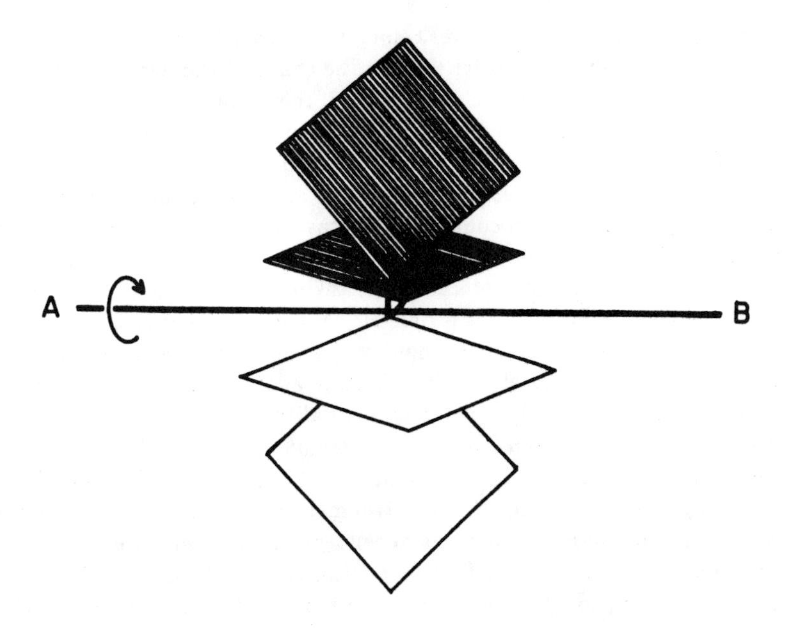

Figure 1: A scheme of the Crookes' radiometer as an example for chiral interaction. The difference in the optical absorption coefficient between black and silver wings generates a temperature difference between the wings when light is shining at the device. The air near the black wings expands as a result, which pushes the wings around the axis in the preferred direction of the black wings. The chirality is built into this device by the distribution of the colors, i.e. the optical absorption coefficient over the wings.

2.1 MACROSCOPIC CHIRAL DEVICES

In principle there exist two forms of CI, namely, mechanical and electric interactions. At first mechanical devices are described, namely, the windmill, the Crookes' radiometer and the rotating water sprinkler. The interacting media are flow or air, EM radiation and water flow, respectively. In all these devices there exists an *interface* separating between the device and the medium and this is where the medium interacts with the device. Chirality is therefore *built into* this interface. It is easy to observe that if, say, the surface of a windmill that interacts with the wind did possess reflection symmetry, the windmill "would not know" in which direction to rotate, clock or anti-clock-wise. Similar arguments hold also for the rotating water sprinkler. In the case of the Crookes' radiometer that interacts with light (see Fig. 1) the chirality is built into the rotating wings via the black and silver colors, that is, via the distribution of the optical absorption coefficient rather than in the shape of the wings. This is an impressive example of a physical chirality. In all these examples the interaction between the device and the medium, i.e. CI, ends up in the rotation of the device around a given axis in one preferred direction out of two possible ones.

Chiral interaction exists also in electric form and the examples for this are the electric cell and the thermocouple. In the case of the electric cell the interacting medium is an electrolyte and there exist interfaces separating the medium from the electrodes where the interaction is taking place. The chirality built into these interfaces is *not* in their geometric shape but rather in the chemical contents of the electrodes, that is in the distribution of the *electrochemical potential* on the surface of the electrodes. This is another fine example for the function of physical rather than geometric chirality. As a result of this CI a current can flow in one preferred out of two possible directions along a conductor connecting the electrodes. In the example of the thermocouple the chirality is built again into the interfaces between two metals by creating a difference in the electrochemical potential due to its dependence on temperature. In view of these examples it can be summarized that CI in electric devices results in the flow of currents along conductors in one preferred direction out of two possible ones.

The analysis of macro-CI is not of much interest *per se* but it may be of help in searching for CI in molecular systems such as proteins. This is done by analogy, in particular to electric chiral devices. For this purpose the main features of macroscopic CI are listed.

CI may occur between devices and suitable media, where chirality is built into an interface separating between the device and the medium. CI is taking place at this interface, which results in energy transfer from the medium to the device in the form of a motion taking place in one preferred direction out of two possible ones. There are two kinds of motion. One is mechanical, or massive rotation, about a given axis and the second is electric, in the form of a massless current around a circuit. This kind of motion violates time reversibility but it does obey PT-invariance (P = parity, T = time), that is, if one neglects entropic effects such as friction and electric resistance. This conclusion is of much interest on microscopic scale since it may provide for a mode of organization on a molecular level.

2.2 MICROSCOPIC CHIRAL INTERACTION

There exists a well recognized mode of interaction for which chirality is a necessary condition, namely, the effect of optical activity and related effects (Barron 1982, Wollmer 1983) that involve the scattering of polarized light from chiral substances. A similar effect exists also for polarized electrons (Blum and Thompson, 1989). These interactions are elastic and do not involve any energy transfer to the molecules, and therefore they can not excite any rotational currents throughout the molecular systems. Moreover, optical activity is a bulk interaction which does not require any interface. In view of these points, such interactions are to be considered as "chiral scatterings" rather than CI. It is interesting to point out that the effect of Second Harmonic Generation (SHG) (Meijer et al. 1990) is becoming also useful for detecting chirality.

The model of CI on molecular scale has been previously described in considerable detail (Gilat, 1991) for protein structures. In view of this it may not be necessary to elaborate on this presently. The main idea behind this model is that free ionic motion within the solvent, in the close vicinity of the biomolecule, can cause an internal perturbation, or current, in the molecule, causing it to flow along helical bonds in one preferred direction out of two possible ones. The structural ingredient of the molecule that is mainly responsible for this effect is the electric dipolar peptide bonding along the α-helix in proteins. It is interesting to notice that all these dipole moments are consistently oriented in a parallel order along the peptide chain, so that the effect is accumulative throughout the α-helix structure. The free ions moving in the close vicinity of the molecule are deflected from their random original tracks by the cumulative dipole-moment of the helical peptide, which induces an electric field along the helix and this results in a perturbation along a specific direction so that, by Lenz rule, it tries to reduce the effect of the induced field. Negative ions are deflected by the dipole moment in the opposite direction when compared to positive ions, but the field induced along the peptide bond is the same for both ions. The presence of an electric-dipole along the peptide H-bonding is the source of the physical chirality that enables the selection of the direction of the motion of the perturbation along the helical chain. The situation for β-sheets, which are also a constituent of proteins, is not treated here since its detailed 3-dimensional structure is not well recognized. Nevertheless, the 2-dimensional structure of β-sheet, in particular, the relative arrangements of the $NH^+ \ldots CO^-$ H-bonds also permits the generation of flow of perturbation around the rings in a compatible fashion, throughout the structure.

Another important ingredient of chiral interaction is the presence of an interface separating between the molecule and the interacting medium, i.e., the solvent. Actually, it is well known (Tschesche, 1983) that all soluble proteins must fold into a globular structure before they can function. The globular structure provides for the microscopic interface surrounding the molecule which is necessary for chiral interaction. This is demonstrated in Fig. 2 where it is shown schematically how CI averages out to zero in the absence of an interface.

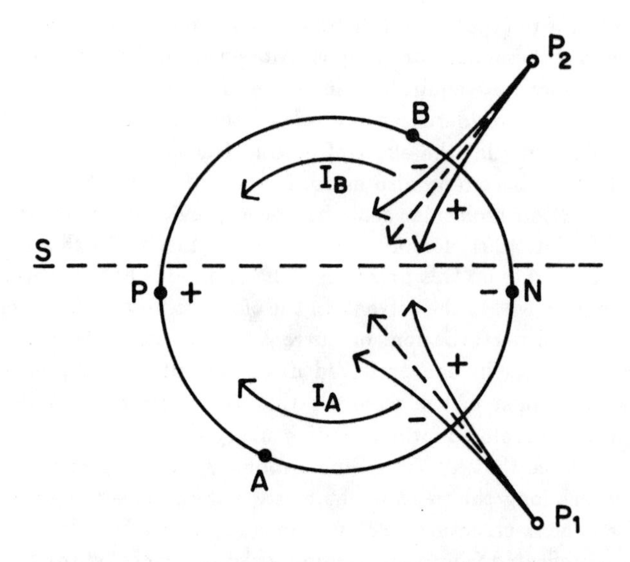

Figure 2. This figure explains schematically the need for the interface in generating microscopic CI. PN is an electric dipole moment attached to a ring of chemical bonds APBN. An ion P_1 approaching the ring from below will be deflected from its original track and will induce a perturbation I_A in a clock-wise direction. By the same token, an ion P_2 approaching from above will induce a similar perturbation I_B in the opposite direction, so that on the average these perturbations will cancel out. In order to obtain CI it is necessary to prevent the ions approaching from one of these direction, say P_2. This is accomplished by the surface marked by S.

In Fig. 3, chiral interaction is presented in a schematic form at the globular surface. The ionic motion in the solvent occurs at one side of the interface whereas the perturbation caused by this motion is taking place along the bonds throughout the molecule on the other side of the interface.

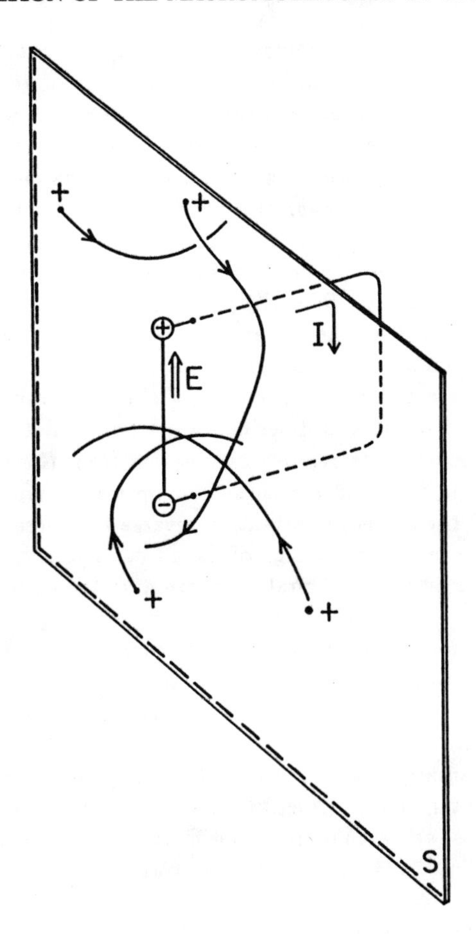

Figure 3: A schematic presentation of CI at the protein globular surface. The electric dipole moment represents the peptide H-bonds along a segment of α-helix. The surface S separates between the solvent and the moleuclar interior. Free ions are moving in the solvent, in front of S, and being deflected from their original tracks they generate an induced time-average field $\vec{E}$ causing the perturbation I to flow in one specific direction out of two possible ones along the closed rings of chemical bonds inside the molecule.

2.3 EXPERIMENTAL VERIFICATION

As yet, chiral interaction is still just a hypothesis awaiting experimental verification. A preliminary attempt to observe it has recently been performed

(Khanna, 1993) and yielded encouraging results, but this must be repeated in a more complete form before it can be considered to be of genuine value. In addition, there exist a few indirect evidences that support the CI hypothesis. Before discussing any possible experiment it may be of importance to indicate two objective difficulties in performing such experiments, namely, the size of the effect of CI, being of rather small magnitude, and the fact of its being an intrinsic molecular perturbation, which limits its accessibility to direct observation.

Two experiments have been proposed to look at chiral interaction. The first one is of direct nature (Gilat, 1990, 1991) and it involves a racemic solution of hydrophobic amino-acid at the water/air interface. In the presence of a magnetic field perpedicular to this interface the solution becomes slightly off-racemic at this interface due to the presence of chiral interaction there. The theoretical background for this is described elsewhere (Gilat 1990, 1991). It therefore becomes necessary to remove monolayers from the interface in the presence of a magnetic field and measure their optical activity elsewhere. If an effect is observed then it is important to reverse the direction of the field which should result in reversing the sign of the effect. The only reason for such an effect can be the existence of chiral interaction at the close vicinity of the water-air interface.

The possibility of modifying the racemic distribution of amino-acids at the water-air interface in the presence of a magnetic field opens the door for an evolutionary process of natural selection of a certain enantiomer L or D. This depends on the direction of the terrestrial magnetic field with respect to the ocean surface where an amino-acid concentration happened to exist in the pre-biotic era. The advantage factor (Avetisov et al., 1991) of one enantiomer with respect to another was rather small, but then there could happen certain amplification processes (Gilat, 1993) that might have further increased the advantage factor.

Another experiment that can support the CI hypothesis is of indirect nature (Gilat, 1991) and it involves a general assumption that CI is a *necessary condition* for enzymatic activity of soluble proteins. Such an assumption arises from the possibility that CI may play a significant role of control mechanism in the autocatalytic activity of enzymes. If so, then certain structural ingredients of proteins which are important for CI, become of biological significance. In the model of CI described above the ionic strength of the solvent plays a significant role for the validity of CI and hence, the assumption of the necessity of CI can be expressed as: "Soluble protein can function only if the ionic strength c of the solvent is larger than a certain threshold concentration c_0" (Gilat 1991). Very little can be said presently about the value of c_0 and this may vary from one protein to another. It is proposed, however, that close to this value the enzymatic activity K will rise sharply as a function of $c \geq c_0$ and then level off, so that for $c >> c_0$ the dependence of K on c will be marginal. Such an experiment can be readily performed and actually there is a certain evidence for such a behaviour in an experiment performed by Record et al. (1985). This is an encouraging evidence but still, being indirect, it cannot provide for an unambiguous support for the hypothesis of CI.

3. CONCLUSIONS AND COMMENTS

The hypothesis of chiral interaction in molecular systems represents a new mode of approach based on dynamical possibilities linked to structural ingredients of molecules, proteins in particular. The combination of chiral structure together with the existence of globular structure, as well as additional elements such as electric dipoles and closed rings of chemical bonds gives rise to intrinsic molecular perturbations that can move along the continuous chemical bonds in a preferred direction. This is due to stochastic free ionic motion in the solvent in the close vicinity of the molecule. It is interesting to notice as well that owing to the highly dielectric nature of water, the effect of the ionic motion is limited to only a few atomic layers around the molecule.

The phenomenon of chiral interaction on a molecular scale, if it can be validated experimentally, may account for several unusual features which may be of biological significance. It has already been pointed out that CI violates time-reversibility but maintains PT-variance. This result leads to the possible violation of the detailed balance principle (Reichl, 1980). Such a mode of interaction may, therefore, be regarded as a mode of organization that can allow for information transfer throughout the molecule. This is the reason why CI may account for the control mechanism of the autocatalytic function of enzymes which can be considered as a microscopic complex machine operation. The details of such a mechanism are still obscure and this may become a novel subject for research.

Apart from violating time reversibility CI has another characteristic of nonergodicity. The very property of selecting one mode of motion out of two equivalent ones prevents the molecular system from reaching thermal equilibrium which is characterized by filling completely the phase space. *Nonergodicity* on a molecular level may be regarded as a *necessary* condition for *molecular evolution*. In other words, a molecular system that can readily reach thermal equilibrium becomes unsuitable for evolution. For this reason most of the known molecules in nature have remained unchanged since their creation. The chiral nature of proteins together with their globular structure, which provides them with an interface, made this molecular systems capable of prebiotic evolution. CI, being a time-irreversible and nonergodic mode of organization may be of vital significance to the function of biomolecules which is related to the biological advantage of chirality mentioned in the Introduction.

In addition to these and in the absence of a complete experiment that can validate CI, it is also important to mention that CI is based largely on symmetry arguments being hard to refute. The consistency and compatibility of many structural elements of globular proteins with the performance of CI contribute also to the viability of the CI hypothesis.

In conclusion, it is necessary to emphasize that the C.I. hypothesis in molecular systems is still in its early stages of development. It still requires a more convincing experimental support and then it will be necessary to obtain more insight of its biological implications such as the possibility of its serving as control mechanism for enzymatic activity as well as for its relevance to molecular evolution and other possibilities and implications.

ACKNOWLEDGEMENT

The author wishes to thank the Ecole Normale Superieure as well as Professor André Rassat for the kind hospitality. This research was supported by the Fund of Promotion of Research at the Technion.

REFERENCES

Avetisov V.A., V.I. Goldanskii and V.V. Kuzmin, 1991, *Physics Today* **44**, 33-41.

Barron, L.D., 1982, *Molecular Light Scattering and Optical Activity*, Cambridge University Press.

Blum K. and D. Thompson, 1989, *J. Phys.* **B22**, 1823-1844.

Gilat G., 1985, *Chem. Phys. Lett.* **121**, 9-12.

Gilat, G., 1986, *Chem. Phys. Lett.* **125**, 129-133.

Gilat, G., 1989, *J. Phys.* **A22**, L545-L550.

Gilat G., 1990, *Chem. Phys.* **140**, 195-198.

Gilat G., 1990, *Found. Phys. Lett.* **3**, 189-196.

Gilat G., 1991, *Mol. Engin.* **1**, 161-178.

Gilat G., 1993, Chiral Interaction and Biomolecular Evolution, in C. Ponnampouma and J. Chela-Flores (Eds.) *Chemical Evolution and Origin of Life*, A. Deepak Publ.

Gilat G., 1994, *J. Math. Chem.* to be published.

Gilat G. and L.S. Schulman, 1985, *Chem. Phys. Lett.* **121**, 13-16.

Khanna R., 1993, Private Communication.

Meijer E.W., E.E. Havinga and G.L.J.A. Rikken, 1990, *Phys. Rev. Lett.* **65**, 37-39.

Record Jr. M.T., C.F. Anderson, P. Mills, M. Mossing and Jung-Hue Roe, 1985, *Adv. Biophys.* **20**, 109-135

Reichl L.E., 1980, *A Modern Course in Statistical Mechanics*, University of Texas Press, Austin.

Tschesche H., 1983, in W. Hoppe, W. Lohmann, H. Markl and H. Ziegler (eds.), *Biophysics*, Springer Verlag, Berlin, p. 37.

Wollmer A., 1983, in W. Hoppe, W. Lohmann, H. Markl and H. Ziegler (eds.), *Biophysics*, Springer Verlag, p. 144.

COSMOLOGICAL SOURCES OF MOLECULAR CHIRALITY

J. Chela-Flores
International Centre for Theoretical Physics, 34100 Trieste, Italy
and
Instituto Internacional de Estudios Avanzados, Apartado 17606 Parque Central, Caracas 1015A, Venezuela
and
N. Kumar
International Centre for Theoretical Physics, 34100 Trieste, Italy
and
Indian Institute of Science, Department of Physics, Bangalore 560012, India.

ABSTRACT

We address the old paradox in chemical evolution that chiral amino acids, rather than racemic mixtures, are abundant on Earth since they are universally present in proteins. Amino acids have still not been observed in interstellar space, in spite of the fact that cosmic organic matter is known to exist. We assume that the electroweak interaction is a truly universal chiral influence that may have induced a bias at a cosmological level of left-handed amino acids . The real question, however, is one of the smallness of the strength of the interaction which is due to its ultra short range. We begin by noting that this was not so in the symmetric phase before the Salam-Weinberg transition. This motivates us to discuss the minimal model, which assumes that supercooling and tunneling may have been possible between the two vacua of the Salam-Weinberg model of the electroweak interactions. We show that the electroweak phase transition may have been delayed due to supercooling and that subsequently tunneling may have occurred from an unbroken SU(2)XU(1) metastable symmetric-vacuum to the stable asymmetric-vacuum state. It is shown that for a certain choice of the parameters (the Higgs mass), the symmetric phase can persist long enough to strongly favour a definite homochirality of the primordial molecular clouds, which may in principle contain amino acids or their precursors. Clearly the model should be tested for consistency with cosmological constraints. One experimentally verifiable consequence is that there may exist in interstellar clouds amino acids, as indeed other asymmetric organic compounds, which are homochiral with preferred sign.

1. INTRODUCTION

The problem of the origin of the preference in biochemistry for chiral molecules has remained an important topic of research for a considerable time. In a seminal paper Frank proposed the well-known general mechanism for the evolution of biomolecular homochirality (Frank, 1953). Perhaps the most persistent theme in the search for the sources of biochirality was initiated after parity was discovered to be violated: the nuclear weak interaction, known to be responsible for this discrete symmetry breakdown, is also responsible for the preference of living organisms for chiral molecules (Bonner, 1979; Ulbricht, 1981; Salam, 1991; Mason, 1992):

The electroweak interaction, being a universal chiral influence (Barron, 1986), is a likely candidate responsible for inducing a bias in an otherwise racemic mixture of amino acids (we do not discuss, for the sake of brevity, biochirality in the molecules of life other than the amino acids).

Ab initio quantum chemistry calculations incorporating the electroweak interactions have suggested the energetic stabilization of the left-handed amino acids over the right-handed ones (Mason and Tranter, 1985). However, the effect is intrinsically weak since the electroweak energy-difference ΔE_{ew} between the enantiomers of a given amino acid is of the order of 10^{-14} J $mole^{-1}$, alternatively $\Delta E_{ew}/k_B T$ is of the order of 10^{-17} for a temperature of 300 K.

However, up to the present two possibilities are still open:

(a) Either the polymerization of chiral amino acids occurred on the early Earth (the standard scenario of chemical evolution), or

(b) A prebiotic abundance of preferentially left-handed amino acids may have been deposited during the Hadean subera and Early Archean. This remote period, before 3.5 Gyr ago, is known to have been characterized by heavy bombardment of meteorites. Estimates suggest that this phenomenon may have been responsible for the deliverance of organics comparable to those produced by other energy sources (Chyba, 1990, Chyba and Sagan, 1992).

The second possibility has received some experimental support based on samples taken from the Murchison meteorite (Engel and Nagy, 1982; Engel *et al.*, 1990; Pillinger, 1982).

We would like to discuss whether the possibility (b) is compatible with our current understanding of the laws of physics, and further we discuss the physical concepts that may amplify the small number $\Delta E_{ew}/k_B T$. We proceed, in Sec. 2, to consider the theoretical basis for the possible interstellar presence of already chirally biased amino acids.

2. A COSMOLOGICAL SCENARIO

From symmetry considerations it is clear that a P-odd (parity-violating) term in the Hamiltonian will, in general, lift the degeneracy between the ground states of the two handed enantiomers of an asymmetric molecule in the absence of tunnelling which can be ignored. Detailed quantum chemical calculations have confirmed that the molecules of L-amino acids and D-sugars are indeed energetically stabilized relative to the other chiral counterparts. This is in conformity with the chiral dominance of these biomolecules as observed in living systems (Mason and Tranter, 1985).

In contrast, a laboratory synthesis invariably yields racemic mixtures. The energy difference is, however, abysmally small as mentioned in Sec. 1. One must then, as has indeed been done, invoke some amplifying mechanism - either a physical cooperative phenomenon giving long-range coherence, or a Darwinian (chemical) kinetics involving autocatalysis, competition, and stabilization by selection.

There is, however, yet another possible prebiotic origin of chiral molecules (one to which we are compelled by the evidence of the Murchison meteorite, as discussed briefly in the Introduction). In the following we consider this. The smallness of the enantiomeric energy splitting is due ultimately to the weakness of the parity-violating term in the Hamiltonian (Mason and Tranter, 1985):

$$H_{pv} = -(G_F/4\sqrt{2}m_e c)\sum_a \sum_i N_a\,[(p_i.\sigma_i)\delta^3(r_i - r_a) + \delta^3(r_i - r_a)\,p_i.\sigma_i] \qquad (1)$$

that couples electrons to the nucleons. Here, G_F is the Fermi constant and N_a denotes the neutron number of the nucleus of atom a; the other symbols have their usual meaning. The summation is over all nuclear electron pairs in the atom a with a second summation over all atoms in the molecule.

In terms of the Standard Model (SM) of the electroweak interactions, the above Hamiltonian represents the neutral-current coupling of the electron to the neutron mediated by the massive neutral vector boson Z. Now the smallness of the parity-violating interaction (effective coupling) is not due to the smallness of the weak charge, but rather it is due to the smallness of the range of the weak interaction, inasmuch as it is mediated by the exchange of a massive vector boson. Indeed, the range

$$\lambda = h / M_B c << 1 \text{ fermi}, \tag{2}$$

where M_B is the vector boson mass.

Now, in the SM the mass of the intermediate vector gauge boson is generated through a spontaneous breaking of the local gauge symmetry SU(2)XU(1). In other words, a certain scalar (Higgs) field ϕ condenses to a non-zero expectation value below a phase transition temperature T_c giving masses to the otherwise massless gauge bosons. Thus, in our present observable universe the true physical vacuum is in the asymmetric phase ($<\phi> \neq 0$).

Above the transition (for $T > T_c$), however, the universe was in the symmetric phase ($<\phi> = 0$) and the vector bosons were massless, like the photon, and the P-odd Hamiltonian was at par with the electromagnetic interaction - one could speak of 'weak' coulomb field of a weak charge having infinite range. In the first approximation this will enhance the coupling strength in Eqn. (1) by $(a / \lambda)^2 \sim 10^{17}$, where $a \sim 10^{-8}$, the chemical bond length. In standard cosmology this Salam-Weinberg phase transition seems to have taken place about 10^{-11}s after the big bang at $T (= T_c) \approx 10^2$ Gev. If, however, the universe supercooled through this temperature and the high-temperature symmetric phase lived long enough, up to the epoch of nucleosynthesis of the light elements (H, C, N, O, P) and galaxy formation, then we could indeed have an episodic view (scenario) where molecules formed in the symmetric (false) metastable vacuum.

While the quantum chemistry in such a symmetric vaccum is not worked out, it is expected that the parity-violating terms in the molecular Hamiltonian will be as effective as the usual electromagnetic terms, and large chiral bias will result. The molecules will be synthesized with handedness (i.e., with preferred chirality). One could expect essentially optically pure primordial molecular clouds. While this *delayed transition* will have important consequences at the low-energy molecular level (i.e., affect chirality), it is not expected to modify other observed cosmological constraints. *This, however, must be checked.* Further, the latent heat of this delayed transition may be used up primarily in the rest masses of various massive particles produced, and thus contribute to entropy without appreciable reheating.

In the following section we atttempt to estimate the bound on the mass of the Higgs scalar that is imposed by the long lifetime of the symmetric vacuum required.

3. TENDENCY TO TUNNEL TO THE STABLE ASYMMETRIC PHASE ON THE TIME SCALE OF THE ORDER OF THE AGE OF THE UNIVERSE

Let us consider a single Higgs scalar field ϕ and the associated effective potential $U(\phi)$ at zero temperature including gauge vector loop corrections (Sher, 1989):

$$U(\phi) = (1/2) \mu^2 \phi^2 - \lambda\phi^4 + B\phi^4 \ln\phi^2/ v^2 \tag{3}$$

where $v = <\phi>$ denotes the expectation value of the Higgs field (the spontaneous symmetry breaking) with the empirical value of $B = 9.7 \times 10^{-5}$ (dimensionless).

Here, v = 248 Gev. $U(\phi)$ has two minima in the case (Sher, 1989 Sec. 4.2.1) when μ^2 is positive, but not too large and

$$m_H < m_{CW}, \tag{4}$$

where the Coleman-Weinberg mass is given by the expression (Guinon *et al.*, 1990, Eqn. 2.159):

$$m_{CW}^2 = 8Bv^2 \tag{5}$$

Under the condition given by Eqn. (4) two minima exist, one symmetric and one asymmetric. This is the case discussed by Frampton (1976) who, however, assumed the stable phase to be symmetric (i.e., when $\lambda < B$). In the present analysis we consider the case when the stable vacuum is the asymmetric phase (i.e., $\lambda > B$), but the universe supercools to persist in the symmetric phase which is, clearly, metastable. We now estimate the life-time of this metastable (false) symmetric vacuum. Following Frampton (1976), the expected number N of tunnelings in the past (backward light-cone) can be shown to be:

$$N = (10^{164}/324) \{ B^2 v^4 (1 - \eta)^4 / [I(\eta)^4] \exp\{ -54\pi^2 / [B(1 - \eta)^3] I(\eta)^4 \} \tag{6}$$

with $B = 9.7\text{x } 10^5$, $\eta = \lambda/B$ and

$$I(\eta) = [(2\eta - 1) x^2 - \eta x^2 + x^4 \ln x^2]^{1/2} dx \tag{7}$$

where x_o (>0) is the zero of the integrand, and $1 \geq \eta \geq 0.5$.

In order to have the metastable symmetric vacuum live long enough (i.e., until the galaxy formation era) we should have N << 1. With the empirical value B = 9.7×10^{-5} and v = 248 Gev, the Higgs mass is unfortunately low. It is clear that this happens over a range of values of η (The latent heat, and therefore, reheating is small - it is a weakly first-order phase transition). This bound may hopefully be relaxed for a more realistic effective potential $U(\phi)$, since in this preliminary scenario we have not taken into account the renormalization group improvement of the effective potential, and the fact that the top quark mass may be very large, altering $U(\phi)$ drastically.

Our result implies a lower bound for the Higgs mass that is not significantly different from the well known Linde-Weinberg lower bound (Guinion *et al.*, 1990; Sher, 1989). We would like to note that, strictly speaking, quarks and leptons are all massless in the symmetric phase and while we do need these charged particles to have non-zero masses, for physical reasons, the actual mass problem is not fully understood at present.

4. DISCUSSION AND CONCLUSIONS

We have discussed a minimal theoretical model in terms of which we have presented a possible scenario for the cosmological prebiotic origin of molecular homochirality of the correct sign. The scenario is compatible with the Standard Model of the electroweak interactions. This implies bounds on the Higgs mass.

As long as the full parameters of the SU(2)XU(1) gauge symmetry model are not experimentally fully known and allow a range of possible values for the Higgs mass (Guinon *et al.*, 1990), the present scenario remains a possibility However, this clearly has to be tested against known cosmological and high energy physics constraints.

The question of the entropy produced in the transition from the symmetric to the asymmetric vacuum deserves some consideration. In our scenario supercooling has delayed the phase transition into the broken symmetry phase of the Salam-Weinberg model as late as the era of galaxy formation; for this reason we assume that the entropy generated will not be sufficient to reheat significantly the substratum (matter and radiation). We expect the latent heat to materialize into particle rest masses. Clearly this aspect of the scenario requires further investigation.

The most relevant changes introduced by the delayed electroweak phase transition, due to supercooling, is at the level of quantum chemistry at the molecular level; the chemical bond will be modified since in the unbroken phase of the electroweak interaction the weak sector is generated by massless intermediate vector bosons. Therefore, the electronic shells will interact with the nucleus through a force other than the coulomb interaction; indeed, we maintain that this is the source of the molecular changes that gives rise to the chirality of amino acids. A related question concerns primordial nucleosynthesis (Weinberg, 1972; Novikov, 1983). The rate of interactions between leptons, protons, and neutrons undergo rapid conversions into one another

$$p + e^- \leftrightarrow n + \nu_e \tag{8}$$

$$p + \nu_e \leftrightarrow n + e^+ \tag{9}$$

These weak processes (8) and (9) are initially in a state of equilibrium between the nucleons. The neutron fraction (n_f) was 50% when the time was approximately 10^{-11} s after the hot big bang. A shifting balance of n_f takes place as the universe continues its expansion and the temperature continues to decrease. Consequently, the rate of these reactions decreases in the standard model and they are practically absent in the first few seconds of the expansion. The final value of n_f is approximately 0.15. In the present scenario the rate of the processes will be affected by the unbroken phase, which is delayed by supercooling, but the ratio n_f is expected to remain unaffected in that the equilibrium is expected to be hastened, but the abundances will be determined by the respective chemical potentials.

We have assumed the concept that the universe is in a single bubble. Some variations of the scenario are possible (Sher, 1989) and should be considered at a later date. In fact, multiple bubbles may occur, all of which would be on equal footing with our own universe. This possibility is consistent with the laws of physics (Coleman, 1977) and would lead to variations of the present scenario.

A verifiable prediction of our proposal would be the detection of optical activity of the interstellar matter assumed to contain not only homochiral amino acids but other homochiral molecules as well. We suggest that one should look for *multiple wavelengths anomalous optical rotation,* which compares the angle between the planes of polarization at two wavelentghs close to an absorption edge.

ACKNOWLEDGMENTS

The authors would like to thank Profs J. Pati, Q. Shafi, and J. Strathdee, for valuable discussions.

REFERENCES

Barron, L.D., 1986: True and false chirality and parity violation. Chem. Phys. Lett. **123,** 423-427..

Bonner, W.A. , 1979: Experiments on the abiotic origin and amplification of optical activity. In: Origins of optical activity in nature. Ed. D.C. Walker. Elsevier: Amsterdam

Chyba, C.F., 1990: Extraterrestial amino acids and terrestial life. Nature **348,** 113-114.

Chyba, C.F. and Sagan, C., 1992: Endogenous production, exogenous delivery and impact-shock synthesis of organic molecules: an inventory for the origins of life. Nature **355,** 125-132.

Coleman, S., 1977: Fate of the false vacuum: Semiclassical theory. Phys. Rev. D **15,** 2929-2936.

Engel, M.H., Macko, S.A., and Silfer, J.A., 1990: Carbon isotope composition of individual amino acids in the Murchinson meteorite. Nature **348,** 47-49.

Engel, M.H. and Nagy, B., 1982: Distribution and enantiomeric composition of amino acids in the Murchison meteorite. Nature **296,** 837-840.

Frampton, P.H., 1976: Vacuum instability and Higgs scalar mass. Phys. Rev. Letters **37,** 1378-1380.

Frank, F.C., 1953: On spontaneous asymmetric synthesis. Biochim. Biophys. Acta **11,** 459-463.

Guinon, J.F., Haber, H.E., Kane, G., and Dawson, S., 1990: The Higgs hunter's guide. Addison-Wesley: Menlo park, California.

Mason, S.F., 1992: Chemical Evolution Origin of the Elements, Molecules, and Living Systems. Clarendon Press: Oxford.

Mason, S.F. and Tranter, G.E., 1985: The electroweak origin of biomolecular handedness. Proc. R. Soc. Lond. A **397,** 45-65.

Novikov, I.D., 1983: Evolution of the universe. Cambridge University Press: London.

Pillinger, C.T., 1982: Not quite the full circle? - non-racemic amino acids in the Murchison meteorite. Nature **296,** 802.

Salam, A., 1991: The role of chirality in the origin of life. J. Mol. Evol., **33,** 105-113.

Sher, M., 1989: Electroweak Higgs potentials and vacuum stability. Phys. Reports **179,** 273-418.

Ulbricht, T.L.V., 1981: Reflections on the origin of optical asymmetry on Earth. Origins of Life **11,** 55-70.

Weinberg S., 1972: Gravitation and cosmology. John Wiley: New York. pp. 545-561.

PART VI
CELEBRATION OF CYRIL PONNAMPERUMA'S 70TH BIRTHDAY

Award of a plaque by
Professor Luciano Bertocchi, Deputy Director of the International Center for Theoretical Physics, to Professor Cyril Ponnamperuma on the occasion of his 70th birthday.

Message for Professor Cyril Ponnamperuma

I had very much hoped to have been able to participate in this Conference on Chemical Evolution and the Origin of Life: Self-Organisation of the Macromolecules of Life, and especially in the celebrations dedicated to the 70th birthday of Professor Cyril Ponnamperuma. Unfortunately, ill health has prevented me from doing so but I would like to send the following message to Cyril.

Our collaboration dates back many years to the time of the CHERAWN Conferences and has developed over the years through Cyril's deep involvement with the activities of the Third World Academy of Sciences, of which I am President. I feel priviledged to have known and worked with Cyril. I have valued his friendship and guidance on numerous occasions. I send him my very best wishes for a happy, healthy life and for many more fruitful years of collaboration.

With my very best wishes.

Abdus Salam

Créteil, October 18, 1993

Dear Professor CHELA-FLORES,

First, I would like to thank you, Professor PONNAMPERUMA and the whole Organising Committee for your kind invitation to the 1993 Conference.

Unfortunately, I will not be able to join you for this very special event and I am deeply sorry. I will try to be at least just a few minutes with you through this letter.

I would have been more than happy to attend, especially because the 93 conference is celebrating the 40 years of dense and fruitful scientific career of a man who has largely contributed to the development of the reseach field of chemical evolution, the origin of life, and exobiology. Professor Cyril Ponnamperuma was not only one of the initiators of this field in the USA, but he largely contributed to its expansion at the international level, from Sri Lanka to Europe, including France.

I remember quite well when 15 years ago, fresh Post-Doc, with a NSF grant (largely obtained with Cyril's help), I came to the University of Maryland and spent one full year at LCE, where Cyril asked me to have in charge the "Planetary Chemistry Group". What a souvenir, what an experience! Not because 1978-79 in the USA was a period of strong events: from the Three Miles Island disaster to the American hostages of the Teheran US Embassy, and the gas shortage problems in USA. But because it was for me a year of great scientific discoveries.

The discovery of what is a well structured - although very cosmopolitan laboratory, with real and fruitful exchanges between its groups, through many seminars, and freequent meetings, many social activities, which help in bringing together the members of a large laboratory, thanks to the Director of LCE and his charming spouse.

The discovery of a field at the interface of chemistry, astrophysics and chemical evolution: planetary chemistry, an attracting field for chemists working on the problem of the origin of life on Earth, and looking for new avenues of research, more associated with real observations, true applications and likely to be financially supported by official research administrations.

The discovery of the chemistry on Jupiter and, more generally, in the outer solar system. It was the time of the Voyager 1 encounter with Jupiter, and a large part of LCE became involved in Jovian researches: Jovian chemistry simulations (UV and electron irradiation), infrared studies of compounds of interest for the Jovian system (in particular of sulphur oxides) to interpret IRIS spectra of Jupiter and its satellites (in particular Io). I will always rememberthe trip of Shubash GUPTA out of Maryland (was it in New Jersey, Cyril?) to bring back a sample of SO3. This was just in the middle of the gas shortage period in the US, and the official plates of Shubash's car (he was driving a Maryland state car) did not help him to get gas without following the strong restrictions and without waiting in the long lines at the gas stations. When he finally came back to LCE, like a Knight of the Round Table with the Holy Grail, he did have the SO3 sample. But some colleagues of us deeply regretted this success, becaus for the next few weeks, the whole building was almost contaminated by this badly-smelling compound (in spite of the several precautions taken by Shubash). Fortunately, the IR spectrometer, which was used by Shubash for studying SO3 (at Professor KHANNA's laboratory) was not affected by this compound and we got nice IR spectra of gaseous SO3, which showed that it is not present in Io's atmosphere.

When I came back to France I did not work on SO3, but I largely used Professor PONNAMPERUMA's view of what should be a laboratory of chemical evolution and developed a group on space organic physical-chemistry: Cyril, at least indirectly, shares the paternity of it.

Thus let me express my gratitude to Professor Cyril PONNAMPERUMA. I wish that this conference will be a unique occasion for all of us to pay homage to him.

With my best regards ans my best wishes for the participants of the conference,

very sincerely yours,

Francois Raulin
(ex) LCE member

PROFESSOR CYRIL PONNAMPERUMA

Bibliography

LIST OF BOOKS

R. Buvet and C. Ponnamperuma, eds., 1971: Molecular Evolution I, North-Holland, London.

C. Ponnamperuma, ed., 1972: Exobiology, A Series of Collected Papers, North-Holland, London.

C. Ponnamperuma, 1972: The Origins of Life, Thames and Hudson, London, E. P. Dutton, New York.

J. Oro, S. Miller, C. Ponnamperuma and R. S. Young, eds., 1974: Cosmochemical Evolution and the Origins of Life, Vol. I, D. Reidel Publishing Company, Dordrecht, Boston.

J. Oro, S. Miller, C. Ponnamperuma and R. S. Young, eds., 1974: Cosmochemical Evolution and the Origins of Life, Vol. II, D. Reidel Publishing Company, Dordrecht, Boston.

C. Ponnamperuma and A. G. W. Cameron, eds., 1974: Interstellar Communication: Scientific Perspectives, Houghton Mifflin, Boston.

C. Ponnamperuma, ed., 1975: Chemical Evolution of the Giant Planets, Proceedings of the College Park Colloquia on Chemical Evolution, October 29 - November 1, 1974, Academic Press, New York.

C. Ponnamperuma, ed., 1976: Chemical Evolution of the Precambrian, Proceedings of the College Park Colloquia on Chemical Evolution, October 29 - November 1, 1975, Academic Press, New York.

C. Ponnamperuma, ed., 1978: Chemical Evolution - Comparative Planetology, Proceedings of the College Park Colloquia on Chemical Evolution, September 29 - October 1, 1976, Academic Press, New York.

G. Field, G. Verschuur and C. Ponnamperuma, 1978: Cosmic Evolution, Houghton Mifflin, Boston.

C. Ponnamperuma and L. Margulis, eds., 1980: Limits of Life, Proceedings of the College Park Colloquia on Chemical Evolution, October 18 - 20, 1978, D. Reidel Publishing Company, Dordrect, Holland.

C. Ponnamperuma, ed., 1981: Comets and the Origin of Life, Proceedings of the College Park Colloquia on Chemical Evolution, October 29 - 31, 1980, D. Reidel Publishing Company, Dordrecht, Holland.

C. Ponnamperuma, ed., 1983: Cosmochemistry and the Origin of Life, Proceedings of the NATO Advanced Study Institute, Maratea, Italy, June 1-2, 1981, D. Reidel Publishing Company, Dordrecht.

F. R. Eirich and C. Ponnamperuma, eds., 1990: Prebiological Self Organization of Matter, Proceedings of the Eighth College Park Colloquia on Chemical Evolution, A. Deepak Publishing Company, Hampton, Virginia.

C. Ponnamperuma and C. G. Gehrke, eds., 1992: A Lunar-Based Chemical Analysis Laboratory, A. Deepak Publishing, Hampton, Virginia, 23666, 296 pages.

C. Ponnamperuma and J. Chela-Flores, eds., 1993: Chemical Evolution: Origin of Life, Proceedings of the Trieste Conference on Chemical Evolution and the Origin of Life, 26-30 October, 1992, A. Deepak Publishing Company, Hampton, Va.

ARTICLES AND REPORTS

1957 - 1989

1. D. G. Arnott and C. Ponnamperuma, 1957: "An Emanating Source for I^{123}," International Journal of Applied Radiation and Isotopes, 2:85-86.

2. C. Ponnamperuma, R. M. Lemmon, E. L. Bennett and M. Calvin, 1961: "Deamination of Adenine by Ionizing Radiation," Science, 134:113.

3. C. Ponnamperuma, R. M. Lemmon and M. Calvin, 1962: "Chemical Effect of Ionizing Radiation on Cytosine," Science, 137:605.

4. C. Ponnamperuma, 1962: "The Radiation Chemistry of Nucleic Acid Constituents," Thesis, Office of Technical Services, U.S. Department of Commerce, Washington, D.C.

5. C. Ponnamperuma, R. M. Lemmon and M. Calvin, 1963: "The Radiation Decomposition of Adenine," Radiation Research, 18:540-551.

6. C. Ponnamperuma, R. M. Lemmon, R. Mariner and M. Calvin, 1963: "Formation of Adenine by Electron Irradiation of Methane, Ammonia, and Water," Proc. of Natl. Acad. of Sci., 49:737-740.

7. C. Ponnamperuma, R. Mariner and C. Sagan, 1963: "Formation of Adenosine by Ultraviolet Irradiation of a Solution of Adenine and Ribose," Nature, 199:1199-1200.

8. C. Ponnamperuma, C. Sagan and C. Mariner, 1963: "Synthesis of Adenosine Triphosphate under Possible Primitive Earth Conditions," Nature, 199:222-226.

9. C. Ponnamperuam, 1964: "Chemical Evolution and the Origin of Life," Nature, 201:337-340.

10. C. Ponnamperuma, P. Kirk, R. Mariner and B. Tyson, 1964: "A Coincidence Technique for Paper Chromatography," Nature, 202:393-394.

11. C. Ponnamperuma, 1964: "The Origin of Life in the Universe," NASA Conference, Science in the Space Age, Los Angeles, California, June 1-4, 1964, Government Printing Office, Washington, D.C., 61-69.

12. C. Ponnamperuma and P. Kirk, 1964: "Synthesis of Deoxyadenosine under Simulated Primitive Earth Conditions," Nature, 203:400.

13. C. Ponnamperuma, 1964: "L'evoluzione Chimica e L'Origine Della Vita," Sapere, 619.

14. C. Ponnamperuma, 1964: "Synthesis of Organic Compounds in Primitive Planetary Environments," University of California in "Horizons in Space Biosciences: Exobiology" series, Spring.

15. C. Ponnamperuma and F. Woeller, 1964: "Differences in the Character of C_6 to C_9 Hydrocarbons from Gaseous Methane in Low-Frequencey Electric Discharges," Nature, 203:272-274.

16. C. Ponnamperuma, R. S. Young, E. F. Muñoz and B. K. McCaw, 1964: "Guanine: Formation During the Thermal Polymerization of Amino Acids," Science, 143:1449-1450.

17. R. S. Young and C. Ponnamperuma, 1964: "Early Evolution of Life," American Institution of Biological Sciences, Biological Sciences Curriculum Study Pamphlet No. 11, Boston, D. C. Heath.

18. R. S. Young and C. Ponnamperuma, 1964: "Life: Origin and Evolution," Science, 143:384-388.

1965

19. C. Ponnamperuma, 1965: "A Biological Synthesis of Some Nucleic Acid Constituents," The Origin of Prebiological Systems and of Their Molecular Matrices, S. W. Fox, ed., Academic Press, New York, 221-242.

20. C. Ponnamperuma, 1965: "The Beginnings of Man-Made Life," Medical Opinion and Review, 1:50-53.

21. C. Ponnamperuma, 1965: "The Chemical Origin of Life," Science Journal, 1:39-45.

22. C. Ponnamperuma, 1965: "Chemical Studies on the Origin of Life," Proceedings of the Conference on Exploration of Mars and Venus, August 23-27, Virginia Polytechnic Institute Engineering Extension Series Circular, No. 5, Chapter VII, 8.

23. C. Ponnamperuma, 1965: "Life in the Universe - Imitations and Implications for Space Science," Astronautics and Aeronautics, 3:66-69.

24. C. Ponnamperuma, 1965: "Primordial Organic Chemistry," Exobiology Summer Study Anthology, Space Science Board, National Academy of Sciences, Washington, D.C.

25. C. Ponnamperuma, 1965: "Primordial Organic Chemistry and the Origin of Life," Proceedings of the Third International Symposium on Bioastronautics and the Exploration of Space, San Antonio, Texas, Theodore C. Bedwill, Jr. and Hubertus Strughold, eds., 117-128.

26. C. Ponnamperuma, 1965: "The Search for Extraterrestrial Life," The Science Teacher, 32:21-26.

27. C. Ponnamperuma and R. Mack, 1965: "Nucleotide Synthesis under Possible Primitive Earth Conditions," Science, 148:1221-1223.

28. C. Ponnamperuma and E. Peterson, 1965: "Peptide Synthesis from Amino Acids in Aqueous Solution," Science, 147:1572-1574.

29. R. S. Young, C. Ponnamperuma and B. K. McCaw, May 1965: "Abiogenic Synthesis on Mars," Life Sciences and Space Research III, A Session of the 5th International Space Symposium, Florence, Italy, M. Florkin, ed., (COSPAR) North-Holland Publishing Company, Amsterdam, 127-138.

1966

30. C. Ponnamperuma, 1966: "The Origin of Life," International Dictionary of Geophysics, vol. 2, S. K. Runcorn, ed., Pergamon Press, Oxford, 799-804.

31. C. Ponnamperuma, 1966: "The Role of Radiation in Primordial Organic Synthesis," Radiation Research, Proceedings of the Third International Congress on Radiation Research, Cortina, Italy, G. Silini, ed., North Holland Publishing Company, Amsterdam, 700-713.

32. C. Ponnamperuma, 1966: "Some Recent Work on Prebiological Synthesis of Organic Compounds," Icarus, 5:450-454.

33. C. Ponnamperuma and K. Pering, 1966: "Possible Abiogenic Origin of Some Naturally Occurring Hydorcarbons," Nature, 209:979-982.

34. C. Ponnamperuma and E. Peterson, 1966: "The Ultraviolet Irradiation of an Aqueous Solution of Tyrosine," Radiation Research, 27:519.

1967

35. W. V. Allen and C. Ponnamperuma, 1967: A Possible PrebioticSynthesis of Monocarboxylic Acids," Currents in Modern Biology, 1:24-28.

36. L. D. Caren and C. Ponnamperuma, September 1967: "A Review of Some Experiments on the Synthesis of Jeewanu," NASA TM X-1439.

37. K. Dose and C. Ponnamperuma, 1967: "The Effect of Ionizing Radiation on n-acetylglycine in the Presence of Ammonia," Radiation Research, 31:650.

38. N. W. Gabel and C. Ponnamperuma, 1967: "Model for Origin of Monosaccharides," Nature, 216:453-455.

39. C. Ponnamperuma, 1967: Book Review of "Bioorganic Mechanisms," Medical Opinion and Review by T. C. Bruice and S. J. Benkovic, 2v, 3:125, 1966.

40. C. Ponnamperuma 1967: "On the Origin of Life," Infectious Diseases, Their Evolution and Eradication, A. Cockburn and C. C. Thomas, eds., Springfield, Illinois, 3.

41. C. Ponnamperuma, 1967: "Une Approche Chimique au Probleme de L'Origine de la vie," La biogenese, Ecole Superieure de Physique et de Chimie Industrielle de la ville de Paris, 63-68.

42. C. Ponnamperuma and K. Pering, 1967: "Aliphatic and Alicyclic Hydorcarbons Isolated from Trinidad Lake Asphalt," Geochimica et Cosmochimica Acta, 31:1350-1354.

43. C. Ponnamperuma and F. Woeller, 1967: " α -Aminonitriles Formed by an Electric Discharge through a Mixture of Anhydrous Methane and Ammonia," Currents in Modern Biology, 1:156-158.

44. C. Ponnamperuma, R. S. Young and L. D. Caren, 1967: "Some Chemical and Microbiological Studies of Surtsey," Surtsey Research Progress Report, The Surtsey Research Society, Reykjavik, Iceland, 3:70-80.

45. C. Reid, L. E. Orgel and C. Ponnamperuma, 1967: "Nucleoside Synthesis under Potentially Prebiotic Conditions," Nature, 216:936.

1968

46. G. W. Hodgson and C. Ponnamperuma, 1968: "Prebiotic Porphyrin Genesis: Porphyrins from Electric Discharge in Methane, Ammonia, and Water Vapor," Proc. of the National Acad. of Sciences, 59:18-22.

47. C. Ponnamperuma, 1968: "Chemical Evolution and the Origin of Life," Arbeistagung úber Extraterrestrische biophysik und Biologie und Raumfahrt Medizin, 2nd Marburg, October 9-10, 1967, Tagungsbericht, 183-195.

48. C. Ponnamperuma, June 1968: "Chemical Evolution and the Origin of Life," Sonderdrúck aus Forschungsbericht W. 68-30 des bundesminesteriums fúr wissenschaftliche Forschung, Seite 183-195.

49. C. Ponnamperuma, 1968: "Ultraviolet Radiation and the Origin of Life," Photophysiology, III, A. C. Giese, ed., Academic Press, 253-267.

50. C. Ponnamperuma and L. Caren, 1968: "Chemical Studies on the Origin of Life," Encyclopedia of Polymer Science and Technology, 9:649-659.

51. C. Ponnamperuma and N. W. Gabel, 1968: "Current Status of Chemical Studies on the Origin of Life," Space Life Sciences, 1:64-96.

52. J. Rabinowitz, S. Chang and C. Ponnamperuma, 1968: "Phosphorylation by Way of Inorganic Phosphate as a Potential Prebiotic Process," Nature, 218:442-443.

53. A. Schwartz and C. Ponnamperuma, 1968: "Phosphorylation of Adenosine with Linear Polyphosphate Salts in Aqueous Solution," Nature, 218:443.

1969

54. S. Chang, J. Flores and C. Ponnamperuma, 1969: "Peptide Formation Mediated by Hydrogen Cyanide Tetramer: A Possible Prebiotic Process," Proc. of National Acad. of Science, 64:1011-1015.

55. W. T. Huntress, J. S. Baldeschwieler and C. Ponnamperuma, 1969: "Ion-Molecule Reactions in Hydrogen Cyanide," Nature, 223:468-471.

56. C. Munday, K. Pering and C. Ponnamperuma, 1969: "Synthesis of Acyclic Isoprenoids by the γ-Irradiation of Isoprene," Nature, 223:867-868.

57. K. Pering and C. Ponnamperuma, 1969: "Alicyclic Hydrocarbons from an Unusual Deposit on Derbyshire, England: A Study in Possible Diagenesis," Geochimica et Cosmochimica Acta, 33:528-532.

58. C. Ponnamperuma, 1969: "Chemical Evolution and the Origin of Life," Proceedings of the Fourth International Symposium on Bioastronautics and the Exploration of Space, San Antonio, Texas, June 1968, 47-61.

59. C. Ponnamperuma, F. Woeller, J. Flores, M. Romiez and W. Allen, 1969: "Synthesis of Organic Compounds by the Action of Electric Discharges in Simulated Primitive Atmospheres," American Chemical Society, Advances in Chemistry, 80:280-288.

60. F. Woeller and C. Ponnamperuma, 1969: "Organic Synthesis in a Simulated Jovian Atmosphere," Icarus, 10:386-392.

1970

61. K. A. Kvenvolden, S. Chang, J. W. Smith, J. Flores, K. Pering, C. Saxinger, F. Woeller, K. Keil, I. Breger and C. Ponnamperuma, 1970: "Carbon Compounds in Lunar Fines from Mare Tranquillitatis - I: Search for Molecules of Biological Significance," Geochimica et Cosmoshimica Acta Supplementa 1: Proceedings of the Apollo 11 Lunar Science Conference, A. A. Levinson, ed., Pergamon Press, 2:1813-1828.

62. S. Chang, J. A. Williams, C. Ponnamperuma and J. Rabinowitz, 1970: "Phosphorylation of Uridine with Inorganic Phosphates," Space Life Sciences, 2:144-150.

63. G. Hodgson, E. Bunnenberg, B. Halpern, E. Peterson, K. Kvenvolden and C. Ponnamperuma, 1970: "Carbon Compounds in Lunar Fines from Mare Tranquillitatis - II: Search for Porphyrins," Geochimica et Cosmochimica Acta Supplement I: Proceedings of the Apollo 11 Lunar Science Conference, A. A. Levinson, ed., Pergamon Press, 2:1845-1956.

64. G. W. Gehrke, R. W. Zumwalt, W. A. Aue, D. L. Stalling, A. Duffield, K. A. Kvenvolden and C. Ponnamperuma, 1970: "Carbon Compounds in Lunar Fines from Mare Tranquilitatis - III: Organosiloxanes in Hydrochloric Acid Hydolysates," Geochimica et Cosmochimica Acta Supplement 1: Proceedings of the Apollo 11 Lunar Science Conference, A. A. Levinson, ed., Pergamon Press, 2:1845-1856.

65. G. W. Hodgson, E. Peterson, K. A. Kvenvolden, E. Bunnenberg, B. Halpern and C. Ponnamperuma, 1970: "Search for Porphyrins in Lunar Dust," Science, 167:763-765.

66. S. Chang, J. W. Smith, I. Kaplan, J. Lawless, K. A. Kvenvolden and C. Ponnamperuma, C., 1970: "Carbon Compounds in Lunar Fines from Mare Tranquillitatis - IV. Evidence for Oxides and Carbides," Geochimica et Cosmochimica Acta Supplement 1: Proceedings of the Apollo 11 Lunar Science Conference, A. A. Levinson, ed., Pergamon Press, 2:1857-1869.

67. K. A. Kvenvolden, J. Lawless, K. Pering, E. Peterson, J. Flores, C. Ponnamperuma, I. R. Kaplan and C. Moore, 1970: "Evidence for Extraterrestrial Amino Acids and Hydrocarbons in the Murchison Meteorite," Nature, 288:923-926.

68. K. A. Kvenvolden and C. Ponnamperuma, eds., 1970: "A Search for Carbon and Its Compounds in Lunar Samples from Mare Tranquillitatis," NASA SP-257.

69. C. Ponnamperuma, 1970: "A La Recherche de la Vie sur la Lune," Atomes, 25:169-174.

70. C. Ponnamperuma, 1970: "Chemical Evolution and the Origin of Life," New York State Journal of Medicine, presented at the 163rd Annual Convention of the Medical Society of the State of New York, February 9, 1969.

71. C. Ponnamperuma, 1970: "Search for Life on the Moon," Priroda, 8:61-65.

72. C. Ponnamperuma, 1970: "Editorial: Space Biology and the Origin of Life," Space Life Sciences, 2:119-120.

73. C. Ponnamperuma, L. Caren and N. Gebel, 1970: "Molecular Differentiation in Primordial Systms," Cell Differentiation, Chapter 2, O. A. Schjeide and J. deVellias, eds., Van Nostrand, New York, 15-30.

74. C. Ponnamperuma and H. P. Klein, 1970: "The Coming Search for Life on Mars," The Quarterly Review of Biology, 45:235-258.

75. C. Ponnamperuma, K. A. Kvenvolden, S. Chang, R. Johnson, G. Pollack, D. Philpott, I. Kaplan, J. Smith, J. W. Schopf, C. W. Gehrke, G. Hodgson, I. A. Breger, B. Halpern, A. Duffiendl, K. Krauskopf, E. Barghoorn, H. Holland and K. Keil, 1970: "Search for Organic Compounds in the Lunar Dust from the Sea of Tranquillity," Science, 167:760-762.

76. M. W. West and C. Ponnamperuma, 1970: "Chemical Evolution and the Origin of Life (A Comprehensive Bibliography)," Space Life Sciences, 2:225-295.

1971

77. M. S. Chadha, J. G. Lawless, J. Flores and C. Ponnamperuma, 1971: "Experiments in Simulated Jovian Atmosphere," Molecular Evolution, R. Buvet and C. Ponnamperuma, eds., 1, North Holland Publishing Company, 143-151.

78. S. Chang, K. A. Kvenvolden, J. Lawless, C. Ponnamperuma, and I. Kaplan, 1971: "Carbon, Carbides, and Methane in an Apollo 12 Sample," Science, 171:474-477.

79. K. A. Kvenvolden, J. G. Lawless and C. Ponnamperuma, 1971: "Non-Protein Amino Acids in the Murchison Meteorite," Proceedings of the National Academy of Sciences, 68:486-490.

80. G. W. Hodgson, E. Bunnenberg, B. Halpern, E. Peterson, K. A. Kvenvolden and C. Ponnamperuma, 1971: "Lunar Pigments: Porphyrin-like Compounds from Apollo 12 Samples," Proceedings of the Second Lunar Science Conference, vol. 2, 1865-1874.

81. C. E. Folsome, J. G. Lawless, M. Romiez and C. Ponnamperuma, 1971: "Heterocyclic Compounds Indigenous to the Murchison Meteorite," Nature, 232:108-109.

82. J. G. Lawless, K. A. Kvenvolden, E. Peterson, C. Ponnamperuma and C. Moore, 1971: "Amino Acids Indigenous to the Murray Meteorite," Science, 173:626-627.

83. H. Noda and C. Ponnamperuma, 1971: "Polymer Formation in a Simulated Jovian Atmosphere," Molecular Evolution, 1, R. Buvet and C. Ponnamperuma, eds., North-Holland Publishing Company, 236-244.

84. K. Pering and C. Ponnamperuma, 1971: "Aromatic Hydrocarbons in the Murchison Meteorite," Science, 173:237-239.

85. C. Ponnamperuma, 1971: "Organic Geochemistry: Methods and Results," Die Naturwissenschaften, Book Review.

86. C. Ponnamperuma and S. Chang, 1971: "The Role of Phosphates in Chemical Evolution," Molecular Evolution, 1, R. Buvet and C. Ponnamperuma, eds., North-Holland Publishing Company, 216-223.

87. P. W. Banda and C. Ponnamperuma, 1971: "Polypeptides from the Condensation of Amino Acid Adenylates," Space Life Science, 3(1):54-62.

88. C. Ponnamperuma and J. Skehan, 1971: "Boston College Environmental Center Summer Institute on Surtsey and Iceland," NASA TM X-62, 009.

89. J. Rabinowitz, S. Chang and C. Ponnamperuma, 1971: "Possible Mechanism for Prebiotic Phosphorylation," Prebiotic and Chemical Evolution, A. P. Kimball and J. Oro, eds., North-Holland Publishing Company, Amsterdam, 70-77.

90. A. Schwartz and C. Ponnamperuma, 1971: "Phosphorylation of Nucleosides by Condensed Phosphates in Aqueous Systems," Prebiotic and Biochemical Evolution, A. P. Kimball and J. Oro, eds., North-Holland Publishing Company, Amsterdam, 78-82.

91. D. Buhl and C. Ponnamperuma, 1971: "Interstellar Molecules and the Origin of Life," Space Life Sciences, 3, 157-164.

92. M. S. Chadha, J. J. Flores, J. G. Lawless and C. Ponnamperuma, 1971: "Organic Synthesis in a Simulated Jovian Atmosphere, II," Icarus, 15, 39-44.

93. M. S. Chadha, L. Replogle, J. Flores and C. Ponnamperuma, 1971: "Possible Role of Aminoacetonitriles in Chemical Evolution," Bioorganic Chemistry, 1, 269-274.

94. K. Dose and C. Ponnamperuma, 1971: "Effect of Ionizing Radiation and UV Light on N-Acetylglycine and the Presence of Ammonia," Biophysik, 7:311-321.

95. C. Ponnamperuma, 1971: "Amino Acids in the Murchison Meteorite - A Summary," Intra-Science Chemical Report, 5:403.

96. C. Ponnamperuma and M. Sweeney, 1971: "The Role of Ionizing Radiation in Primordial Organic Synthesis," Theory and Experiment in Exobiology, A. Schwartz, ed., Wolters-Nordhoff, 1:1-40.

97. M. W. West, E. C. Gill and C. Ponnamperuma, 1971: "Chemical Evolution and the Origin of Life," Space Life Sciences, 3:293-304.

98. C. Ponnamperuma, 1971: "Chemical Evolution and the Origin of Life," Graduate School Chronicle, University of Maryland, 1:4-6.

99. C. Ponnamperuma, 1971: "Primordial Organic Chemistry and the Origin of Life," Quarterly Review of Biophysics, 4:77-106.

100. W. C. Saxinger and C. Ponnamperuma, 1971: "Experimental Investigation on the Origin of the Genetic Code," Journal of Molecular Evolution, 1:63-73.

101. W. C. Saxinger, C. Ponnamperuma and C. Woese, 1971: "Evidence for Interaction of Nucleotides with Immobilized Amino Acids and Its Significance for the Origin of the Genetic Code," Nature, 234:172-174.

1972

102. J. Flores and C. Ponnamperuma, 1972: "Ploymerization of Amino Acids under Primitive Earth Conditions," Journal of Molecular Evolution, 2:1-9.

103. C. Gehrke, R. Zumwalt, K. Kuo, W. Aue, D. Stalling, K. A. Kvenvolden and C. Ponnamperuma, 1972: "Amino Acid Analysis of Apollo 14 Samples," Proceedings of the Third Lunar Science Conference, Geochimica et Cosmochimica Acta, Supplement 3, The MIT Press, 2:2119-2129.

104. C. Gehrke, R. Zumwalt, K. Kuo, J. Rash, W. Aue, D. Stalling, K. Kvenvolden and C. Ponnamperuma, 1972: "Research for Amino Acids in Lunar Samples," Space Life Sciences, 3:439-499.

105. G. W. Hodgson, K. Kvenvolden, E. Peterson and C. Ponnamperuma, 1972: "A Quest of Phoryphyrins in Lunar Soil: Samples from Apollo 11, 12, and 14," Space Life Sciences, 3:419-424.

106. J. G. Lawless, K. Kvenvolden, E. Peterson, C. Ponnamperuma and E. Jarosewich, 1972: "Evidence for Amino Acids of Extraterrestrial Origin in the Orgueil Meteorite," Nature, 236 (5341), 66-67.

107. P. Molton and C. Ponnamperuma, 1972: "The Survival of Common Terrestrial Microorganisms under Simulated Jovian Conditions," Nature, 238:217-218.

108. C. Ponnamperuma, 1972: "A la Recherche de la vie Extraterrestre," Sciences, Revue de la Civilisation Scientifique, 76:49-55.

109. C. Ponnamperuma, 1972: "Lunar Organic Analysis: Implications for Chemical Evolution," Space Life Sciences, 3:493-496.

110. C. Ponnamperuma, 1972: "Organic Compounds in the Murchison Meteorite," Annals New York Academy of Sciences, presented at the Conference on Interstellar Molecules and Cosmochemistry, New York, June 16-18, 1971, 194:56-70.

111. C. Ponnamperuma, 1972: Editorial, Space Life Sciences, vol. 3, no. 4.

112. C. Ponnamperuma and N. Gabel, 1972: "Prebiological Synthesis of Organic Compounds," Chemistry and Space Research, A. Rembaum and R. Landel, eds., Elsevier, New York, 45-82.

113. C. Ponnamperuma and N. Gabel, 1972: "Primordial Organic Chemistry," Exobiology, C. Ponnamperuma, ed., North Holland, Amsterdam, 1:81-116.

114. C. Ponnamperuma and P. Molton, 1972: "Astrochimica et Esobiologia," Enciclopedia della Scienze e della Tecnica, A. Mondadori, ed., 89-100.

115. C. Saxinger, C. Ponnamperuma and D. Gillespie, 1972: "Nucleic Acid Hybridization with RNA Immobilized on Filter Paper," Proceedings of the National Academy of Sciences, 69:2975-2978.

1973

116. C. E. Folsome, J. G. Lawless, M. Romiez and C. Ponnamperuma, 1973: "Heterocyclic Compounds Recovered from Carbonaceous Chondrites," Geochimica et Cosmochimica Acta, 37:455-465.

117. C. Gehrke, R. Zumwalt, K. Kuo, C. Ponnamperuma, C. N. Cheng, and A. Shimoyama, 1973: "Extractable Organic Compounds in Apollo 15 and 16 Fines," Proceedings of the Fourth Lunar Science Conference, 2249-2259.

118. P. Molton, J. Williams and C. Ponnamperuma, 1973: "The Survival of Common Bacteria in a Liquid Culture under Carbon Dioxide at High Temperatures," Nature, 243:242-243.

119. A. Mondadori, 1973: A Biography of Cyril Ponnamperuma, Biographic Encyclopedia of Scientists and Technologists of the 20th Century, Milan, Italy.

120. C. Ponnamperuma, 1973: "Origin of Life," Chemical Geology, Book Review, 11:233-234.

121. C. Ponnamperuma, 1973: "Interstellar Molecules: Significance for Prebiotic Chemistry," Molecules in the Galactic Environment, M. Gordon and L. Snyder, eds., John Wiley & Sons, New York. Presented at the Symposium on Interstellar Molecules, National Radio Astronomical Observatory, Charlottesville, Virginia, October 4-6, 1971.

122. C. Ponnamperuma and P. Molton, 1973: "Life on Jupiter?" New Scientist, 692-693.

123. C. Ponnamperuma and P. Molton, 1973: "Prospect of Life on Jupiter," Space Life Sciences, 4:32-44.

124. R. S. Young and C. Ponnamperuma, March 1973: "Early Evolution of Life," BSCS Pamphlet, revised.

1974

125. M. S. Chadha, P. Molton and C. Ponnamperuma, 1974: "Amino-nitriles: Possible Role in Chemical Evolution," Origins of Life,6:127-136, 1975. Also in Cosmochemical Evolution and the Origins of Life, Vol. II, ed. by J. Oro et al., D. Reidel Publishing Company, Dordrecht, Boston.

126. P. Molton and C. Ponnamperuam, 1974: "Organic Synthesis in a Simulated Jovian Atmosphere. III. Synthesis of Amino-nitriles," Icarus, 21:166-174.

127. W. K. Park, A. Hochstim and C. Ponnamperuma, 1974: "Organic Synthesis by Quench Reactions," Origins of Life, 6:99-107, 1975. Also in Cosmochemical Evolution and the Origins of Life, Vol. II, ed. by Oro et al., D. Reidel Publishing Company, Dordrecht and Boston.

128. C. Ponnamperuma and N. W. Gabel, 1974: "The Precellular Evolution and Organization of Molecules," Symposia of the Society for General Microbiology. XXIV. Evolution in the Microbial World, 393-413.

129. C. Ponnamperuma, 1974: "The Chemical Basis of Extraterrestrial Life," in Interstellar Communication: Scientific Perspectives, ed. by C. Ponnamperuma and A. G. W. Cameron, Houghton Mifflin, Boston, 45-58.

130. C. Saxinger and C. Ponnamperuma, 1974: "Interactions between and Nucleotides in the Prebiotic Milieu," Origins of Life, 5:189-200. Also in Cosmochemical Evolution and the Origins of Life, Vol. I, ed. by Oro et al., D. Reidel Publishing Company, Dordrecht and Boston, 189-200.

131. C. Ponnamperuma, 1974: A book review, "De Vitae Origine," Nature, 247(5438):241.

132. C-N. Cheng and C. Ponnamperuma, 1974: "Extraction of Amino Acids from Soils and Sediments with Superheated Water," Geochimica et Cosmochimica Acta, 38:1843-1848.

133. M. Gay, P. Molton and C. Ponnamperuma, 1974: "Electron Impact Induced Fragmentation of Some Substituted 3-Amino-propionitriles," Organic Mass Spectrometry, 9:1124.

134. C. Ponnamperuma, 1974: "How Life Began," Maryland, Vol. II, No. 4, 7-9.

135. C. Ponnamperuma, 1974: "Chemistry - The Origins of Life," A.A.A.S. Science Books, Vol. 9, No. 4, III, 1974.

1975

136. C. Ponnamperuma, April 17, 1975: "The Search for Extraterrestrial Life," Chancellor's Lecture 1975, University of Maryland, College Park, Maryland, April 17.

137. C. Ponnamperuma, 1975: A book review, "Carbonaceous Meteorites," Chemical Geology, 16(4):315.

138. C. Gehrke, R. Zumwalt, K. Kuo, C. Ponnamperuma and A. Shimoyama, 1975: "Search for Amino Acids in Returned Lunar Soil," Origins of Life, 6:541-550.

139. C. Ponnamperuma, 1975: A book review, "The Origin of Life and Evolutionary Biochemistry," Bioscience, 669.

1976

140. C. Ponnamperuma, 1976: "The Indian Science Congress, 1976 - An Agenda for Action," Science, 192(45):2.

141. C. Ponnamperuma, 1976: "The Organic Chemistry and Biology of the Atmosphere of the Planet Jupiter," Icarus 29:321-328.

142. C. Ponnamperuma, 1976: Editorial, "International Seminar on the Origin of Life," Origins of Life, 7(3):91.

143. A. Negron-Mendoza and C. Ponnamperuma, 1976: "Formation of Biologically Relevant Carboxylic Acids during the Gamma Irradiation of Acetic Acid," Origins of Life, 7(3):191-196.

144. J. Hulshof and C. Ponnamperuma, 1976: "Prebiotic Condensation Reactions in an Aqueous Medium: A Review of Condensing Agents," Origins of Life, 7(3):197-224.

145. C. Ponnamperuma, 1976: "Organic Synthesis in a Simulated Jovian Atmosphere," The Chemical Evolution of the Giant Planets, ed. by C. Ponnamperuma, Academic Press, New York. 221-231.

146. C. Ponnamperuma, November 1976: "Life beyond the Earth," Astronautics and Aeronautics, 50-55.

147. A. M. Sweeney, A. Toste and C. Ponnamperuma, 1976: "Formation of Amino Acids by Cobalt-60 Irradiation of Hydrogen Cyanide Solutions," Origins of Life, 7(3):187-189.

1977

148. E. Griffith, C. Ponnamperuma and N. Gabel, 1977: "Phosphorus, A Key to Life on the Primitive Earth," Origins of Life, 8(2):71-85.

149. H. Mizutani and C. Ponnamperuma, 1977: "The Evolution of the Protein Synthesis System, I," Origins of Life, 8(3):183-220.

150. C. Ponnamperuma, A. Shimoyana, T. Hobo, M. Yamada and R. Pal, 1977: "Possible Surface Reactions on Mars: Implications for Viking Biology Results," Science, 197, 455-457.

151. R. Hanel, B. Conrath, D. Gautier, P. Gierasch, S. Kumar, V. Kunde, P. Lowman, W. Maguire, J. Pearl, J. Pirraglio, C. Ponnamperuma and R. Samuelson, 1977: "The Voyager Infrared Spectroscopy and Radiometry Investigation," Space Science Review, 21:129-157.

152. Ponnamperuma, C., January 19, 1977: "Are We Alone in the Universe?" Paper presented at the Special Premier of "The Loneliness Factor," Buhl Planetarium, Pittsburgh, Pennsylvania.

153. C. Ponnamperuma, 1977: "Cosmo-chemistry and the Origins of Life," in Origins of Life, the Earth and the Universe, ed. by Mitsubishi-Kasei Institute of Life Sciences, Heibonsha, Tokyo, 35-70.

1978

154. I. Draganic, Z. Draganic, A. Shimoyama and C. Ponnamperuma, 1978: "Evidence for Amino Acids in Hydrolysates of Compounds Formed by Ionizing Radiation: I. Aqueous Solutions of HCN, NH_4CN, and NaCN," Origin of Life, ed. by H. Noda, Center for Academic Publications, Tokyo, 129-134.

155. Z. Draganic, I. Draganic, A. Shimoyama and C. Ponnamperuma, 1978: "Evidence for Amino Acids in Hydrolysates of Compounds Formed by Ionizing Radiation: II. Aqueous Solutions of CH_3CN and C_2H_5 CN," Origin of Life, ed. by H. Noda, Center for Academic Publications,Tokyo, 83-88.

156. T. Hobo, C. Ponnamperuma, A. G. Hook and B. Donn, 1978: "Lower Molecular Weight Hydrocarbon Formation in an Open Flow System by Fischer Tropsch Reaction," Origin of Life, ed. by H. Noda, Center for Academic Publications, Tokyo, 89-94.

157. H. Mizutani and C. Ponnamperuma, 1978: "The Effect of Polynucleotides on the Dimerization of Glycine," Origin of Life, ed. by H. Noda, Center for Academic Publications, Tokyo, 273-277.

158. A. Negron-Mendoza and C. Ponnamperuma, 1978: "Interconversion of Biologically Important Carboxylic Acids by Radiation," Origin of Life, ed. by H. Noda. Center for Academic Publications, Tokyo, 101-104.

159. C. Ponnamperuma, 1978: "Prebiotic Molecular Evolution," Origin of Life, ed. by H. Noda, Center for Academic Publications, Tokyo, 67-81.

160. C. Ponnamperuma, 1978: "The Origin of Life in the Universe," EPOCA, 1437:48.

161. C. Ponnamperuma, 1978: "Cosmochemistry, and the Origin of Life," Proceedings of the Robert A. Welch Foundation Conferences on Chemical Research. XXI. Cosmochemistry, ed. by W. O. Milligan, Houston, 137-197.

162. C. Ponnamperuma, 1978: "Our Most Remote Ancestors," Chemistry, 51(9):6-12.

163. C. Ponnamperuma, A. Shimoyama, M. Yamada, T. Hobo and R. Pal, 1978: "Possible Surface Reactions on Mars: II. Implication for Viking Labeled Release Results," Origin of Life, ed. by H. Noda, Center for Academic Publications, Tokyo, 45-49.

164. A. Shimoyama, N. Blair and C. Ponnamperuma, 1978: "Synthesis of Amino Acids under Primitive Earth Conditions in the Presence of Clay," Origin of Life, ed. by H. Noda, Center for Academic Publications, Tokyo, 95-100.

1979

165. R. Hanel, B. Conrath, M. Flasar, V. Kunde, P. Lowman, W. Maguire, J. Pearl, J. Pirraglia, R. Samuelson, D. Gautier, P. Gierasch, S. Kumar and C. Ponnamperuma, 1979: "Infrared Observations of the Jovian System from Voyager I," Science, 204(4396):972-76.

166. R. Hanel, B. Conrath, M. Flasar, L. Herath, V. Kunde, P. Lowman, W. Maguire, J. Pearl, J. Pirraglia, R. Samuelson, D. Gautier, P. Gierasch, L. Horn, S. Kumar and C. Ponnamperuma, 1979: "Infrared Observations of the Jovian System from Voyager 2," Science, 206(4421):952-56.

167. R. K. Kotra, A. Shimoyama, C. Ponnamperuma and P. E. Hare, 1979: "Amino Acids in a Carbonaceous Chondrite from Antarctica," Journal of Molecular Evolution, 13(3):179-184.

168. J. Pearl, R. Hanel, V. Kunde, W. Maguire, K. Fox, S. Gupta, C. Ponnamperuma and F. Raulin, 1979: "Io: Identification of Gaseous SO_2 and New Upper Limits for other Gases on IO," Nature, 286:755-58.

169. C. Ponnamperuma, 1979: "The Emergence of Life," J. Royal Swaziland Soc. Sci. Tech., 2(1):9-13.

170. C. Ponnamperuma, 1979: "Primordial Organic Chemistry," Chemistry in Britain, 15(11):560-68.

171. F. Raulin, A. Bossard, G. Toupance and C. Ponnamperuma, 1979: "Abundance of Organic Compounds Photochemically Produced in the Atmosphere of the Outer Planets," Icarus, 38:358-366.

172. A. Shimoyama, C. Ponnamperuma and K. Yanai, 1979: "Amino Acids in the Yamato Carbonaceous Chondrite from Antractica," Nature, 282:394-96.

173. A. Shimoyama, C. Ponnamperuma and K. Yanai, 1979: "Amino Acids in the Yamato Meteorite 74662, an Antractic Carbonaceous Chrondrite," Mem. Natl. Inst. of Polar Resarch, Special Issue 15, 196-204.

1980

174. E. Friebele, A. Shimoyama and C. Ponnamperuma, 1980: "Possible Selective Adsorption of Enantiomers by Na-Montmorillonite," Proceedings, ISSOL Conference, 1980, D. Reidel Publishing, Dordrecht, 1981.

175. E. Friebele, A. Shimoyama and C. Ponnamperuma, 1980: "Adsorption of Protein and Non-Protein Amino Acidson a Clay Mineral: A Possible Role of Selection in Chemical Evolution," Journal of Molecular Evolution, 16(314):269-278.

176. C. N. Karkhanis, A. N. Goodwin and C. Ponnamperuma, January 1980: "Paleobiology and Organic Geochemistry of Archean Helen Iron Formation, Michipicoten Area," Journal of the Geological Society of India, 21(1):1-9.

177. H. Kristjansson and C. Ponnamperuma, 1980: "Purification and Properties of Malate Dehydrogenase from the Extreme Thermophile, Bacillus Caldolyticus," Limits of Life, C. Ponnamperuma and L. Margulis, eds., D. Reidel Publishing, Dordrecht, 47-54.

178. H. Kristjansson and C. Ponnamperuma, 1980: "Purification and Properties of Malate Dehydrogenase from the Extreme Thermophile, Bacillus Caldolyticus," Origins of Life, 10(2):185-192.

179. H. Mizutani and C. Ponnamperuma, March 1980: "The Evolution of the Protein Synthesis System II: From Chemical Evolution to Biological Evolution," Origins of Life, 10(1):31-38.

180. A. Negron-Mendoza, C. Ponnamperuma and R. L. Graff, 1980: "γ-Irradiation of Malic Acid in Aqueous Solution," Origins of Life, 10(4).

181. T. Owen, J. Caldwell, A. R. Rivolo, A. L. Lane, C. Sagan, G. Hunt and C. Ponnamperuma, 1980: "Observations of the Spectrum of Jupiter from 1500 to 2000°A with IUE: Evidence for C_2H_2 and P_2," Astrophysical Journal, 236:L39-L42.

182. L. G. Pleasant and C. Ponnamperuma, 1980: "Chemical Evolution and the Origin of Life: Bibliography Supplement, 1977," Origins of Life, 10(1):69-87.

183. L. G. Pleasant and C. Ponnamperuma, 1980: "Chemical Evolution and the Origin of Life: Bibliography Supplement, 1978," Origins of Life, 10(4):379-404.

184. C. Ponnamperuma, 1980: Book Review, "Life Sciences and Space Research," Space Science Reviews, Volume 16, 25:84.

185. C. Ponnamperuma, Winter 1980: "Primordial Cosmochemistry and the Origin of Life," Science/Technology and the Humanities (STTH), 3(1):1-11.

186. C. Ponnamperuma, 1980: "Uncontaminated Carbonaceous Chondrites from the Antarctic," Antractic Journal of the United States, 14(5):42-44.

187. F. Raulin, P. Price and C. Ponnamperuma, 1980: "Quantitative Gas Chromatographic Analysis of Volatile Amines in the Presence of Low Molecular Weight Hydrocarbons and Nitriles," American Laboratory, 121(10):45.

188. A. Shimoyama and C. Ponnamperuma, 1980: "Adsorption of Some Amino Acids on Na-Montmorillonite: Implications of the Adsorption for Chemical Evolution," Biogeochemistry of Amino Acids, P. E. Hare, ed., John Wiley & Sons, New York, 145-152.

1981

189. S. Gupta, E. Ochiai and C. Ponnamperuma, 1981: "Organic Synthesis in the Atmosphere of Titan," Nature, 293:5835.

190. R. Hanel, B. Conrath, F. M. Flasar, V. Kunde, W. Maguire, J. Pearl, J. Pirraglia, R. Samuelson, L. Herath, M. Allison, D. Cruikshank, D. Gautier, P. Gierasch, L. Horn, R. Koppany and C. Ponnamperuma, C., 1981: "Infrared Observations of the Saturnian System from Voyager 1," Science, 212(4491):192-200.

191. T. Hobo, Y. Masaaki, S. Suzuki, Sh. Araki, A. Shimoyama and C. Ponnamperuma, 1981: "Recognition of Amino Acid Chirality in Gas Chromatography by Sequential Use of Enantiomeric Stationary Phases," in Bunseki Kagaku, 30(6):T71-T76 (in Japanese).

192. R. K. Kotra, A. Shimoyama, C. Ponnamperuma, P. E. Hare and K. Yanai, 1981: "Organic Analysis of the Antractic Carbonaceous Chondrites," Origin of Life, Y. Wolman, ed., D. Reidel Publishing Co., Dordrecht, Holland, 51-57.

193. F. Raulin and C. Ponnamperuma, 1981: "Possible Role of Phosphine in Chemical Evolution," Origin of Life, Y. Wolman, ed., D. Reidel Publishing Co., Dordrect, Holland, 107-114.

194. F. Stoetzel, A. Shimoyama, C. Ponnamperuma and R. Buvet, 1981: "The Role of Analytical Proceduresin the Formation of Biochemicals from Experiments Simulating the Chemical Evolution of Primeval Earth," Origin of Life, Y. Wolman, ed., D. Reidel Publishing Co., Dordrect, Holland, 197-199.

195. E. Friebele, A. Shimoyama and C. Ponnamperuma, 1981: "Possible Selective Adsorption of Enantiomers by Na-Montmorrillionite," Origin of Life, Y. Wolman, ed., 1980, D. Reidel Publishing, Co., Dordrect, Holland.

196. C. Walters, A. Shimoyama and C. Ponnamperuma, 1981: "Organic Geochemistry of the Isua Supracrystals," Origins of Life, Y. Wolman, ed., D. Reidel Publishing Co., Dordrect, Holland, 473-479.

197. H. Mizutani and C. Ponnamperuma, 1981: "Effect of Polynucleotides on the Dimerization of Glycine," Origins of Life, 11(3):237-242.

198. L. G. Pleasant and C. Ponnamperuma, 1981: "Chemical Evolution and the Origin of Life, Bibliography Supplement, 1979," Origins of Life, 11(3):273-288.

199. C. Ponnamperuma, 1981: "Harold Clayton Urey: 1893-1981," Origins of Life, 11(3):269-270.

200. C. Ponnamperuma, May 1981: "Harold Clayton Urey: Chemist of the Cosmos," Sky and Telescope, 61(5):397.

201. C. Ponnamperuma, September 7, 1981: Book Review, "Life Beyond Earth -- The Intelligent Earthling's Guide to Life in the Universe," Chemical and Engineering News, 78-79.

202. C. Ponnamperuma, 1981: "The Quickening of Life," The Fire of Life, Smithsonian Exposition Books.

1982

203. M. Akiyama, A. Shimoyama and C. Ponnamperuma, 1982: "Amino Acids from the Late Precambrian Thule Group, Greenland," Origins of Life, 12, 215-227.

204. R. Hanel, B. Conrath, F. M. Flasar, V. Kunde, W. Maguire, J. Pearl, J. Pirraglia, R. Samuelson, D. Cruikshank, D. Goutier, P. Gierasch, L. Horn and C. Ponnamperuma, 1982: "Infrared Observations of the Saturnian System from Voyager 2, Science, 215, No. 4534.

205. A. Negron-Mendoza and C. Ponnamperuma, 1982: "Prebiotic Formation of Higher Molecular Weight Compounds from the Phyotolysis of Aqueous Acetic Acid," Photobiology, 36, 595-597.

206. H. Okihana and C. Ponnamperuma, 1982: "A Protective Function of the Coacervates Against UV Light on the Primitive Earth, "Nature, 299, 347-349.

207. C. Ponnamperuma, 1982: "Genetic Takeover," New Scientist, 96, book review, 453-454.

208. C. Ponnamperuam, 1982: "Life in the Universe" and "Cosmic Dawn," Physics Today, 35, 59.

209. C. Ponnamperuma, 1982: "Life Itself," New Scientist, 94, 435.

210. C. Ponnamperuma, A. Shimoyama and E. Friebelle, 1982: "Clay and the Origins of Life," Origins of Life, 12, no. 4, p. --.

211. C. Ponnamperuma, E. Friebelle and A. Shimoyama, 1982: "The Role of Clay Minerals in Prebiotic Protein Synthesis," Japan Scientific Societies Press, 15-30.

212. C. Ponnamperuma, 1982: "Evidence for the Earliest Life on Earth," McGraw-Hill Yearbook of Science and Technology.

213. C. Ponnamperuma, April 1982: "Chemical Evolution of Life," Science Today, 43-47.

214. C. Ponnamperuma and S. Gupta, 1982: "Organic Synthesis in the Atmosphere of Titan," Origins of Life, 12, 266.

215. C. Ponnamperuma, 1982: "Cosmochemistry and the Origins of Life, Basic Research as an Integrated Component of a Self-Reliant Base of Science and Technology," Proceedings of the Indian Research Council Association, M. Menon and A. Sharma, eds., 111-128.

216. E. Ochiai and C. Ponnamperuma, 1982: "Inorganic Chemistry of Earliest Sediments -- Bioorganic Chemical Aspects of the Origins and Evolution of Life," Cosmochemistry and the Origin of Life, Proceedings of the NATO Advanced Study Institute Meeting, Maratea, Italy.

217. C. Ponnamperuma, November 3-7, 1982: "Cosmochemistry and the Origin of Life," Proceedings of the XVI Mexican Congress of Pure and Applied Chemistry, Morelia, Mexico.

218. C. Ponnamperuma, 1982: "Chemical Evolution: Historical BAckground," NASA Special Publication, James Lawaless, ed.

219. C. Ponnamperuma, 1982: "Exobiology," Funk and Wagnall's New Encyclopedia, vol. --, pgs. --.

220. C. Ponnamperuma, 1982: "Fossil Organic Matter," 1982 McGraw-Hill Yearbook of Science and Technology, vol. --, p. --.

221. C. Ponnamperuma and E. Friebele, 1982: "The Antiquity of Carbon," Nuclear and Chemical Dating Techniques, American Chemical Society Symposium, Series no. 1760.

222. C. Ponnamperuma and E. Ochiai, 1982: "Comets and the Origin of Life," Comets, L. L. Wilkening, ed., University of Arizona Press, 696-703.

223. C. Ponnamperuma, 1982: "The Winged-Bean and the World Food Crisis," Proceedings of the Second International Symposium on the Winged Bean, W. Herath, ed., Colombo, Sri Lanka.

224. C. Ponnamperuma, 1982: "Cosmochemistry," Proceedings of the NATA Advanced Study Institute Meeting, Moratea, Italy, June 1-12, 1981, D. Reidel Publishing Co., Dordrech, Holland.

225. C. Ponnamperuma, March 21-29, 1982: "Geochemical Aspects of the Origin of Life," Proceedings of the 7th Annual Meetings of the Japan Society for the Study of the Origin of Life, N. Morimoto, ed., Tokuhama, Japan.

226. C. Ponnamperuma, 1982: "Vita della Origine," Encylopedia Italiana, G. Bedeski, ed., vol. --, p. --.

227. C. Ponnamperuma, A. Shimoyama and E. Friebelle, 1982: "Clay and the Origins of Life," Origins of Life, 12, 9-40.

228. C. Ponnamperuma, December 1982: "Cosmic Dawn: The Origin of Matter and Life," Physics Today, vol. --, p. --.

229. C. Ponnamperuma, 1982: "Cosmochemistry and the Origin of Life in the Universe," Proceedings of the Anaheim Symposium, Texas University Press, Arizona, vol. --, p. --.

230. C. Ponnamperuma, 1982: "De Origine Vitae," Europeo, vol. -- , p. 54-62.

231. C. Ponnamperuma, January 3-8, 1982: "Chemistry in Our Lives, Basic Research as an Integral Component of a Self-Reliant Base of Science and Technology," Proceedings of the Indian Science Congress Association Meeting, Mysoir, India, 54-62.

232. C. Ponnamperuma, 1982: "Cosmochemistry and the Origin of Life", J. Chem. Educ., 59 (2) 89-90.

233. A. Negron-Mendoza and C. Ponnamperuma, 1982: "Role of Succinic Acid in Chemical Evolution", Origins of Life, 12, pp. 427-431.

234. H. Okihana and C. Ponnamperuma, 1982: "Functions of the Coacervate Droplets", Origins of Life, 12, pp. 347-353.

235. L. G. Pleasant and C. Ponnamperuma, 1982: "Chemical Evolution and the Origin of Life: Bibliography Supplement 1980," Origins of Life, 12, D. Reidel Publishing Co., Dordrecht, Holland, 93-118.

236. C. Ponnamperuma and M. K. Hobish, 1982: "The Galapagos Hydrothermal Vent Ecologies: Possibilities for Neoabiogenesis?" in the First Symposium on Chemical Evolution and the Origins and Evolution of Life, NASA Conference Publication 2276, 77.

237. M. K. Hobish, B. J. Cherayil and C. Ponnamperuma, 1982: "NMR Studies of Interactions between Nucleic Acids and Proteins," in the First Symposium on Chemical Evolution and the Origins and Evolution of Life, NASA Conference Publication 2276, 60.

136. C. Ponnamperuma, 1982: "Genetic Takeover," New Scientist, 96, book review, 453-454.

137. C. Ponnamperuam, 1982: "Life in the Universe" and "Cosmic Dawn," Physics Today, 35, 59.

138. C. Ponnamperuma, 1982: "Life Itself," New Scientist, 94, 435.

139. C. Ponnamperuma, A. Shimoyama and E. Friebelle, 1982: "Clay and the Origins of Life," Origins of Life, 12, no. 4.

140. C. Ponnamperuma, E. Friebelle and A. Shimoyama, 1982: "The Role of Clay Minerals in Prebiotic Protein Synthesis," Japan Scientific Societies Press, 15-30.

141. C. Ponnamperuma, 1982: "Evidence for the Earliest Life on Earth," McGraw-Hill Yearbook of Science and Technology.

142. C. Ponnamperuma, April 1982: "Chemical Evolution of Life," Science Today, 43-47.

143. C. Ponnamperuma and S. Gupta, 1982: "Organic Synthesis in the Atmosphere of Titan," Origins of Life, 12, 266.

144. C. Ponnamperuma, 1982: "Cosmochemistry and the Origins of Life, Basic Research as an Integrated Component of a Self-Reliant Base of Science and Technology," Proceedings of the Indian Research Council Association, M. Menon and A. Sharma, eds., 111-128.

1983

238. C. Ponnamperuma, 1983: "The Geochemical Approach to the Study of the Origin of Life," reprinted from the Proceedings of the Indo-U.S. Workshop on "Precambrians of South India," Hyderabad, January 12-14, 1982, 61-70.

239. K. Kobayashi and C. Ponnamperuma, 1983: "Chemical Evolution and Trace Elements", Kagaku no Ryoiki, 37 (8), pp. 578-583 (in Japanese).

240. L. G. Pleasant and C. Ponnamperuma, 1983: "Chemical Evolution and the Origin of Life. Bibliography Supplement 1981", Origins of Life, 13, 61-80.

241. C. Ponnamperuma, 1983: "Cosmochemistry and the Origin of Life", in Cosmochemistry and the Origin of Life (ed. by C. Ponnamperuma), D. Reidel Publishing Co., Dordrecht, pp. 1-34.

242. C. Ponnamperuma (ed.), 1983: "Cosmochemistry and the Origin of Life", Proceedings of the NATO Advanced Study Institute,

Maratea, Italy, June 1-12, 1981, D. Reidel Publishing Co., Dordrecht, p. 386.

243. M. K. Hobish and C. Ponnamperuma, 1983: "Structural Requirements for Associations between Phenylalanine and Nucleotides Comprising Its Genetic Code Sequences," Biochemistry 22, 33a.

244. B. J. Cherayil, M. K. Hobish and C. Ponnamperuma, 1983: "Structural Requirements for Associations between Mononucleotides and Dipeptides," Biochemistry 22, 34a.

245. X. Liang, M. K. Hobish and C. Ponnamperuma, 1983: "Chirality of Amino Acids formed by Electric Discharge on a Model Primitive Earth Atmosphere," Biochemistry 22, 34a.

1984

246. C. Ponnamperuma and M. Hobish, 1984: "The Stereochemical Approach to Studies of the Origin of the Genetic Code," in Molecular Evolution and Protobiology, K. Matsuno, K. Dose, K. Harada and D. Rohlfing, eds., Plenum Publishing Corporation, New York, 295.

247. W. Wen-Qing, K. Kobayashi and C. Ponnamperuma, 1984: "Prebiotic Synthesis in a Mixture of Methane, Mitrogen, Water and Phosphine," in Acta Scientiarum Naturalum, Univeristy of Peking, 6, 36-44, (in Chinese)

248. L. G. Pleasant and C. Ponnamperuma, 1984: "Chemical Evolution and the Origin of Life, Bibliography Supplement, 1982," in Origins of Life, 15, D. Reidel Publishing Company, 55-69.

249. W. Altekar, H. Kristjansson, C. Ponnamperuma and L. Hochstein, 1984: "On Archaebacterial ATPase from Halobacterium saccharovorum", Origins of Life, 14, pp. 733-738.

250. L. G. Pleasant and C. Ponnamperuma, 1984: "Chemical Evolution and the Origin of Life. Bibliography Supplement 1982", Origins of Life, 15 (1), pp. 55-69.

251. C. Ponnamperuma and M.K. Hobish, 1984: "The Stereochemical Approach to Studies of the Origin of the Genetic Code", in Molecular Evolution and Protobiology, (ed. by K. Matsuno, K. Dose, K. Harada, and D.L. Rohlfing), Plenum Press, New York, pp. 295-312.

252. W. Wang, K. Kobayashi, and C. Ponnamperuma, 1984: "Prebiotic Synthesis in a Mixture of Methane, Nitrogen and Phosphine", Acta Science Nat. University Peking, 6, pp. 36-44 (in Chinese).

1985

253. C. Ponnamperuma, 1985: "Louis Pasteur and the Origin of Life," World's Debt to Pasteur, Alann R. Liss, Inc., 117-130.

254. C. Ponnamperuma, April 1985: "The Track of Extraterrestrial Life Grows Warmer," Aerospace America, the American Institute of Aeronautics and Astronautics, 62-65.

255. V. Navale, R. Suresh and C. Ponnamperuma, December 2, 1985: "Precambrian of India - Possible Source of Evidence for Early Life," Proceedings of Indian National Science Academy, 51, A, No. 6, 1033-1053.

256. M. A. Corigliano-Murphy, X. Liang, C. Ponnamperuma, D. Dalzoppo, A. Fontana, T. Kanmera and I. Chaiken, 1985: "Synthesis and Properties of an All-D Model Ribonuclease S-Peptide," International Journal Peptide Protein Res., 25, 225-231.

257. C. Ponnamperuma, 1985: "Synthesis and Analysis in Chemical Evolution," The Search for Extraterrestrial Life: Recent Developments, M. D. Papagiannis, ed., 185-197.

258. K. Kobayashi and C. Ponnamperuma, 1985: "Trace Elements in Chemical Evolution, I," Origins of Life, 16, D. Reidel Publishing Company, 41-55.

259. K. Kobayashi and C. Ponnamperuma, 1985: "Trace Elements in Chemical Evolution II: Synthesis of Amino Acids under Simulated Primitive Earth Conditions in the Presence of Trace Elements," Origins of Life, 16, D. Reidel Publishing Company, 57-67.

260. M. K. Hobish and C. Ponnamperuma, eds., 1985: "Global Habitability and Chemical Evolution," Special Issue of Origins of Life, 15(4), D. Reidel Publishing Co.

1986

261. K. Kobayashi and C. Ponnamperuma, 1986: "Abiotic Synthesis of Compounds of Biological Imporrtance by Electric Discharge in Simuated Paleoatmosphere," Viva Origino, 14, 42-54 (in Japanese).

262. R. Navarro-Gonzalez, A. Negron-Mendoza and C. Ponnamperuma, 1986: "Methane as a Chemical Dosímeter in Prebiotic Experiments, I. Electrical Discharges, Heat and Shock Waves", Origins of Life, 16 (1986) pp. 301-302.

263. A. Negron-Mendoza, R. Navarro-Gonzalez and C. Ponnamperuma, 1986: "Influence of Na-Montmorillonite in the Gamma Radiolysis of Acetic Acid. Implications in Prebiotic Chemistry", Origins of Life, 16 (1986) pp. 303-304.

264. K. Kobayashi, L. Hua, P. E. Hare, M. K. Hobish and C. Ponnamperuma, 1986: "Abiotic Synthesis of Nucleosides by Electric Discharge in a Simulated Primitive Earth Atmosphere," Origins of Life, 16(3/4), 277.

1987

265. L. G. Pleasant, R. Wade and C. Ponnamperuma, 1987: "Chemical Evolution and the Origin of Life, Bibliography Supplement, 1983," Origins of Life, D. Reidel Publishing Company, 171-184.

266. R. Wade, J. Powers and C. Ponnamperuma, 1987: "Chemical Evolution and the Origin of Life, Bibliography Supplement 1984," Origins of Life, 17, D. Reidel Publishing Company, 185-206.

1988

267. C. Ponnamperuma and K. S. Kumar, September 25-29, 1988: "The CHEMRAWN II Recommendations for Increasing the World Food Production - Biotechnology a Priority," Workshop on "Advanced Technologies for Increased Agricultural Production." UNDP Program and the Italian Association for International Development (AISI), Santa Margherita Ligure, Italy.

268. C. Ponnamperuma and K. S. Kumar, September 25-29, 1988: "Winged Bean - Potential for Cloning Improved Varieties," Workshop on "Advanced Technologies for Increased Agricultural Production." UNDP Program and the Italian Association for International Development (AISI), Santa Margherita Ligure, Italy.

1989

269. Y. Honda, R. Navarro-Gonzalez and C. Ponnamperuma, 1989: "Chemical Yields of Biologically Important Compounds from Electric Discharges", Radiat. Phys. Chemistry, 33 (1989) p. 287.

270. R. Navarro-Gonzales, A. Negron-Mendoza, Y. Honda and C. Ponnamperuma, 1989: "The γ - Radiolysis of Aqueous Solution of Urea, Implications for Chemical Evolution", Radiat. Phys. Chem., 33:287.

271. R. Navarro-Gonzalez, A. Negron-Mendoza, M. E. Aguirre-Calderon and C. Ponnamperuma, 1989: The γ-irradiation of Aqueous Hydrogen Cyanide in the Presence of Ferrocyanide or Ferricyanide, Implication to Prebiotic Chemistry," Advances in Space Research, Vol. 9 (No. 6), pp. (6) 57-(6) 61.

272. Y. Honda, R. Navarro-Gonzalez and C. Ponnamperuma, 1989: "A Quantitative Assay of Biologically Important Compounds in Simulated Primitive Earth Experiments," Advances in Space Research, Vol. 9 (No. 6), pp. (6) 63-(6) 66.

273. R. Navarro-Gonzalez and C. Ponnamperuma, 1989: "The Mechanism of Bacterial Bioluminescence", Rev. Soc. Quim. Mexico, Journal of the Mexican Chemical Society, 33 (1989), 54-60 (in Spanish).

274. C. Ponnamperuma, 1989: "Experimental Studies on the Origin of Life", J. British Interplanetary Society, 42, 397-400.

275. R. Navarro-Gonzalez, "The Role of Hydrogen Cyanide in Chemical Evolution.", Doctoral Dissertation, Graduate School of the University of Maryland, College Park, Studies done under the guidance and supervision of C. Ponnamperuma, April 26, 1989.

276. K. Kobayashi, P. E. Hare and C. Ponnamperuma, 1989: "Analysis of Sugars in the Products of Spark Discharge in Simulated Primitive Atmospheres by GC-MS.", Bunseki Kagaku, Vol. 38, No. 11, 608-612 (in Japanese).

277. K. Kobayashi, H. Yanagawa, L. Hua, C. Ponnamperuma, and T. Oshima, 1989: "Abiotic Synthesis in Simulated Primitive Planetary Atmospheres.", Institute of Space and Astronautical Science, pp. 112-113.

278. R. C. Wade, J. V. Powers and C. Ponnamperuma, 1989: "Chemical Evolution and the Origin of Life, Bibliography Supplement 1985," Origins of Life, 19, 199-220.

279. Y. Su, Y. Honda, P. E. Hare and C. Ponnamperuma, 1989: "Search of Peptide-like Materials in Electric Discharge Experiments," Origins of Life, 19, 237-238.

1990

280. S. Tyagi and C. Ponnamperuma, 1990: "Nonrandomness in Prebiotic Peptide Synthesis," Journal of Molecular Evolution, 30:391-399.

281. J. V. Powers and C. Ponnamperuma, 1990: "Chemical Evolution and the Origin of Life, Bibliography Supplement 1986," Origins of Life, 20, Kluwer Academic Publishers, 55-86.

282. N. Senaratne, M. K. Hobish and C. Ponnamperuma, 1990: "Direct Interactions between Amino Acids and Nucleotides as a Possible Explanation for the Origin of the Genetic Code," in Prebiological Self Organization of Matter, C. Ponnamperuma and F. Eirich, eds., A. Deepak Publishing, Hampton, Virginia, p. 148.

283. S. Tyagi and C. Ponnamperuma, 1990: "A Study of Peptide Synthesis by Amino Amyl Nucleotide Anhydrides in Presence of Complementary Homopolynucleotides," in Prebiological Self Organization of Matter, C. Ponnamperuma and F. Eirich, eds., A. Deepak Publishing, Hampton, Virginia, pp. 197-210.

284. C. Ponnamperuma, Y. Honda and R. Navarro-Gonzalez, 1990: "Asymmetry and Origin of Life," in Symmetries in Science IV, B. Gruber and J. H. Yopp, eds., Plenum Press, New York, pp. 193-203.

285. R. Navarro-Gonzalez, A. Negron-Mendoza, S. Ramos, and C. Ponnamperuma, 1990: "Radiolysis of Aqueous Solutions of Acetic Acid in the Presence of Na-Montmorillonite," Sciences Geologiques, Proceedings of the 9th International Clay Conference, V.C. Farmer and Y. Tardy, eds, 55-65.

286. Honda, Y., R. Navarro-González and C. Ponnamperuma, 1990: "The Electrolysis of a Simulated Primitive Atmosphere: CH_4-N_2-H_2O. I. Development of Dosimetric Methods," Abstracts of Papers for the Thirty-Eight Annual Meeting of the Radiation Research Society, New Orleans, Louisiana, April 7-12, 142.

287. Navarro-González, R., Y. Honda and C. Ponnamperuma, 1990: "The Electrolysis of a Simulated Primitive Atmosphere: CH_4-N_2-H_2O. II. An Investigation of the Initial Products," Abstracts of Papers for the Thirty-Eight Annual Meeting of the Radiation Research Society, New Orleans, Louisiana, April 7-12, 142.

LIST OF PARTICIPANTS

A. Allegrini, Universita di Torino, Torino, ITALY
A. Altstein, Russian Academy of Sciences, Institute of Gene Biology, Vavilov Str. 34/5, 117334 Moscow, RUSSIA
A.A. Bakasov, Joint Institute for Nuclear Research, Frank Laboratory of Neutron Physics, P.O. Box 79, Moscow Region, 141980 Dubna, RUSSIA
H. Baltscheffsky, University of Stockholm, Arrhenius Laboratory Department of Biochemistry, S-106 91 Stockholm, SWEDEN
C. Blomberg, Royal Institute of Technology, Department of Theoretical Physics, Lindstedsvagen 15, S-10044 Stockholm, SWEDEN
O. Carelse, University of Zimbabwe, Department of Biochemistry, P.O. Box Mp 167, Mount Pleasant, Harare, ZIMBABWE
M.S. Chadha, Bhabha Atomic Research Centre, Bio-Organic Division, 400 085 Bombay, INDIA
J. Chela-Flores, International Centre for Theoretical Physics, P.O. Box 586, Miramare, (34100) Trieste, ITALY
C. Cosmovici, Consiglio Nazionale Delle Ricerche, Istituto di Fisica Dello Spazio Interplanetario, Via G. Galilei, Casella Postale 27, 00044 Frascati, ITALY
G. De Sarrazin, Universidad de Los Andes, Departamento de Matematicas, Facultad de Ciencias, La Hechicera, Merida, VENEZUELA
J. Doskocil, Academy of Sciences of the Czech Republic, Institute of Molecular Genetics, Flemingovo Nam. 2, 166 37 Prague, CZECH REPUBLIC
F.R. Eirich, Polytechnic University of New York, 333 Jay Street, Brooklyn, New York 11201, USA
M.O. Eze, University of Nigeria, Department of Biochemistry, Nsukka, NIGERIA
A. Falaschi, International Centre for Genetic Engineering & Biotechnology, Padriciano 99, 34100 Trieste, ITALY
O. Famurewa, Ondo State University, Department Applied Microbiology, Ondo State, Ado-Ekiti, NIGERIA
A.R. Figureau, Institut de Physique Nucleaire de Lyon, Université Claude Bernard, 43. Bldv Du 11 Novembre 1918, 69622 Villeurbanne, FRANCE
G. Gilat, Israel Institute of Technology, Technion, Department of Physics, Technion City, 32 000 Haifa, ISRAEL
V.I. Goldanskii, Russian Academy of Sciences, Institute of Chemical Physics, Ulitsa Kosygina 4, V-334, SU-117977 SZD Moscow, RUSSIA
M. Hack, Osservatorio Astronomico di Trieste, Via G.B. Tiepolo 11, 34131 Trieste, ITALY
R. Haynes, University of Toronto, Department of Biology, Toronto, CANADA
W. Hebel, Directorate General for Science, Research & Development, Commission of the European Communities, Rue de la Loi 200, B-1049 Brussels, BELGIUM
B. Heinz, Palomar College, 1140 West Mission Road, San Marcos, California 92069, USA
M.K. Hobish, Consortium for Analysis in Space Exploration, 5606 Rockspring Road, Baltimore, Maryland 21209-4318, USA
R.K. Khanna, University of Maryland at College Park, Laboratory of Chemical Evolution, Department of Chemistry, College Park, Maryland 20742, USA
I.S. Kulaev, Russian Academy of Sciences, Institute of Biochemistry, Pustchino On Onka, 142292 Moscow, RUSSIA
N. Kumr, Indian Institute of Science, Department of Physics, 560 012 Bangalore, INDIA
G. Lenaz, Università degli Studi Bologna, Dipartimento de Biochimica, Facoltà di Medicina, Via Irnerio 48, 40126 Bologna, ITALY

G. Longo, Università degli Studi di Bologna, Dipartimento di Fisica, Via Irnerio 46, 40126 Bologna, ITALY

A.J. MacDermott, University of Oxford, Physical Chemistry Laboratory, South Parks Road, OX1 3QZ Oxford, UNITED KINGDOM

C. Matthews, University of Illinois at Chicago, Department of Chemistry, Box 4348, Chicago, Illinois 60680, USA

T. Mbungu, Université de Kinshasa, Departement de Physique, Faculté Des Sciences, B.P. 190, 11 Kinshasa, ZAIRE

R. Mistretta, Università di Torino, Dipartimento de Fisica Via Pietro Giuria 1, (10125) Torino, ITALY

R. Mohan, Max-Planck-Institut für Biophysikalische Chemie, AM Fassberg 11, Postfach 2841, 3400-Nikolausberg, Gottingen, GERMANY

E.D. Mshelia, Abubakar Tafawa Balewa University, Department of Physics, P.M.B. 0248, Bauchi, NIGERIA

A. Negron-Mendoza, Universidad Nacional Autonoma de Mexico, Instituto de Ciencias Nucleares, Apdo. Postal 70-543, Circuito Exterior. C.U., 04510 DF Mexico City, MEXICO

T. Oshima, Tokyo Institute of Technology, Tokyo, JAPAN

M.K. Pasha, University of Chittagong, Department of Botany, University Post Office, Chittagong, BANGLADESH

C. Ponnamperuma, University of Maryland at College Park, Laboratory of Chemical Evolution, Department of Chemistry, College Park, Maryland, 20742, USA

B. Prieur, Université de Paris VI, Paris, FRANCE

K.S. Rao, International Centre for Theoretical Physics, P.O. Box 586, Miramare, (34100) Trieste, ITALY

S. Rauch-Wojciechows, Linkoping University, Department of Mathematics, 581 83 Linkoping, SWEDEN

M. Rizzotti, Università degli Studi di Padova, Dipartimento de Biologia, Via Trieste 75, 35121 Padova, ITALY

H.N. Rosu, Universidad de Guanajuato, Istituto de Fisica, Apartdo Postal E-143, Prov. Guanajuato, Leon, MEXICO

T. Saito, University of Tokyo, Institute for Cosmic Ray Research, 3-2-1 Midori-Cho, Tanashi-Shi (188) Tokyo, JAPAN

M. Schidlowski, Max-Planck-Institut für Chemie, Postfach 30603, D 55020 Mainz, GERMANY

O.T. Scott, California State University, Department of Biology, 1811 Nordhoff, Northridge, California 91330, USA

P.L. Selvelli, Osservatorio Astronomico di Trieste, Via G.B. Tiepolo 11, 34131 Trieste, ITALY

T.K. Shah, International Centre for Theoretical Physics, P.O. Box 586, Miramare, 34100 Trieste, ITALY

I. Simon, Hungarian Academy of Sciences, Institute of Enzymology, Pob 7, H-1518 Budapest, HUNGARY

P. Stadler, Santa Fe Institute, Santa Fe, New Mexico 87505, USA

H. Stavliotis, University of Patras, Department of Physics, Faculty of Science, 26110 Patras, GREECE

W.H. Thiemann, Universität Bremen, Fachbereich 2-Chemie, Postfach 33 04 40, Loebenerstrasse Nw 2, 2800 Bremen, GERMANY

G. Vitiello, Universita di Salerno, Dipartimento di Fisica, Via S.Allende, Baronissi, 84081 Salerno, ITALY

W. Wang, Max-Planck-Institut für Chemie, Postfach 30603, D 55020 Mainz, GERMANY

W. Wang, Beijing University {University of Peking}, Department of Technical Physics, 100871 Beijing, CHINA